Fundamentals of Semiconductor Fabrication

Fundamentals of Semiconductor Fabrication

GARY S. MAY

Motorola Foundation Professor
School of Electrical and Computer Engineering
Georgia Institute of Technology
Atlanta, Georgia

SIMON M. SZE

UMC Chair Professor
National Chiao Tung University
National Nano Device Laboratories
Hsinchu, Taiwan

WILEY

JOHN WILEY & SONS, INC.

Senior Acquisitions Editor *Bill Zobrist*
Production Editor *Sandra Dumas*
Senior Marketing Manager *Katherine Hepburn*
Senior Designer *Kevin Murphy*
Production Management Services *Argosy*

Photo Credit: Nicole Capello/Georgia Institute of Technology
Cover Description: An eight-inch silicon wafer containing Intel Pentium processors.

This book was typeset in 10/12 *New Caledonia* (*NC*) by Argosy and printed and bound by *Hamilton Printing Company.* The cover was printed by *Phoenix Color Corp.*

The paper in this book was manufactured by a mill whose forest management programs include sustained yield harvesting of its timberlands. Sustained yield harvesting principles ensure that the number of trees cut each year does not exceed the amount of new growth.

This book is printed on acid-free paper. ∞

May, Gary, S., Simon M. Sze
Fundamentals of Semiconductor Fabrication

ISBN 0-471-23279-3
Wiley International Edition ISBN 0-471-45238-6
Printed in the United States of America

10 9 8 7 6 5 4 3 2 1

To LeShelle and Therese,

who enable, uplift, and sustain us.

Preface

This book provides an introduction to semiconductor fabrication technology, from crystal growth to integrated devices and circuits. It covers theoretical and practical aspects of all major steps in the fabrication sequence. It is intended as a textbook for senior undergraduates or first-year graduate students in physics, chemistry, electrical engineering, chemical engineering, and materials science. The book can be used conveniently in a semester-length course on integrated circuit fabrication. Such a course may or may not be accompanied by a corequisite laboratory. The text can also serve as a reference for practicing engineers and scientists in the semiconductor industry.

Chapter 1 gives a brief historical overview of major semiconductor devices and key technology developments, as well as an introduction to basic fabrication steps. Chapter 2 deals with crystal growth techniques. The next several chapters are organized according to a typical fabrication sequence. Chapter 3 presents silicon oxidation. Photolithography and etching are discussed in Chapters 4 and 5, respectively. Chapters 6 and 7 present the primary techniques for the introduction of dopants: diffusion and ion implantation. The final chapter on individual process steps, Chapter 8, covers various methods of thin film deposition. The final three chapters focus on broad, summative topics. Chapter 9 ties the individual process steps together by presenting the process flows for critical process technologies, integrated devices, and microelectrical mechanical systems (MEMS). Chapter 10 introduces high-level integrated circuit manufacturing issues, including electrical testing, packaging, process control, and yield. Finally, Chapter 11 discusses the future outlook and challenges for the semiconductor industry.

Each chapter begins with an introduction and a list of learning goals and concludes with a summary of important concepts. Solved example problems are provided throughout, and suggested homework problems appear at the end of the chapter. The concept of process simulation is presented in several chapters, using the popular SUPREM and PROLITH software packages as application vehicles. Mastery of this software is intended to supplement, but not replace, learning the fundamental concepts associated with microelectronics processing.

A complete set of detailed solutions to all end-of-chapter problems has been prepared. This instructor's manual is available to all adopting faculty. The figures in the text are also available, in electronic format, from the publisher at the following website: http://www.wiley.com/college/may.

Acknowledgments

The authors wish to thank Drs. T. C. Chang, T. S. Chao, M. C. Chiang, F. H. Ko, M. C. Liaw, and S. C. Wu of the National Nano Device Laboratories, and Prof. T. L. Li of the National Chia-Yi University for their helpful suggestions and discussions. We are further indebted to Mr. N. Erdos for technical editing of the manuscript, Ms. Iris Lin for typing the many revisions of the draft, and Ms. Y. G. Yang of the Semiconductor Laboratory, National Chiao Tung University who furnished the hundreds of technical illustrations used in the book.

At John Wiley and Sons, we wish to thank Mr. W. Zobrist, who encouraged us to undertake the project. S. M. Sze wishes also to acknowledge the Spring Foundation of the National Chiao Tung University for its financial support, and the United Microelectronic Corporation for the UMC Chair Professorship grant that provided the environment to work on this book. G. S. May would like to similarly acknowledge the Motorola Foundation Professorship.

Contents

CHAPTER 1

Introduction

Semiconductor devices are the foundation of the electronics industry, which is the largest industry in the world, with global sales of over one trillion dollars since 1998. Figure 1.1 shows the sales volume of the semiconductor device-based electronics industry in the past 20 years and projects sales to the year 2010. Also shown are the gross world product (GWP) and the sales volumes of the automobile, steel, and semiconductor industries.[1,2] Note that the electronics industry surpassed the automobile industry in 1998. If current trends continue, the sales volume of the electronics industry will reach three trillion dollars and will constitute about 10% of GWP by 2010. The semiconductor industry, a subset of the electronics industry, will grow at an even higher rate to surpass the steel industry in the early twenty-first century and to constitute 25% of the electronics industry in 2010.

The multitrillion dollar electronics industry is fundamentally dependent on the manufacture of semiconductor integrated circuits (ICs). The solid-state computing, telecommunications, aerospace, automotive, and consumer electronics industries all rely heavily on these devices. A basic knowledge of semiconductor *materials*, *devices*, and *processes* is thus essential to the understanding of modern electronics. Although this text deals primarily with the basic processes involved in IC fabrication, a brief historical review of each of these three topics is warranted.

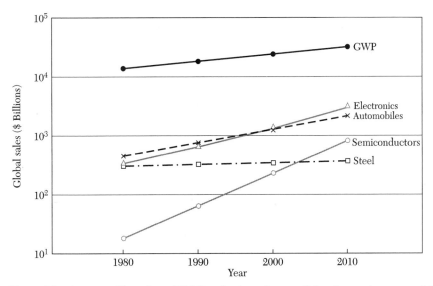

Figure 1.1   Gross world product (GWP) and sales volumes of the electronics, automobile, semiconductor, and steel industries from 1980 to 2000 and projected to 2010.[1,2]

▶ 1.1 SEMICONDUCTOR MATERIALS

Germanium was one of the first materials used in semiconductor device fabrication. In fact, the first transistor, developed by Bardeen, Brattain, and Shockley in 1947, was made of this element. However, germanium was rapidly replaced by silicon in the early 1960s. Silicon became the dominant material because it has several advantages. First, silicon can be easily oxidized to form a high-quality silicon dioxide (SiO_2) insulator, and SiO_2 is an excellent barrier layer for the selective diffusion steps needed in IC fabrication. Silicon also has a wider bandgap than germanium, which means that silicon devices can operate at a higher temperature than their germanium counterparts. Finally, and perhaps most important, as a primary constituent in ordinary sand, silicon is a very inexpensive and abundant element in nature. Thus, in addition to its processing advantages, silicon provides a very low-cost source material.

The next most popular material for IC fabrication is gallium arsenide (GaAs). Although GaAs has a higher electron mobility than silicon, it also possesses severe processing limitations, including less stability during thermal processing, a poor native oxide, high cost, and much higher defect densities. Silicon is therefore the material of choice in ICs and is emphasized more thoroughly in this text. GaAs is used for circuits that operate at very high speeds (in excess of 1 GHz) but with low to moderate levels of integration.

▶ 1.2 SEMICONDUCTOR DEVICES

The unique properties of semiconductor materials have enabled the development of a wide variety of ingenious devices that have literally changed our world. These devices have been studied for over 125 years.[3] To date, there are about 60 major devices, with over 100 device variations related to them.[4] Some major semiconductor devices are listed in Table 1.1 in chronological order.

The earliest systematic study of semiconductor devices (metal-semiconductor contacts) is generally attributed to Braun,[5] who in 1874 discovered that the resistance of contacts between metals and metal sulfides (e.g., copper pyrite) depended on the magnitude and polarity of the applied voltage. The electroluminescence phenomenon (for the light-emitting diode) was discovered by Round[6] in 1907. He observed the generation of yellowish light from a crystal of carborundum when he applied a potential of 10 V between two points on the crystals.

In 1947, the point-contact transistor was invented by Bardeen and Brattain.[7] This was followed by Shockley's[8] classic paper on p–n junctions and bipolar transistors in 1949. Figure 1.2 shows the first transistor. The two point contacts at the bottom of the triangular quartz crystal were made from two stripes of gold foil separated by about 50 μm ($1 \ \mu m = 10^{-4}$ cm) and pressed onto a semiconductor surface. The material used was germanium. With one gold contact forward biased (i.e., positive voltage with respect to the third terminal) and the other reverse biased, transistor action was observed; that is, the input signal was amplified. The bipolar transistor is a key semiconductor device and has ushered in the modern electronic era.

In 1952, Ebers[9] developed the basic model for the thyristor, which is an extremely versatile switching device. The solar cell was developed by Chapin et al.[10] in 1954 using a silicon p–n junction. The solar cell is a major candidate for obtaining energy from the sun because it can convert sunlight directly to electricity and is environmentally benign. In 1957, Kroemer[11] proposed the heterojunction bipolar transistor to improve transistor performance. This device is potentially one of the fastest semiconductor devices. In 1958, Esaki[12] observed negative resistance characteristics in a heavily doped p–n junction, which led to the discovery of the tunnel diode. The tunnel diode and its associated tunneling phenomenon are important for ohmic contacts and carrier transport through thin layers.

TABLE 1.1 Major Semiconductor Devices

Year	Semiconductor Device	Author(s)/Inventor(s)	Ref.
1874	Metal-semiconductor contact[a]	Braun	5
1907	Light emitting diode[a]	Round	6
1947	Bipolar transistor	Bardeen and Brattain; Shockley	7, 8
1949	p–n junction[a]	Shockley	8
1952	Thyristor	Ebers	9
1954	Solar cell[a]	Chapin, Fuller, and Pearson	10
1957	Heterojunction bipolar transistor	Kroemer	11
1958	Tunnel diode[a]	Esaki	12
1960	MOSFET	Kahng and Atalla	13
1962	Laser[a]	Hall et al.	15
1963	Heterostructure laser[a]	Kroemer; Alferov and Kazarinov	16, 17
1963	Transferred-electron diode[a]	Gunn	18
1965	IMPATT diode[a]	Johnston, DeLoach, and Cohen	19
1966	MESFET	Mead	20
1967	Nonvolatile semiconductor memory	Kahng and Sze	21
1970	Charge-coupled device	Boyle and Smith	23
1974	Resonant tunneling diode[a]	Chang, Esaki, and Tsu	24
1980	MODFET	Mimura et al.	25
1994	Room-temperature single-electron memory cell	Yano et al.	22
2001	15-nm MOSFET	Yu et al.	14

MOSFET, metal-oxide-semiconductor field-effect transistor; MESFET, metal-semiconductor field-effect transistor; MODFET, modulation-doped field-effect transistor.

[a]Denotes a two-terminal device; otherwise, it is a three- or four-terminal device.

Figure 1.2 The first transistor.[7] (Photograph courtesy of Bell Laboratories.)

The most important device for advanced integrated circuits is the MOSFET (metal-oxide-semiconductor field-effect transistor), which was reported by Kahng and Atalla[13] in 1960. Figure 1.3 shows the first device using a thermally oxidized silicon substrate. The device has a gate length of 20 μm and a gate oxide thickness of 100 nm (1 nm = 10^{-7} cm). The two keyholes are the source and drain contacts, and the top elongated area is the aluminum gate evaporated through a metal mask. Although present-day MOSFETs have been scaled down to the deep-submicron regime, the choice of silicon and thermally grown silicon dioxide used in the first MOSFET remains the most important combination of materials. The MOSFET and related integrated circuits now constitute about 90% of the semiconductor device market. An ultrasmall MOSFET with a channel length of 15 nm has been demonstrated recently.[14] This device can serve as the basis for the most advanced integrated circuit chips containing over one trillion (>10^{12}) devices.

In 1962, Hall et al.[15] first achieved lasing in semiconductors. In 1963, Kroemer[16] and Alferov and Kazarinov[17] proposed the heterostructure laser. These proposals laid the foundation for modern laser diodes, which can be operated continuously at room temperature. Laser diodes are the key components for a wide range of applications, including digital video disks, optical-fiber communication, laser printing, and atmospheric pollution monitoring.

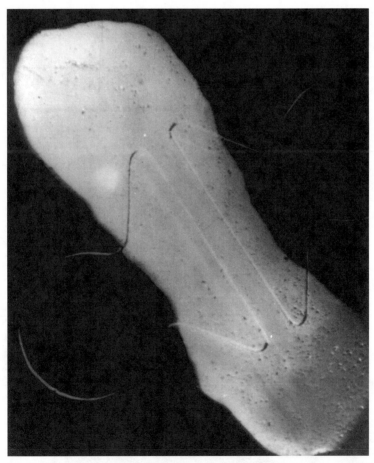

Figure 1.3 The first metal-oxide-semiconductor field-effect transistor.[13] (Photograph courtesy of Bell Laboratories.)

Three important microwave devices were invented or realized in the next three years. The first device was the transferred-electron diode (TED; also called Gunn diode), invented by Gunn[18] in 1963. The TED is used extensively in such millimeter-wave applications as detection systems, remote controls, and microwave test instruments. The second device is the IMPATT diode; its operation was first observed by Johnston et al.[19] in 1965. IMPATT diodes can generate the highest continuous-wave (CW) power at millimeter-wave frequencies of all semiconductor devices. They are used in radar systems and alarm systems. The third device is the MESFET (metal-semiconductor field-effect transitor), invented by Mead[20] in 1966. It is a key device for monolithic microwave integrated circuits (MMIC).

An important semiconductor memory device was invented by Kahng and Sze[21] in 1967. This is the nonvolatile semiconductor memory (NVSM), which can retain its stored information when the power supply is switched off. Although it is similar to a conventional MOSFET, the major difference is the addition of a "floating gate" in which semipermanent charge storage is possible. Because of its attributes of nonvolatility, high device density, low power consumption, and electrical rewritability (i.e., the stored charge can be removed by applying voltage to the control gate), the NVSM has become the dominant memory for portable electronic systems such as the cellular phone, notebook computer, digital camera, and smart card.

A limiting case of the floating-gate nonvolatile memory is the single-electron memory cell (SEMC), which is obtained by reducing the length of the floating gate to ultrasmall dimensions (e.g., 10 nm). At this dimension, when an electron moves into the floating gate, the potential of the gate will be altered so that it will prevent the entrance of another electron. The SEMC is the ultimate floating-gate memory cell, since we need only one electron for information storage. The operation of a SEMC at room temperature was first demonstrated by Yano et al.[22] in 1994. The SEMC can serve as the basis for the most advanced semiconductor memories, which can contain over one trillion bits.

The charge-coupled device (CCD) was invented by Boyle and Smith[23] in 1970. CCD is used extensively in video cameras and in optical sensing applications. The resonant tunneling diode (RTD) was first studied by Chang et al.[24] in 1974. RTD is the basis for most quantum-effect devices, which offer extremely high density, ultrahigh speed, and enhanced functionality, because it permits a greatly reduced number of devices to perform a given circuit function. In 1980, Mimura et al.[25] developed the MODFET (modulation-doped field-effect transistor). With the proper selection of heterojunction materials, the MODFET is the fastest field-effect transistor.

Since the invention of the bipolar transistor in 1947, the number and variety of semiconductor devices have increased tremendously as advanced technology, new materials, and broadened comprehension have been applied to the creation of new devices. However, one compelling question remains: What processes are required to construct these wondrous devices from basic semiconductor materials?

▶ 1.3 SEMICONDUCTOR PROCESS TECHNOLOGY

1.3.1 Key Semiconductor Technologies

Many important semiconductor technologies have been derived from processes invented centuries ago. For example, the growth of metallic crystals in a furnace was pioneered by Africans living on the western shores of Lake Victoria more than 2000 years ago.[26] This process was used to produce carbon steel in preheated forced-draft furnaces. Another example is the lithography process, which was invented in 1798. In this original process, the pattern, or image, was transferred from a stone plate (*litho*).[27] This section considers the milestones of technologies that were applied for the first time to semiconductor processing or developed specifically for semiconductor device fabrication.

Some key semiconductor technologies are listed in Table 1.2 in chronological order. In 1918, Czochralski[28] developed a liquid–solid monocomponent growth technique. Czochralski growth is the process used to grow most of the crystals from which silicon wafers are produced. Another growth technique was developed by Bridgman[29] in 1925. The Bridgman technique has been used extensively for the growth of gallium arsenide and related compound semiconductor crystals. Although the semiconductor properties of silicon have been widely studied since early 1940, the study of semiconductor compounds was neglected for a long time. In 1952, Welker[30] noted that gallium arsenide and related III-V compounds were semiconductors. He was able to predict their characteristics and to prove them experimentally. Technology and devices using these compounds have since been actively studied.

The diffusion of impurity atoms in semiconductors is important for device processing. Basic diffusion theory was considered by Fick[31] in 1855. The idea of using diffusion techniques to alter the type of conductivity in silicon was disclosed in a patent in 1952 by Pfann.[32] In 1957, the ancient lithography process was applied to semiconductor device fabrication by Andrus.[33] He used photosensitive, etch-resistant polymers (photoresist) for pattern transfer. Lithography is a key technology for the semiconductor industry. The continued growth of the industry has been the direct result of improved lithographic technology. Lithography is also a significant economic factor, currently representing over 35% of IC manufacturing costs.

TABLE 1.2   Key Semiconductor Technologies

Year	Technology	Author(s)/Inventor(s)	Ref.
1918	Czochralski crystal growth	Czochralski	28
1925	Bridgman crystal growth	Bridgman	29
1952	III-V compounds	Welker	30
1952	Diffusion	Pfann	32
1957	Lithographic photoresist	Andrus	33
1957	Oxide masking	Frosch and Derrick	34
1957	Epitaxial CVD growth	Sheftal, Kokorish, and Krasilov	35
1958	Ion implantation	Shockley	36
1959	Hybrid integrated circuit	Kilby	37
1959	Monolithic integrated circuit	Noyce	38
1960	Planar process	Hoerni	39
1963	CMOS	Wanlass and Sah	40
1967	DRAM	Dennard	41
1969	Polysilicon self-aligned gate	Kerwin, Klein, and Sarace	42
1969	MOCVD	Manasevit and Simpson	43
1971	Dry etching	Irving, Lemons, and Bobos	44
1971	Molecular beam epitaxy	Cho	45
1971	Microprocessor (4004)	Hoff et al.	46
1982	Trench isolation	Rung, Momose, and Nagakubo	47
1989	Chemical mechanical polishing	Davari et al.	48
1993	Copper interconnect	Paraszczak et al.	49

CVD, chemical vapor deposition; CMOS, complementary metal-oxide-semiconductor field-effect transistor; DRAM, dynamic random access memory; MOCVD, metalorganic CVD.

The oxide masking method was developed by Frosch and Derrick[34] in 1957. They found that an oxide layer can prevent most impurity atoms from diffusing through it. In the same year, the epitaxial growth process based on the chemical vapor deposition technique was developed by Sheftal et al.[35] Epitaxy, derived from the Greek words *epi*, meaning "on," and *taxis*, meaning "arrangement," describes a technique of crystal growth to form a thin layer of semiconductor materials on the surface of a crystal that has a lattice structure identical to that of the crystal. This method is important for the improvement of device performance and the creation of novel device structures. In 1959, a rudimentary integrated circuit was made by Kilby.[37] It contained one bipolar transistor, three resistors, and one capacitor, all made in germanium and connected by wire bonding—a hybrid circuit. Also in 1959, Noyce[38] proposed the monolithic IC by fabricating all devices in a single semiconductor substrate (*monolith* means "single stone") and connecting the devices by aluminum metallization. Figure 1.4 shows the first monolithic IC of a flip-flop circuit containing six devices. The aluminum interconnection lines were obtained by etching evaporated aluminum layer over the entire oxide surface using the lithographic technique. These inventions laid the foundation for the rapid growth of the microelectronics industry.

The planar process was developed by Hoerni[39] in 1960. In this process, an oxide layer is formed on a semiconductor surface. With the help of a lithography process, portions of the oxide can be removed and windows cut in the oxide. Impurity atoms will diffuse only through the exposed semiconductor surface, and p–n junctions will form in the oxide window areas.

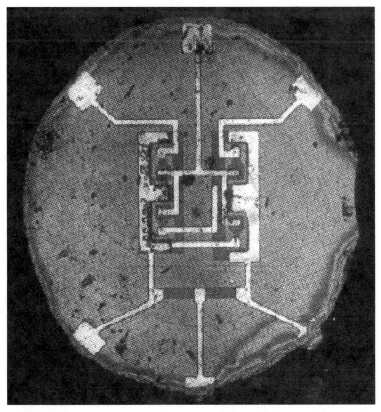

Figure 1.4 The first monolithic integrated circuit.[37] (Photograph courtesy of Dr. G. Moore.)

As the complexity of the IC increased, technology has moved from NMOS (n-channel MOSFET) to CMOS (complementary MOSFET), which employs both NMOS and PMOS (p-channel MOSFET) to form the logic elements. The CMOS concept was proposed by Wanlass and Sah[40] in 1963. The advantage of CMOS technology is that logic elements draw significant current only during the transition from one state to another (e.g., from 0 to 1) and draw very little current between transitions, allowing power consumption to be minimized. CMOS technology is the dominant technology for advanced ICs.

In 1967, an important two-element circuit, the dynamic random access memory (DRAM), was invented by Dennard.[41] The memory cell contains one MOSFET and one charge-storage capacitor. The MOSFET serves as a switch to charge or discharge the capacitor. Although a DRAM is volatile and consumes relatively high power, we expect that DRAMs will continue to be the first choice among various semiconductor memories for nonportable electronic systems in the foreseeable future.

To improve device performance, the polysilicon self-aligned gate process was proposed by Kerwin et al.[42] in 1969. This process not only improved device reliability but also reduced parasitic capacitances. Also in 1969, the metalorganic chemical vapor deposition (MOCVD) method was developed by Manasevit and Simpson.[43] This is a very important epitaxial growth technique for compound semiconductors such as GaAs.

As the device dimensions were reduced, the dry etching technique was developed to replace wet chemical etching for high-fidelity pattern transfer. This technique was initiated by Irving et al.[44] in 1971 using a CF_4/O_2 gas mixture to etch silicon wafers. Another important technique developed in the same year by Cho[45] was molecular beam epitaxy. This technique has the advantage of near-perfect vertical control of composition and doping down to atomic dimensions. It is responsible for the creation of numerous photonic devices and quantum-effect devices.

In 1971, the first microprocessor was made by Hoff et al.,[46] who put the entire central processing unit (CPU) of a simple computer on one chip. It was a four-bit microprocessor (Intel 4004), shown in Figure 1.5, with a chip size of 3 mm by 4 mm, and it contained 2300 MOSFETs. It was fabricated by a p-channel polysilicon gate process using an 8-μm design rule. This microprocessor performed as well as those in \$300,000 IBM computers of the early 1960s—each of which needed a CPU the size of a large desk. This was a major breakthrough for the semiconductor industry. Currently, microprocessors constitute the largest segment of the industry.

Since the early 1980s, many new technologies have been developed to meet the requirements of ever-shrinking minimum feature lengths. Three key technologies are trench isolation, chemical mechanical polishing, and copper interconnect. Trench isolation technology was introduced by Rung et al.[47] in 1982 to isolate CMOS devices. This approach eventually replaced all other isolation methods. In 1989, the chemical mechanical polishing method was developed by Davari et al.[48] for global planarization of the interlayer dielectrics. This is a key process for multilevel metallization. At submicron dimensions, a widely known failure mechanism is electromigration, which is the transport of metal ions through a conductor due to the passage of an electrical current. Although aluminum has been used since the early 1960s as interconnect material, it suffers from electromigration at high electrical current. Copper interconnect was introduced in 1993 by Paraszczak et al.[49] to replace aluminum for minimum feature lengths approaching 100 nm. This book considers all the technologies listed in Table 1.2.

1.3.2 Technology Trends

Since the beginning of the microelectronics era, the smallest linewidth (or the minimum feature length) of an integrated circuit has been reduced at a rate of about 13% per year.[50]

Figure 1.5 The first microprocessor.[46] (Photograph courtesy of Intel Corp.)

At that rate, the minimum feature length will shrink to about 50 nm in the year 2010. Device miniaturization results in reduced unit cost per circuit function. For example, the cost per bit of memory chips has halved every 2 years for successive generations of DRAM circuits. As device dimensions decrease, the intrinsic switching time decreases. Device speed has improved by four orders of magnitude since 1959. Higher speeds lead to expanded IC functional throughput rates. In the future, digital ICs will be able to perform data processing and numerical computation at terabit-per-second rates. As devices becomes smaller, they consume less power. Therefore, device miniaturization also reduces the energy used for each switching operation. The energy dissipated per logic gate has decreased by over one million times since 1959.

Figure 1.6 shows the exponential increase of the actual DRAM density versus the year of first production from 1978 to 2000. The density increases by a factor of 2 every 18 months. If trends continue, DRAM density will increase to 8 Gb in the year 2005 and to 64 Gb around the year 2012. Figure 1.7 shows the exponential increase of microprocessor computational power. Computational power also increases by a factor of 2 every 18 months. Currently, a Pentium-based personal computer has the same computational power as that of a CRAY 1 supercomputer of the late 1960s, yet it is three orders of magnitude smaller. If these trends continue, we will reach 100 GIP (billion instructions per second) in the year 2010.

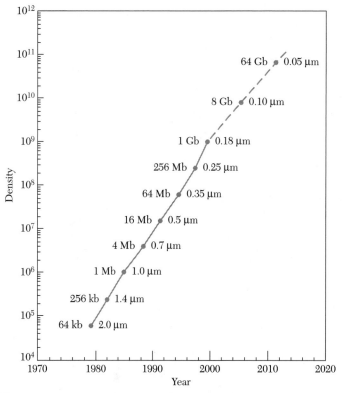

Figure 1.6 Exponential increase of dynamic random access memory density versus year based on the Semiconductor Industry Association roadmap.[50]

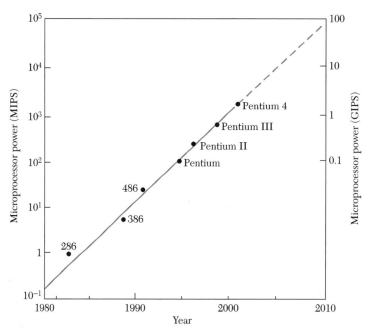

Figure 1.7 Exponential increase of microprocessor computational power versus year.

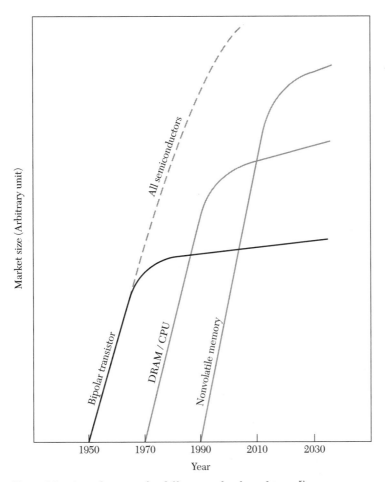

Figure 1.8 Growth curves for different technology drivers.[51]

Figure 1.8 illustrates the growth curves for different technology drivers.[51] At the beginning of the modern electronic era (1950–1970), the bipolar transistor was the technology driver. From 1970 to 1990, the DRAM and the microprocessor based on MOS devices were the technology drivers because of the rapid growth of personal computers and advanced electronic systems. Since 1990, nonvolatile semiconductor memory has been the technology driver, mainly because of the rapid growth of portable electronic systems.

► 1.4 BASIC FABRICATION STEPS

Today, planar technology is used extensively for IC fabrication. Figures 1.9 and 1.10 show the major steps of a planar process. These steps include oxidation, photolithography, etching, ion implanation, and metallization. This section describes these steps briefly. More detailed discussions can be found in Chapters 3 to 8; Chapter 9 describes the integration of these unit process steps to form semiconductor devices.

1.4.1 Oxidation

The development of a high-quality silicon dioxide (SiO_2) has helped to establish the dominance of Si in the production of commercial ICs. Generally, SiO_2 functions as an insulator

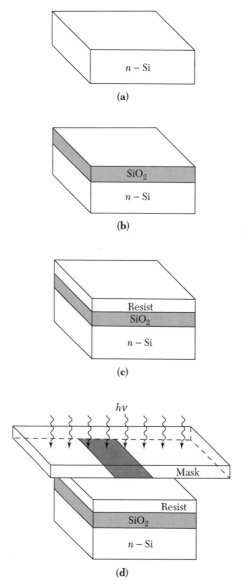

Figure 1.9 (*a*) A bare *n*-type Si wafer. (*b*) An oxidized Si wafer by dry or wet oxidation. (*c*) Application of resist. (*d*) Resist exposure through the mask.

in a number of device structures or as a barrier to diffusion or implantation during device fabrication. In the fabrication of a *p–n* junction (Fig. 1.9), the SiO_2 film is used to define the junction area.

There are two SiO_2 growth methods, dry and wet oxidation, depending on whether dry oxygen or water vapor is used. Dry oxidation is usually used to form thin oxides in a device structure because of its good Si–SiO_2 interface characteristics, whereas wet oxidation is used for thicker layers because of its higher growth rate. Figure 1.9*a* shows a section of a bare Si wafer ready for oxidation. After the oxidation process, a SiO_2 layer is formed all over the wafer surface. For simplicity, Figure 1.9*b* shows only the upper surface of an oxidized wafer. More details on oxidation may be found in Chapter 3.

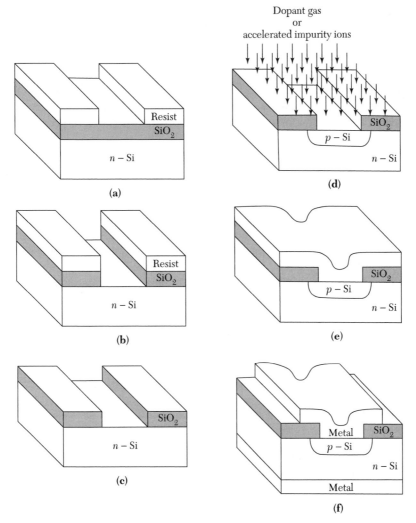

Figure 1.10 (a) The wafer after the development. (b) The wafer after SiO_2 removal. (c) The final result after a complete lithographic process. (d) A p–n junction is formed in the diffusion or implantation process. (e) The wafer after metallization. (f) A p–n junction after the complete processes.

1.4.2 Photolithography and Etching

Another technology, called *photolithography*, is used to define the geometry of the p–n junction. After the formation of SiO_2, the wafer is coated with an ultraviolet (UV) light–sensitive material called a *photoresist*, which is spun on the wafer surface by a high-speed spinner. Afterward (Fig. 1.9c), the wafer is baked at about 80°C to 100°C to drive the solvent out of the resist and to harden the resist for improved adhesion.

Figure 1.9d shows the next step, which is to expose the wafer through a patterned mask using a UV light source. The exposed region of the photoresist-coated wafer undergoes a chemical reaction depending on the type of resist. The area exposed to light becomes polymerized and difficult to remove in an etchant. The polymerized region remains when the wafer is placed in a developer, whereas the unexposed region (under the opaque area) dissolves and washes away.

Figure 1.10a shows the wafer after the development. The wafer is again baked to 120°C to 180°C for 20 minutes to enhance the adhesion and improve the resistance to the subsequent etching process. Then, an etch using buffered hydrofluoric acid (HF) removes the unprotected SiO$_2$ surface (Fig. 1.10b). Lastly, the resist is stripped away by a chemical solution or an oxygen plasma system. Figure 1.10c shows the final result of a region without oxide (a window) after the lithography process. The wafer is now ready for forming the p–n junction by a diffusion or ion implantation process. Photolithography and etching are described more thoroughly in Chapters 4 and 5, respectively.

1.4.3 Diffusion and Ion Implantation

In the diffusion method, the semiconductor surface not protected by the oxide is exposed to a source with a high concentration of opposite-type impurity. The impurity moves into the semiconductor crystal by solid-state diffusion. In ion implantation, the intended impurity is introduced into the semiconductor by accelerating the impurity ions to a high energy level and then implanting the ions in the semiconductor. The SiO$_2$ layer serves as a barrier to impurity diffusion or ion implantation. After the diffusion or implantation process, the p–n junction is formed, as shown in Figure 1.10d. Due to lateral diffusion of impurities or lateral straggle of implanted ions, the width of the p-region is slightly wider than the window opening. Diffusion and ion implantation are discussed in Chapters 6 and 7, respectively.

1.4.4 Metallization

After diffusion or ion implantation, a metallization process is used to form ohmic contacts and interconnections (Fig. 1.10e). Metal films can be formed by physical vapor deposition or chemical vapor deposition. The photolithography process is again used to define the front contact, which is shown in Figure 1.10f. A similar metallization step is performed on the back contact without using a lithography process. Normally, a low-temperature (≤ 500°C) anneal would also be performed to promote low-resistance contacts between the metal layers and the semiconductor. Metallization is discussed in more detail in Chapter 8.

▶ 1.5 SUMMARY

Semiconductor devices have an enormous impact on our society and the global economy because they serve as the foundation of the largest industry in the world—the electronics industry.

This introductory chapter has presented a historical review of major semiconductor devices, from the first study of the metal-semiconductor contact in 1874 to the fabrication of an ultrasmall 15-nm MOSFET in 2001. Of particular importance are the invention of the bipolar transistor in 1947, which ushered in the modern electronics era; the development of the MOSFET in 1960, which is the most important device for integrated circuits; and the invention of the nonvolatile semiconductor memory in 1967, which has been the technology driver of the electronics industry since 1990.

This chapter also described key semiconductor technologies. The origins of these technologies can be traced back as far as two millennia. Of particular importance are the development of the lithographic photoresist in 1957, which established the basic pattern transfer process for semiconductor devices; the invention of the integrated circuit in 1959, which was seminal to the rapid growth of the microelectronics industry; and the developments of the DRAM in 1967 and the microprocessor in 1971, which constitute the two largest segments of the semiconductor industry.

In this book, each chapter deals with a key IC fabrication process step or sequence of steps. Each chapter is presented in a clear and coherent fashion without heavy reliance on the original literature. However, a few important papers are listed at the end of each chapter for reference and for further reading.

▶ REFERENCES

1. *2000 Electronic Market Data Book*, Electron. Ind. Assoc., Washington, DC, 2000.

2. *2000 Semiconductor Industry Report*, Ind. Technol. Res. Inst., Hsinchu, Taiwan, 2000.

3. Most of the classic device papers are collected in S. M. Sze, Ed., *Semiconductor Devices: Pioneering Papers*, World Sci., Singapore, 1991.

4. K. K. Ng, *Complete Guide to Semiconductor Devices*, McGraw-Hill, New York, 1995.

5. F. Braun, "Uber die Stromleitung durch Schwefelmetalle," *Ann. Phys. Chem.*, **153,** 556 (1874).

6. H. J. Round, "A Note on Carborundum," *Electron. World*, **19,** 309 (1907).

7. J. Bardeen and W. H. Brattain, "The Transistor, a Semiconductor Triode," *Phys. Rev.*, **71,** 230 (1948).

8. W. Shockley, "The Theory of p–n Junction in Semiconductors and p–n Junction Transistors," *Bell Syst. Tech. J.*, **28,** 435 (1949).

9. J. Ebers, "Four Terminal p–n–p–n Transistors," *Proc. IRE*, **40,** 1361 (1952).

10. D. M. Chapin, C. S. Fuller, and G. L. Pearson, "A New Silicon p–n Junction Photocell for Converting Solar Radiation into Electrical Power," *J. Appl. Phys.*, **25,** 676 (1954).

11. H. Kroemer, "Theory of a Wide-Gap Emitter for Transistors," *Proc. IRE*, **45,** 1535 (1957).

12. L. Esaki, "New Phenomenon in Narrow Germanium p–n Junctions," *Phys. Rev.*, **109,** 603 (1958).

13. D. Kahng and M. M. Atalla, "Silicon-Silicon Dioxide Surface Device," in *IRE Device Research Conference*, Pittsburgh, 1960. (The paper can be found in Ref. 3.)

14. B. Yu, et al., "15 nm Gate Length Planar CMOS Transistor," *IEEE IEDM Technical Digest*, Washington, DC, p. 937 (2001).

15. R. N. Hall, et al., "Coherent Light Emission from GaAs Junctions," *Phys. Rev. Lett.*, **9,** 366 (1962).

16. H. Kroemer, "A Proposed Class of Heterojunction Injection Lasers," *Proc. IEEE*, **51,** 1782 (1963).

17. I. Alferov and R. F. Kazarinov, "Semiconductor Laser with Electrical Pumping," U.S.S.R. Patent 181, 737 (1963).

18. J. B. Gunn, "Microwave Oscillations of Current in III-V Semiconductors," *Solid State Commun.*, **1,** 88 (1963).

19. R. L. Johnston, B. C. DeLoach, Jr., and B. G. Cohen, "A Silicon Diode Microwave Oscillator," *Bell Syst. Tech. J.*, **44,** 369 (1965).

20. C. A. Mead, "Schottky Barrier Gate Field Effect Transistor," *Proc. IEEE*, **54,** 307 (1966).

21. D. Kahng and S. M. Sze, "A Floating Gate and Its Application to Memory Devices," *Bell Syst. Tech. J.*, **46,** 1283 (1967).

22. K. Yano, et al. "Room Temperature Single-Electron Memory," *IEEE Trans. Electron Devices*, **41,** 1628 (1994).

23. W. S. Boyle and G. E. Smith, "Charge Coupled Semiconductor Devices," *Bell Syst. Tech. J.*, **49,** 587 (1970).

24. L. L. Chang, L. Esaki, and R. Tsu, "Resonant Tunneling in Semiconductor Double Barriers," *Appl. Phys. Lett.*, **24,** 593 (1974).

25. T. Mimura, et al., "A New Field-Effect Transistor with Selectively Doped GaAs/n–Al_xGa_{1-x} as Heterojunction," *Jpn. J. Appl. Phys.*, **19,** L225 (1980).

26. D. Shore, "Steel-Making in Ancient Africa," in I. Van Sertima, Ed., *Blacks in Science: Ancient and Modern*, New Brunswick, NJ: Transaction Books, 157 (1986).

27. M. Hepher, "The Photoresist Story," *J. Photo. Sci.*, **12,** 181 (1964).

28. J. Czochralski, "Ein neues Verfahren zur Messung der Kristallisationsgeschwindigkeit der Metalle," *Z. Phys. Chem.*, **92**, 219 (1918).

29. P. W. Bridgman, "Certain Physical Properties of Single Crystals of Tungsten, Antimony, Bismuth, Tellurium, Cadmium, Zinc, and Tin," *Proc. Am. Acad. Arts Sci.*, **60**, 303 (1925).

30. H. Welker, "Über Neue Halbleitende Verbindungen," *Z. Naturforsch.*, **7a**, 744 (1952).

31. A. Fick, "Ueber Diffusion," *Ann. Phys. Lpz.*, **170**, 59 (1855).

32. W. G. Pfann, "Semiconductor Signal Translating Device," U.S. Patent 2,597,028 (1952).

33. J. Andrus, "Fabrication of Semiconductor Devices," U.S. Patent 3,122,817 (filed 1957; granted 1964).

34. C. J. Frosch and L. Derrick, "Surface Protection and Selective Masking during Diffusion in Silicon," *J. Electrochem. Soc.*, **104**, 547 (1957).

35. N. N. Sheftal, N. P. Kokorish, and A. V. Krasilov, "Growth of Single-Crystal Layers of Silicon and Germanium from the Vapor Phase," *Bull. Acad. Sci. U.S.S.R., Phys. Ser.*, **21**, 140 (1957).

36. W. Shockley, "Forming Semiconductor Device by Ionic Bombardment," U.S. Patent 2,787,564 (1958).

37. J. S. Kilby, "Invention of the Integrated Circuit," *IEEE Trans. Electron Devices*, **ED-23**, 648 (1976); U.S. Patent 3,138,743 (filed 1959; granted 1964).

38. R. N. Noyce, "Semiconductor Device-and-Lead Structure," U.S. Patent 2,981,877 (filed 1959; granted 1961).

39. J. A. Hoemi, "Planar Silicon Transistors and Diodes," *IRE Int. Electron Devices Meet.*, Washington, DC (1960).

40. F. M. Wanlass and C. T. Sah, "Nanowatt Logics Using Field-Effect Metal-Oxide Semiconductor Triodes," *Tech. Dig. IEEE Int. Solid-State Circuit Conf.*, p. 32 (1963).

41. R. M. Dennard, "Field Effect Transistor Memory," U.S. Patent 3,387,286 (filed 1967; granted 1968).

42. R. E. Kerwin, D. L. Klein, and J. C. Sarace, "Method for Making MIS Structure," U.S. Patent 3,475,234 (1969).

43. H. M. Manasevit and W. I. Simpson, "The Use of Metal-Organic in the Preparation of Semiconductor Materials. I. Epitaxial Gallium-V Compounds," *J. Electrochem. Soc.*, **116**, 1725 (1969).

44. S. M. Irving, K. E. Lemons, and G. E. Bobos, "Gas Plasma Vapor Etching Process," U.S. Patent 3,615,956 (1971).

45. A. Y. Cho, "Film Deposition by Molecular Beam Technique," *J. Vac. Sci. Technol.*, **8**, S31 (1971).

46. The inventors of the microprocessor are M. E. Hoff, F. Faggin, S. Mazor, and M. Shima. For a profile of M. E. Hoff, see *Portraits in Silicon* by R. Slater, p. 175, MIT Press, Cambridge, 1987.

47. R. Rung, H. Momose, and Y. Nagakubo, "Deep Trench Isolated CMOS Devices," *Tech. Dig. IEEE Int. Electron Devices Meet.*, p. 237 (1982).

48. B. Davari, et al., "A New Planarization Technique, Using a Combination of RIE and Chemical Mechanical Polish (CMP)," *Tech. Dig. IEEE Int. Electron Devices Meet.*, p. 61 (1989).

49. J. Paraszczak, et al., "High Performance Dielectrics and Processes for ULSI Interconnection Technologies," *Tech. Dig. IEEE Int. Electron Devices Meet.*, p. 261 (1993).

50. *The International Technology Roadmap for Semiconductor,* Semiconductor Ind. Assoc., San Jose, 1999.

51. F. Masuoka, "Flash Memory Technology," *Proc. Int. Electron Devices Mater. Symp.*, 83, Hsinchu, Taiwan (1996).

2

Crystal Growth

The two most important semiconductors for discrete devices and integrated circuits are silicon and gallium arsenide. This chapter describes the common techniques for growing single crystals of these two semiconductors. The basic process flow from starting materials to polished wafers is shown in Figure 2.1. The starting materials—silicon dioxide for a silicon wafer, and gallium and arsenic for a gallium arsenide wafer—are chemically processed to form a high-purity polycrystalline semiconductor from which single crystals are grown. The single-crystal ingots are shaped to define the diameter of the material and are sawed into wafers. These wafers are etched and polished to provide smooth, specular surfaces upon which devices will be made. This chapter covers the following topics:

- Basic techniques to grow silicon and GaAs single-crystal ingots
- Wafer-shaping steps from ingots to polished wafers
- Wafer characterization in terms of its electrical and mechanical properties

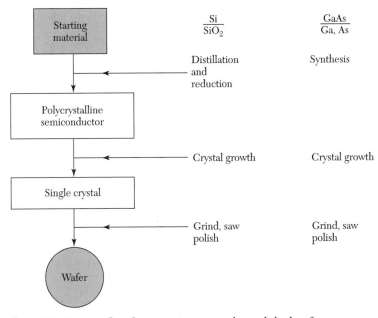

Figure 2.1   Process flow from starting material to polished wafer.

▶ 2.1 SILICON CRYSTAL GROWTH FROM THE MELT

The basic technique for silicon crystal growth from the melt, which is material in liquid form, is the *Czochralski technique*. A substantial percentage (>90%) of silicon crystals for the semiconductor industry are prepared by the Czochralski technique, and so is virtually all the silicon used for fabricating integrated circuits.

2.1.1 Starting Material

The starting material for silicon is a relatively pure form of sand (SiO_2) called *quartzite.* This is placed in a furnace with various forms of carbon (coal, coke, and wood chips). Although a number of reactions take place in the furnace, the overall reaction is

$$SiC(solid) + SiO_2(solid) \rightarrow Si(solid) + SiO(gas) + CO(gas) \tag{1}$$

This process produces metallurgical-grade silicon with a purity of about 98%. Next, the silicon is pulverized and treated with hydrogen chloride (HCl) at 300°C to form trichlorosilane ($SiHCl_3$):

$$Si(solid) + 3HCl(gas) \rightarrow SiHCl_3(gas) + H_2(gas) \tag{2}$$

The trichlorosilane is a liquid at room temperature (it has a boiling point of 32°C). Fractional distillation of the liquid removes the unwanted impurities. The purified $SiHCl_3$ is then used in a hydrogen reduction reaction to prepare the electronic-grade silicon (EGS):

$$SiHCl_3(gas) + H_2(gas) \rightarrow Si(solid) + 3HCl(gas) \tag{3}$$

This reaction takes place in a reactor containing a resistance-heated silicon rod, which serves as the nucleation point for the deposition of silicon. EGS, a polycrystalline material of high purity, is the raw material used to prepare device-quality, single-crystal silicon. Pure EGS generally has impurity concentrations in the parts-per-billion range.[1]

2.1.2 The Czochralski Technique

The Czochralski technique uses an apparatus called a *crystal puller*. A simplified version of this device is shown in Figure 2.2. The puller has three main components: (a) a furnace, which includes a fused-silicon (SiO_2) crucible, a graphite susceptor, a rotation mechanism (clockwise as shown), a heating element, and a power supply; (b) a crystal-pulling mechanism, which includes a seed holder and a rotation mechanism (counterclockwise); and (c) an ambient control, which includes a gas source (such as argon), a flow control, and an exhaust system. In addition, the puller has an overall microprocessor-based control system to control process parameters such as temperature, crystal diameter, pull rate, and rotation speeds, as well as to permit programmed process steps. Various sensors and feedback loops allow the control system to respond automatically, reducing operator intervention.

In the crystal-growing process, polycrystalline silicon (EGS) is placed in the crucible, and the furnace is heated above the melting temperature of silicon. A suitably oriented seed crystal (e.g., <111>) is suspended over the crucible in a seed holder. The seed is inserted into the melt. Part of it melts, but the tip of the remaining seed crystal still touches the liquid surface. It is then slowly withdrawn. Progressive freezing at the solid–liquid interface yields a large, single crystal. A typical pull rate is a few millimeters per minute. For large-diameter silicon ingots, an external magnetic field is applied to the basic Czochralski puller. The purpose of the external magnetic field is to control the concentration of defects, impurities, and oxygen.[2] Figure 2.3 shows a 300-mm (12 in.) and a 400-mm (16 in.) Czochralski-grown silicon ingot.

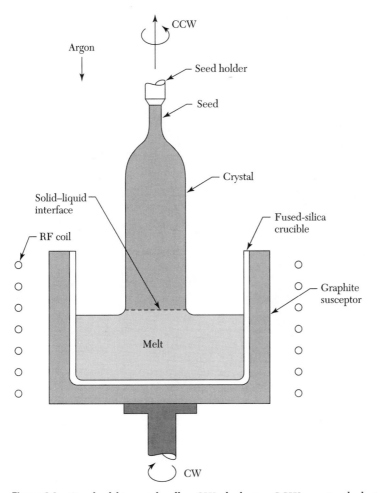

Figure 2.2 Czochralski crystal puller. CW, clockwise; CCW, counterclockwise.

2.1.3   Distribution of Dopant

In crystal growth, a known amount of dopant is added to the melt to obtain the desired doping concentration in the grown crystal. For silicon, boron and phosphorus are the most common dopants for p- and n-type materials, respectively.

As a crystal is pulled from the melt, the doping concentration incorporated into the crystal (solid) is usually different from the doping concentration of the melt (liquid) at the interface. The ratio of these two concentrations is defined as the *equilibrium segregation coefficient, k_0*:

$$k_0 \equiv \frac{C_s}{C_l} \tag{4}$$

where C_s and C_l are, respectively, the equilibrium concentrations of the dopant in the solid and liquid near the interface. Table 2.1 lists values of k_0 for the commonly used dopants for silicon. Note that most values are below 1, which means that during growth the dopants are rejected into the melt. Consequently, the melt becomes progressively enriched with the dopant as the crystal grows.

Figure 2.3 300-mm (12 in.) and 400-mm (16 in.) Czochralski-grown silicon ingots. (Photo courtesy of Shin-Etsu Handotai Co., Tokyo.)

TABLE 2.1 Equilibrium Segregation Coefficients for Dopants in Silicon

Dopant	k_0	Type	Dopant	k_0	Type
B	8×10^{-1}	p	As	3.0×10^{-1}	n
Al	2×10^{-3}	p	Sb	2.3×10^{-2}	n
Ga	8×10^{-3}	p	Te	2.0×10^{-4}	n
In	4×10^{-4}	p	Li	1.0×10^{-2}	n
O	1.25	n	Cu	4.0×10^{-4}	—[a]
C	7×10^{-2}	n	Au	2.5×10^{-5}	—[a]
P	0.35	n			

[a]Deep-lying impurity level.

Consider a crystal being grown from a melt having an initial weight M_0 with an initial doping concentration C_0 in the melt (i.e., the weight of the dopant per 1 g of melt). At a given point of growth when a crystal of weight M has been grown, the amount of dopant remaining in the melt (by weight) is S. For an incremental amount of the crystal with weight dM, the corresponding reduction of the dopant $(-dS)$ from the melt is $C_s dM$, where C_s is the doping concentration in the crystal (by weight):

$$-dS = C_s dM \tag{5}$$

Now, the remaining weight of the melt is $M_0 - M$, and the doping concentration in the liquid (by weight), C_l, is given by

$$C_l = \frac{S}{M_0 - M} \tag{6}$$

Combining Eqs. 5 and 6 and substituting $C_s/C_l = k_0$ yields

$$\frac{dS}{S} = -k_0 \left(\frac{dM}{M_0 - M} \right) \tag{7}$$

Given the initial weight of the dopant, $C_0 M_0$, we can integrate Eq. 7:

$$\int_{C_0 M_0}^{S} \frac{dS}{S} = k_0 \int_{0}^{M} \frac{-dM}{M_0 - M} \tag{8}$$

Solving Eq. 8 and combining with Eq. 6 gives

$$C_s = k_0 C_0 \left(1 - \frac{M}{M_0} \right)^{k_0 - 1} \tag{9}$$

Figure 2.4 illustrates the doping distribution as a function of the fraction solidified (M/M_0) for several segregation coefficients.[3,4] As crystal growth progresses, the composition initially at $k_0 C_0$ will increase continually for $k_0 < 1$ and decrease continually for $k_0 > 1$. When $k_0 \cong 1$, a uniform impurity distribution can be obtained.

▶ **EXAMPLE 1**

A silicon ingot, which should contain 10^{16} boron atoms/cm^3, is to be grown by the Czochralski technique. What concentration of boron atoms should be in the melt to give the required concentration in the ingot? If the initial load of silicon in the crucible is 60 kg, how many grams of boron (atomic weight 10.8) should be added? The density of molten silicon is 2.53 g/cm^3.

SOLUTION   Table 2.1 shows that the segregation coefficient k_0 for boron is 0.8. We assume that $C_s = k_0 C_l$ throughout the growth. Thus, the initial concentration of boron in the melt should be

$$\frac{10^{16}}{0.8} = 1.25 \times 10^{16} \text{ boron atoms/cm}^3$$

Since the amount of boron concentration is so small, the volume of melt can be calculated from the weight of silicon. Therefore, the volume of 60 kg of silicon is

$$\frac{60 \times 10^3}{2.53} = 2.37 \times 10^4 \text{ cm}^3$$

The total number of boron atoms in the melt is

$$1.25 \times 10^{16} \text{ atoms/cm}^3 \times 2.37 \times 10^4 \text{ cm}^3 = 2.96 \times 10^{20} \text{ boron atoms}$$

so that

$$\frac{2.96 \times 10^{20} \text{ atoms} \times 10.8 \text{ g/mol}}{6.02 \times 10^{23} \text{ atoms/mol}} = 5.31 \times 10^{-3} \text{ g of boron} = 5.31 \text{ mg of boron}$$

Note the small amount of boron needed to dope such a large load of silicon.   ◀

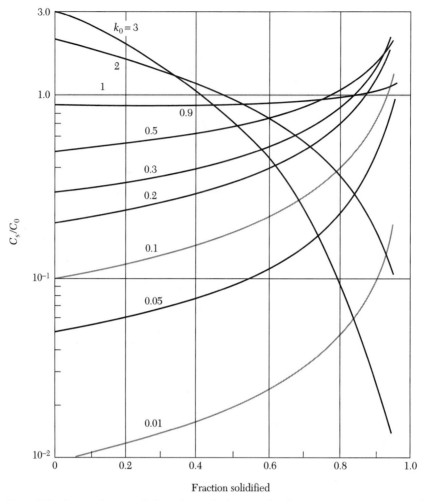

Figure 2.4 Curves for growth from the melt showing the doping concentration in a solid as a function of the fraction solidified.[4]

2.1.4 Effective Segregation Coefficient

While the crystal is growing, dopants are constantly being rejected into the melt (for $k_0 > 1$). If the rejection rate is higher than the rate at which the dopant can be transported away by diffusion or stirring, then a concentration gradient will develop at the interface, as illustrated in Figure 2.5. The segregation coefficient (given in Section 2.1.3) is $k_0 = C_s/C_1(0)$. We can define an effective segregation coefficient k_e, which is the ratio of C_s and the impurity concentration far away from the interface:

$$k_e \equiv \frac{C_s}{C_1} \tag{10}$$

Consider a small, virtually stagnant layer of melt with width δ in which the only flow is that required to replace the crystal being withdrawn from the melt. Outside this stagnant layer, the doping concentration has a constant value C_1. Inside the layer, the doping concentration can be described by the steady-state continuity equation:

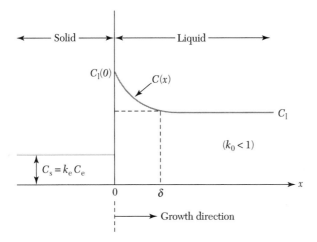

Figure 2.5 Doping distribution near the solid–melt interface.

$$0 = v\frac{dC}{dx} + D\frac{d^2C}{dx^2} \tag{11}$$

where D is the dopant diffusion coefficient in the melt, v is the crystal growth velocity, and C is the doping concentration in the melt.

The solution of Eq. 11 is

$$C = A_1 e^{-vx/D} + A_2 \tag{12}$$

where A_1 and A_2 are constants to be determined by the boundary conditions. The first boundary condition is that $C = C_1(0)$ at $x = 0$. The second boundary condition is the conservation of the total number of dopants; that is, the sum of the dopant fluxes at the interface must be zero. By considering the diffusion of dopant atoms in the melt (neglecting diffusion in the solid), we have

$$D\left(\frac{dC}{dx}\right)_{x=0} + \left[C_1(0) - C_s\right]v = 0 \tag{13}$$

Substituting these boundary conditions into Eq. 12 and noting that $C = C_1$ at $x = \delta$ gives

$$e^{-v\delta/D} = \frac{C_1 - C_s}{C_1(0) - C_s} \tag{14}$$

Therefore,

$$k_e \equiv \frac{C_s}{C_1} = \frac{k_0}{k_0 + (1 - k_0)e^{-v\delta/D}} \tag{15}$$

The doping distribution in the crystal is given by the same expression as in Eq. 9, except that k_0 is replaced by k_e. Values of k_e are larger than those of k_0 and can approach 1 for large values of the growth parameter $v\delta/D$. Uniform doping distribution ($k_e \to 1$) in the crystal can be obtained by employing a high pull rate and a low rotation speed (since δ is inversely proportional to the rotation speed). Another approach to achieve uniform doping is to add ultrapure polycrystalline silicon continuously to the melt so that the initial doping concentration is maintained.

▶ 2.2 SILICON FLOAT-ZONE PROCESS

The *float-zone process* can be used to grow silicon that has lower contamination than that normally obtained from the Czochralski technique. A schematic setup of the float-zone process is shown in Figure 2.6*a*. A high-purity polycrystalline rod with a seed crystal at the bottom is held in a vertical position and rotated. The rod is enclosed in a quartz envelope within which an inert atmosphere (argon) is maintained. During the operation, a small zone (a few centimeters in length) of the crystal is kept molten by a radio-frequency (RF) heater, which is moved from the seed upward so that this *floating zone* traverses the length of the rod. The molten silicon is retained by surface tension between the melting and growing solid-silicon faces. As the floating zone moves upward, single-crystal silicon freezes at the zone's retreating end and grows as an extension of the seed crystal. Materials with higher resistivities can be obtained from the float-zone process than from the Czochralski process because the former can be used to purify the crystal more easily. Furthermore, since no crucible is used in the float-zone process, there is no contamination from the crucible (as with Czochralski growth). At the present time, float-zone crystals are used mainly for high-power, high-voltage devices, where high-resistivity materials are required.

To evaluate the doping distribution of a float-zone process, consider a simplified model, as shown in Figure 2.6*b*. The initial uniform doping concentration in the rod is C_0 (by weight). L is the length of the molten zone at a distance x along the rod, A is the cross-sectional area of the rod, ρ_d is the specific density of silicon, and S is the amount of dopant present in the molten zone. As the zone traverses a distance dx, the amount of dopant added to it at its advancing end is $C_0\rho_d A\, dx$, whereas the amount of dopant removed from it at the retreating end is $k_e(S\, dx/L)$, where k_e is the effective segregation coefficient. Thus,

$$dS = C_0\rho_d A\, dx - \frac{k_e S}{L} dx = \left(C_0\rho_d A - \frac{k_e S}{L} \right) dx \qquad (16)$$

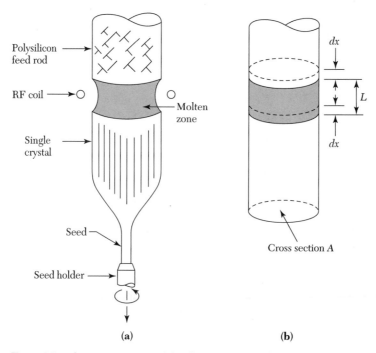

(a)   (b)

Polysilicon feed rod

RF coil

Molten zone

Single crystal

Seed

Seed holder

dx

L

dx

Cross section A

Figure 2.6 Float-zone process. (*a*) Schematic setup. (*b*) Simple model for doping evaluation.

so that

$$\int_0^x dx = \int_{S_0}^S \frac{dS}{C_0\rho_d A - (k_e S/L)} \tag{16a}$$

where $S_0 = C_0\rho_d AL$ is the amount of dopant in the zone when it was first formed at the front end of the rod. From Eq. 16a, we obtain

$$\exp\left(\frac{k_e x}{L}\right) = \frac{C_0\rho_d A - (k_e S_0/L)}{C_0\rho_d A - (k_e S/L)} \tag{17}$$

or

$$S = \frac{C_0 A\rho_d L}{k_e}\left[1 - (1 - k_e)^{-k_e x/L}\right] \tag{17a}$$

Since C_s (the doping concentration in the crystal at the retreating end) is given by $C_s = k_e(S/A\rho_d L)$, then

$$C_s = C_0\left[1 - 1(1 - k_e)^{-k_e x/L}\right] \tag{18}$$

Figure 2.7 shows the doping concentration versus the solidified zone length for various values of k_e.

These two crystal growth techniques can also be used to remove impurities. A comparison of Figure 2.7 with Figure 2.4 shows that a single pass in the float-zone process does not produce as much purification as a single Czochralski growth. For example, for $k_0 = k_e = 0.1$, C_s/C_0 is smaller over most of the solidified ingot made by the Czochralski growth. However, multiple float-zone passes can be performed on a rod much more easily than a crystal can be grown, the end region cropped off, and regrown from the melt. Figure 2.8 shows the impurity distribution for an element with $k_e = 0.1$ after a number of successive passes of the zone along the length of the rod.[4] Note that there is a substantial reduction of impurity concentration in the rod after each pass. Therefore, the float-zone process is ideally suited for crystal purification. This process is also called the *zone-refining* technique, which can provide a very high purity level of the raw material.

If it is desirable to dope the rod rather than purify it, consider the case in which all the dopants are introduced in the first zone ($S_0 = C_1 A\rho_d L$) and the initial concentration C_0 is negligibly small. Equation 17 gives

$$S_0 = S\exp\left(\frac{k_e x}{L}\right) \tag{19}$$

Since $C_s = k_e(S/A\rho_d L)$, we obtain the following from Eq. 19:

$$C_s = k_e C_1 e^{-k_e x/L} \tag{20}$$

Therefore, if $k_e x/L$ is small, C_s will remain nearly constant with distance except at the end that is last to solidify.

For certain switching devices, such as high-voltage thyristors, large chip areas are used—frequently an entire wafer for a single device. This size imposes stringent requirements on the uniformity of the starting material. To obtain homogeneous distribution of dopants, a float-zone silicon slice that has an average doping concentration well below the required amount is used. The slice is then irradiated with thermal neutrons. This process, called *neutron irradiation*, gives rise to fractional transmutation of silicon into phosphorus and dopes the silicon *n*-type:

$$\mathrm{Si}_{14}^{30} + \text{neutron} \rightarrow \mathrm{Si}_{14}^{31} + \gamma\,\text{ray} \rightarrow \mathrm{P}_{15}^{31} + \beta\,\text{ray} \tag{21}$$

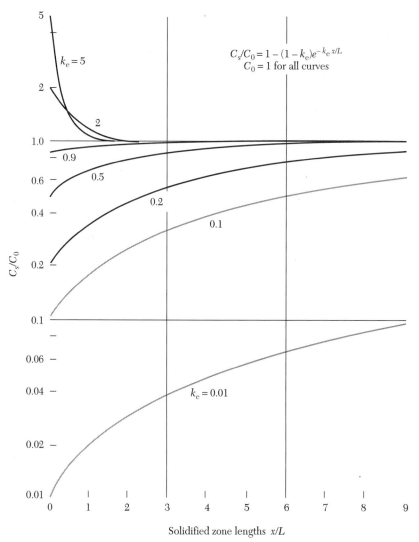

Figure 2.7 Curves for the float-zone process showing doping concentration in the solid as a function of solidified zone lengths.[4]

The half-life of the intermediate element Si_{14}^{31} is 2.62 hours. Because the penetration depth of neutrons in silicon is about 100 cm, doping is very uniform throughout the slice. Figure 2.9 compares the lateral resistivity distributions in conventionally doped silicon and in silicon doped by neutron irradiation.[5] Note that the resistivity variations for the neutron-irradiated silicon are much smaller than those for the conventionally doped silicon.

▶ 2.3 GaAs CRYSTAL GROWTH TECHNIQUES

2.3.1 Starting Materials

The starting materials for the synthesis of polycrystalline gallium arsenide are the elemental, chemically pure gallium and arsenic. Because gallium arsenide is a combination of two materials, its behavior is different from that of a single material such as silicon. The behavior of a combination can be described by a phase diagram. A *phase* is a state

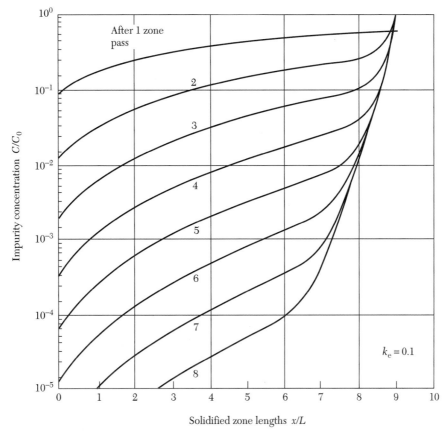

Figure 2.8 Relative impurity concentration versus zone length for a number of passes. L denotes the zone length.[4]

(e.g., solid, liquid, or gaseous) in which a material may exist. A *phase diagram* shows the relationship between the two components, gallium and arsenic, as a function of temperature.

Figure 2.10 shows the phase diagram of the gallium–arsenic system. The abscissa represents various compositions of the two components in terms of atomic percent (lower scale) or weight percent (upper scale).[6,7] Consider a melt that is initially of composition x (e.g., 85 atomic percent arsenic, as shown in Fig. 2.10). When the temperature is lowered, its composition will remain fixed until the liquidus line is reached. At the point (T_l, x), material of 50 atomic percent arsenic (i.e., gallium arsenide) will begin to solidify.

▶ **EXAMPLE 2**

In Figure 2.10, consider a melt of initial composition C_m (weight percent scale) that is cooled from T_a (on the liquidus line) to T_b. Find the fraction of the melt that will be solidified.

SOLUTION At T_b, M_l is the weight of the liquid, M_s is the weight of the solid (i.e., GaAs), and C_l and C_s are the concentrations of dopant in the liquid and the solid, respectively. Therefore, the weights of arsenic in the liquid and solid are $M_l C_l$ and $M_s C_s$, respectively. Because the total arsenic weight is $(M_l + M_s)C_m$, we have

$$M_l C_l + M_s C_s = (M_l + M_s)C_m$$

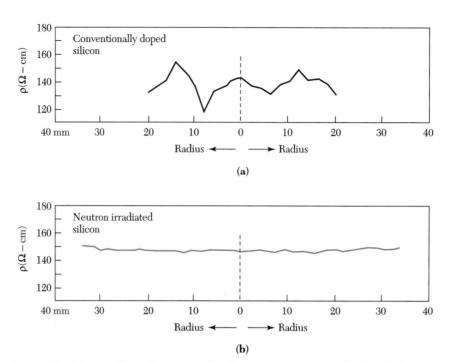

Figure 2.9 (*a*) Typical lateral resistivity distribution in a conventionally doped silicon. (*b*) Silicon doped by neutron irradiation.[5]

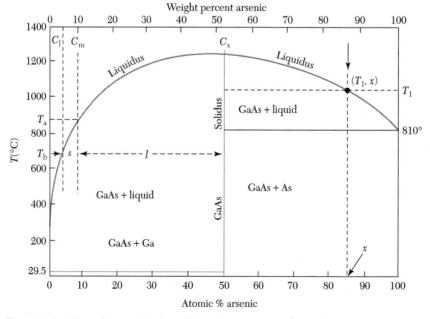

Figure 2.10   Phase diagram for the gallium–arsenic system.[6]

or

$$\frac{M_s}{M_l} = \frac{\text{Weight of GaAs at } T_b}{\text{Weight of liquid at } T_b} = \frac{C_m - C_l}{C_s - C_m} = \frac{s}{l}$$

where s and l are the lengths of the two lines measured from C_m to the liquidus and solidus line, respectively. As can be seen from Figure 2.10, about 10% of the melt is solidified. ◀

Unlike silicon, which has a relatively low vapor pressure at its melting point (approximately 10^{-6} atm at 1412°C), arsenic has much higher vapor pressures at the melting point of gallium arsenide (1240°C). In its vapor phase, arsenic has As_2 and As_4 as its major species. Figure 2.11 shows the vapor pressures of gallium and arsenic along the liquidus curve.[8] Also shown for comparison is the vapor pressure of silicon. The vapor pressure curves for gallium arsenide are double valued. The dashed curves are for arsenic-rich gallium arsenide melt (right side of liquidus line in Fig. 2.10), and the solid curves are for gallium-rich gallium arsenide melt (left side of liquidus line in Fig. 2.10). Because there is a larger amount of arsenic in an arsenic-rich melt than in a gallium-rich melt, more arsenic (As_2 and As_4)

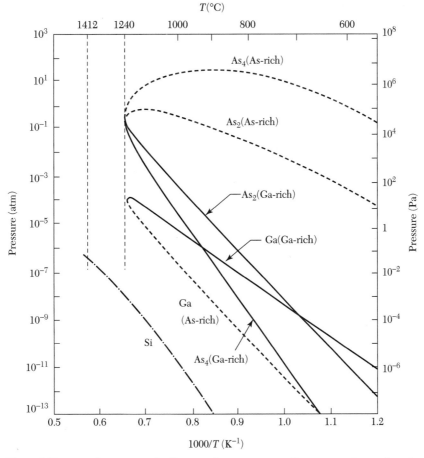

Figure 2.11 Partial pressure of gallium and arsenic over gallium arsenide as a function of temperature.[8] Also shown is the partial pressure of silicon.

will be vaporized from the arsenic-rich melt, thus resulting in a higher vapor pressure. A similar argument can explain the higher vapor pressure of gallium in a gallium-rich melt. Note that long before the melting point is reached, the surface layers of liquid gallium arsenide may decompose into gallium and arsenic. Because the vapor pressures of gallium and arsenic are different, there is a preferential loss of the more volatile arsenic species, and the liquid becomes gallium rich.

To synthesize gallium arsenide, an evacuated, sealed quartz tube system with a two-temperature furnace is commonly used. The high-purity arsenic is placed in a graphite boat and heated to 610°C to 620°C, whereas the high-purity gallium is placed in another graphite boat and heated to slightly above the gallium arsenide melting temperature (1240–1260°C). Under these conditions, an overpressure of arsenic is established (a) to cause the transport of arsenic vapor to the gallium melt, converting it into gallium arsenide; and (b) to prevent decomposition of the gallium arsenide while it is being formed in the furnace. When the melt cools, a high-purity polycrystalline gallium arsenide results. This serves as the raw material to grow single-crystal gallium arsenide.[7]

2.3.2 Crystal Growth Techniques

There are two techniques for GaAs crystal growth: the Czochralski technique and the *Bridgman technique*. Most gallium arsenide is grown by the Bridgman technique. However, the Czochralski technique is more popular for the growth of larger-diameter GaAs ingots.

For Czochralski growth of gallium arsenide, the basic puller is identical to that for silicon. However, to prevent decomposition of the melt during crystal growth, a liquid encapsulation method is employed. The liquid encapsulant is a molten boron trioxide (B_2O_3) layer about 1 cm thick. Molten boron trioxide is inert to the gallium arsenide surface and serves as a cap to cover the melt. This cap prevents decomposition of the gallium arsenide as long as the pressure on its surface is higher than 1 atm (760 Torr). Because boron trioxide can dissolve silicon dioxide, the fused-silica crucible is replaced with a graphite crucible.

To obtain the desired doping concentration in the grown crystal of GaAs, cadmium and zinc are commonly used for p-type materials, whereas selenium, silicon, and tellurium are used for n-type materials. For semiinsulating GaAs, the material is undoped. The equilibrium segregation coefficients for dopants in GaAs are listed in Table 2.2. Similar to those in Si, most of the segregation coefficients are less than 1. The expressions derived previously for Si (Eqs. 4 to 15) are equally applicable to GaAs.

TABLE 2.2   Equilibrium Segregation Coefficients for Dopants in GaAs

Dopant	k_0	Type
Be	3	p
Mg	0.1	p
Zn	4×10^{-1}	p
C	0.8	n/p
Si	1.85×10^{-1}	n/p
Ge	2.8×10^{-2}	n/p
S	0.5	n
Se	5.0×10^{-1}	n
Sn	5.2×10^{-2}	n
Te	6.8×10^{-2}	n
Cr	1.03×10^{-4}	Semiinsulating
Fe	1.0×10^{-3}	Semiinsulating

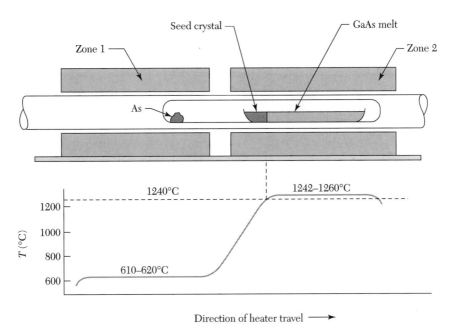

Figure 2.12 Bridgman technique for growing single-crystal gallium arsenide, and a temperature profile of the furnace.

Figure 2.12 shows a Bridgman system in which a two-zone furnace is used for growing single-crystal gallium arsenide. The left-hand zone is held at a temperature (approximately 610°C) to maintain the required overpressure of arsenic, whereas the right-hand zone is held just above the melting point of gallium arsenide (1240°C). The sealed tube is made of quartz, and the boat is made of graphite. In operation, the boat is loaded with a charge of polycrystalline gallium arsenide, with the arsenic kept at the other end of the tube.

As the furnace is moved toward the right, the melt cools at one end. Usually, there is a seed placed at the left end of the boat to establish a specific crystal orientation. The gradual freezing (solidification) of the melt allows a single crystal to propagate at the liquid–solid interface. Eventually, a single crystal of gallium arsenide is grown. The impurity distribution can be described essentially by Eqs. 9 and 15, where the growth rate is given by the traversing speed of the furnace.

▶ 2.4 MATERIAL CHARACTERIZATION

2.4.1 Wafer Shaping

After a crystal is grown, the first shaping operation is to remove the seed and the other end of the ingot, which is last to solidify.[1] The next operation is to grind the surface so that the diameter of the material is defined. After that, one or more flat regions are ground along the length of the ingot. These regions, or *flats*, mark the specific crystal orientation of the ingot and the conductivity type of the material. The largest flat, the *primary flat*, allows a mechanical locator in automatic processing equipment to position the wafer and to orient the devices relative to the crystal. Other smaller flats, called *secondary flats*, are ground to identify the orientation and conductivity type of the crystal, as shown in Figure 2.13. For crystals with diameters equal to or larger than 200 mm, no flats are ground. Instead, a small groove is ground along the length of the ingot.

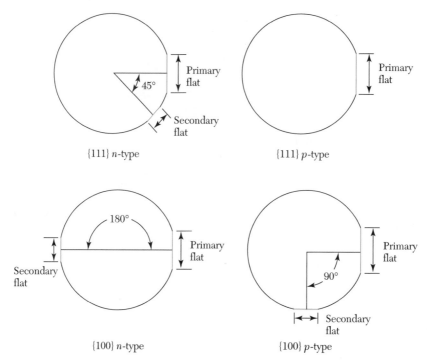

Figure 2.13 Identifying flats on a semiconductor wafer.

The ingot is then ready to be sliced by diamond saw into wafers. Slicing determines four wafer parameters: *surface orientation* (e.g., <111> or <100>); *thickness* (e.g., 0.5–0.7 mm, depending on wafer diameter); *taper*, which is the wafer thickness variations from one end to another; and *bow*, which is the surface curvature of the wafer, measured from the center of the wafer to its edge.

After slicing, both sides of the wafer are lapped using a mixture of Al_2O_3 and glycerine to produce a typical flatness uniformity within 2 μm. The lapping operation usually leaves the surface and edges of the wafer damaged and contaminated. The damaged and contaminated regions can be removed by chemical etching (see Chapter 5). The final step of wafer shaping is polishing. Its purpose is to provide a smooth, specular surface where device features can be defined by photolithographic processes (see Chapter 4). Figure 2.14 shows 200-mm (8 in.) and 400-mm (16 in.) polished silicon wafers in cassettes. Table 2.3 shows the specifications for 125-, 150-, 200-, and 300-mm diameter polished silicon wafers from the Semiconductor Equipment and Materials Institute (SEMI). As mentioned previously, for large crystals (≥ 200 mm diameter), no flats are ground; instead, a groove is made on the edge of the wafer for positioning and orientation purposes.

Gallium arsenide is a more fragile material than silicon. Although the basic shaping operation of gallium arsenide is essentially the same as that for silicon, greater care must be exercised in gallium arsenide wafer preparation. The state of gallium arsenide technology is relatively primitive compared with that of silicon. However, the technology of group III-V compounds has advanced partly because of the advances in silicon technology.

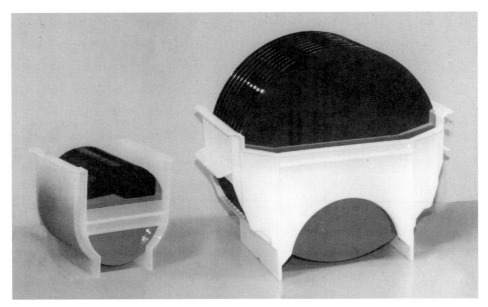

Figure 2.14 200-mm (8 in.) and 400-mm (16 in.) polished silicon wafers in cassettes. (Photo courtesy of Shin-Etsu Handotai Co., Tokyo.)

TABLE 2.3 Specifications for Polished Monocrystalline Silicon Wafers

Parameter	125 mm	150 mm	200 mm	300 mm
Diameter (mm)	125 ± 1	150 ± 1	200 ± 1	300 ± 1
Thickness (mm)	0.6–0.65	0.65–0.7	0.715–0.735	0.755–0.775
Primary flat length (mm)	40–45	55–60	NA	NA
Secondary flat length (mm)	25–30	35–40	NA	NA
Bow (µm)	70	60	30	< 30
Total thickness variation (µm)	65	50	10	< 10
Surface orientation	(100) ± 1°	Same	Same	Same
	(111) ± 1°	Same	Same	Same

NA, not available.

2.4.2 Crystal Characterization

Crystal Defects

A real crystal (such as a silicon wafer) differs from the ideal crystal in important ways. It is finite; thus, surface atoms are incompletely bonded. Furthermore, it has defects, which strongly influence the electrical, mechanical, and optical properties of the semiconductor. There are four categories of defects: point defects, line defects, area defects, and volume defects.

Figure 2.15 shows several forms of *point defects*.[1,9] Any foreign atom incorporated into the lattice at either a substitutional site [i.e., at a regular lattice site (Fig. 2.15*a*)] or an interstitial site [i.e., between regular lattice sites (Fig. 2.15*b*)] is a point defect. A missing atom in the lattice creates a vacancy, also considered a point defect (Fig. 2.15*c*).

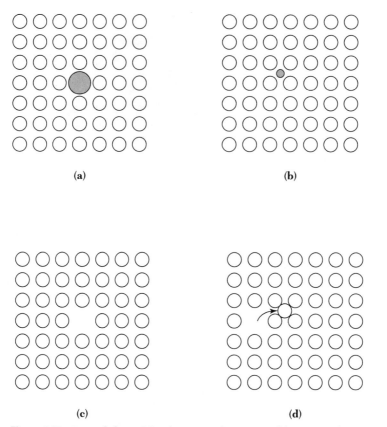

Figure 2.15 Point defects. (*a*) Substitutional impurity. (*b*) Interstitial impurity. (*c*) Lattice vacancy. (*d*) Frenkel-type defect.[9]

A host atom that is situated between regular lattice sites and adjacent to a vacancy is called a *Frenkel defect* (Fig. 2.15*d*). Point defects are particularly important subjects in the kinetics of oxidation and diffusion processes. These topics are considered in Chapters 3 and 6, respectively.

The next class of defects is the *line defect*, also called a *dislocation*.[10] There are two types of dislocations: the edge and screw types. Figure 2.16*a* is a schematic representation of an edge dislocation in a cubic lattice. An extra plane of atoms *AB* is inserted into the lattice. The line of the dislocation would be perpendicular to the plane of the page. The screw dislocation may be considered as being produced by cutting the crystal partway through and pushing the upper part one lattice spacing over, as shown in Figure 2.16*b*. Line defects in devices are undesirable because they act as precipitation sites for metallic impurities, which may degrade device performance.

Area defects represent a large-area discontinuity in the lattice. Typical defects are twins and grain boundaries. *Twinning* represents a change in the crystal orientation across a plane. A grain boundary is a transition between crystals having no particular orientational relationship to one another. Such defects appear during crystal growth. Another area defect is the *stacking fault*.[9] In this defect, the stacking sequence of an atomic layer is interrupted. In Figure 2.17 the sequence of atoms in the stack is *ABCABC* .... If a part of layer *C* is missing, the defect is called an intrinsic stacking fault (Fig. 2.17*a*). If an extra plane *A* is inserted between layers *B* and *C*, it is an extrinsic stacking fault (Fig. 2.17*b*). Such defects may appear during crystal growth. Crystals having these area defects are not usable for integrated circuit manufacture and are discarded.

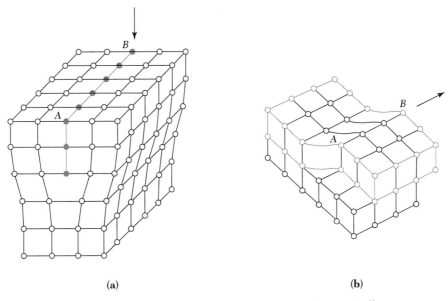

Figure 2.16 (a) Edge and (b) screw dislocation formation in cubic crystals.[10]

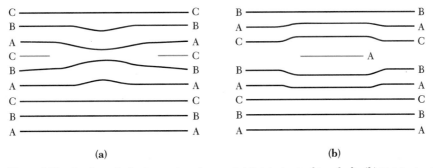

Figure 2.17 Stacking faults in semiconductor. (a) Intrinsic stacking fault. (b) Extrinsic stacking fault.[9]

Precipitates of impurities or dopant atoms make up the fourth class of defects: *volume defects*. These defects arise because of the inherent solubility of the impurity in the host lattice. There is a specific concentration of impurity that the host lattice can accept in a solid solution of itself and the impurity. Figure 2.18 shows solubility versus temperature for a variety of elements in silicon.[11] The solubility of most impurities decreases with decreasing temperature. Thus, if an impurity is introduced to the maximum concentration allowed by its solubility at a given temperature, and the crystal is then cooled to a lower temperature, the crystal can only achieve an equilibrium state by precipitating the impurity atoms in excess of the solubility level. However, the volume mismatch between the host lattice and the precipitates results in dislocations.

Material Properties

Table 2.4 compares silicon characteristics and the requirements for integrated circuit technology having more than 10^7 components, which is referred to as *ultralarge-scale integration* (ULSI).[12,13] The semiconductor material properties listed in Table 2.4 can be measured by various methods. The resistivity is measured by the *four-point probe* method,[14]

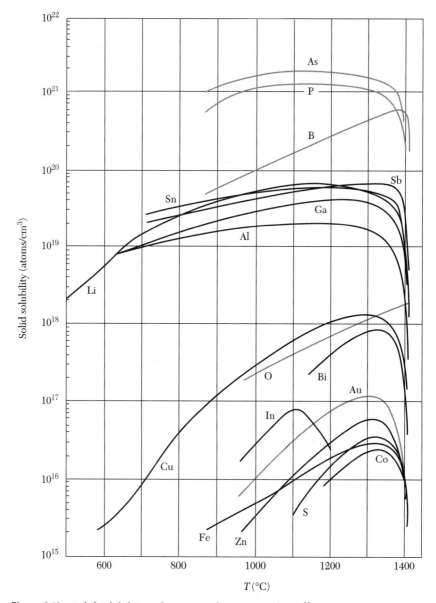

Figure 2.18 Solid solubilities of impurity elements in silicon.[11]

and trace impurities such as oxygen and carbon in silicon can be analyzed by the secondary ion mass spectroscopic (SIMS) technique described in Chapter 6. Note that although current capabilities can meet most of the wafer specifications listed in Table 2.3, many improvements are needed to satisfy the stringent requirements for ULSI technology.[13]

The oxygen and carbon concentrations are substantially higher in Czochralski crystals than in float-zone crystals because of the dissolution of oxygen from the silica crucible and transport of carbon to the melt from the graphite susceptor during crystal growth. Typical carbon concentrations range from 10^{16} to about 10^{17} atoms/cm^3, and carbon atoms in silicon occupy substitutional lattice sites. The presence of carbon is undesirable because

TABLE 2.4 **Comparison of Silicon Material Characteristics and Requirements for ULSI**

Property	Characteristics		Requirements for ULSI
	Czochralski	Float Zone	
Resistivity (phosphorus) n-type (ohm-cm)	1–50	1–300 and up	5–50 and up
Resistivity (antimony) n-type (ohm-cm)	0.005–10	—	0.001–0.02
Resistivity (boron) p-type (ohm-cm)	0.005–50	1–300	5–50 and up
Resistivity gradient (four-point probe) (%)	5–10	20	< 1
Minority carrier lifetime (μs)	30–300	50–500	300–1000
Oxygen (ppma)	5–25	Not detected	Uniform and controlled
Carbon (ppma)	1–5	0.1–1	< 0.1
Dislocation (before processing) (per cm^2)	≤ 500	≤ 500	≤ 1
Diameter (mm)	Up to 200	Up to 100	Up to 300
Slice bow (μm)	≤ 25	≤ 25	< 5
Slice taper (μm)	≤ 15	≤ 15	< 5
Surface flatness (μm)	≤ 5	≤ 5	< 1
Heavy-metal impurities (ppba)	≤ 1	≤ 0.01	< 0.001

ppma, parts per million atoms; ppba, parts per billion atoms.

it aids the formation of defects. Typical oxygen concentrations range from 10^{17} to 10^{18} atoms/cm^3. Oxygen, however, has both deleterious and beneficial effects. It can act as a donor, distorting the resistivity of the crystal caused by intentional doping. On the other hand, oxygen in an interstitial lattice site can increase the yield strength of silicon.

In addition, the precipitates of oxygen due to the solubility effect can be used for gettering. *Gettering* is a general term meaning a process that removes harmful impurities or defects from the region in a wafer where devices are fabricated. When the wafer is subjected to high-temperature treatment (e.g., 1050°C in N$_2$), oxygen evaporates from the surface. This lowers the oxygen content near the surface. The treatment creates a defect-free (or *denuded*) zone for device fabrication, as shown in the inset of Figure 2.19.[1] Additional thermal cycles can be used to promote the formation of oxygen precipitates in the interior of the wafer for gettering of impurities. The depth of the defect-free zone depends on the time and temperature of the thermal cycle and on the diffusivity of oxygen in silicon. Measured results for the denuded zone are shown in Figure 2.19.[1] It is possible to obtain Czochralski crystals of silicon that are virtually free of dislocations.

Commercial melt-grown materials of gallium arsenide are heavily contaminated by the crucible. However, for photonic applications, most requirements call for heavily doped materials (between 10^{17} and 10^{18} cm^{-3}). For integrated circuits or for discrete MESFET devices, undoped gallium arsenide can be used as the starting material with a resistivity of 10^9 Ω-cm. Oxygen is an undesirable impurity in GaAs because it can form a deep donor level, which contributes to a trapping charge in the bulk of the substrate and increases its resistivity. Oxygen contamination can be minimized by using graphite crucibles for melt growth. The dislocation content for Czochralski-grown gallium arsenide crystals is about two orders of magnitude higher than that for silicon. For Bridgman GaAs crystals, the dislocation density is about an order of magnitude lower than that for Czochralski-grown GaAs crystals.

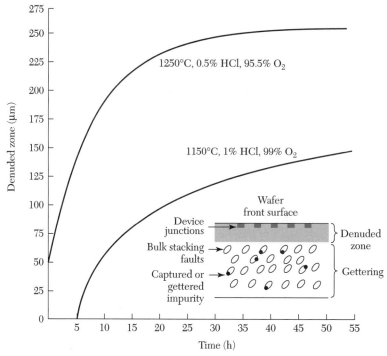

Figure 2.19   Denuded zone width for two sets of processing conditions. Inset shows a schematic of the denuded zone and gettering sites in a wafer cross section.[1]

▶ 2.5 SUMMARY

Several techniques are available to grow single crystals of silicon and gallium arsenide. For silicon crystals, sand (SiO_2) is used to produce polycrystalline silicon, which then serves as the raw material in a Czochralski puller. A seed crystal with the desired orientation is used to grow a large ingot from the melt. Over 90% of silicon crystals are prepared by this technique. During crystal growth, the dopant in the crystal will redistribute. A key parameter is the segregation coefficient, that is, the ratio of the dopant concentration in the solid to that in the melt. Since most of the coefficients are less than 1, the melt becomes progressively enriched with the dopant as the crystal grows.

Another growth technique for silicon is the float-zone process. It offers lower contamination than that normally obtained from the Czochralski technique. Float-zone crystals are used mainly for high-power, high-voltage devices where high-resistivity materials are required.

To make GaAs, chemically pure gallium and arsenic are used as the starting materials that are synthesized to form polycrystalline GaAs. Single crystals of GaAs can be grown by the Czochralski technique. However, a liquid encapsulant (e.g., B_2O_3) is required to prevent decomposition of GaAs at the growth temperature. Another technique is the Bridgman process, which uses a two-zone furnace for gradual solidification of the melt.

After a crystal is grown, it usually goes through wafer-shaping operations to give an end product of highly polished wafers with a specified diameter, thickness, and surface orientation. For example, 200-mm silicon wafers for a MOSFET fabrication line should have a diameter of 200 ± 1 mm, a thickness of 0.725 ± 0.01 mm, and a surface orientation of (100) ± 1°. Wafers with diameters larger than 200 mm are being manufactured for future integrated circuits. Their specifications are listed in Table 2.3.

A real crystal has defects that influence the electrical, mechanical, and optical properties of the semiconductor. These defects are point defects, line defects, area defects, and volume defects. This chapter also discussed means to minimize such defects. For the more demanding ULSI applications, the dislocation density must be less than 1 per square centimeter. Other important requirements are listed in Table 2.4.

▶ REFERENCES

1. C. W. Pearce, "Crystal Growth and Wafer Preparation" and "Epitaxy," in S. M. Sze, Ed., *VLSI Technology*, McGraw-Hill, New York, 1983.

2. T. Abe, "Silicon Crystals for Giga-Bit Scale Integration," in T. S. Moss, Ed., *Handbook on Semiconductors*, Vol. 3, Elsevier Science B. V., Amsterdam/New York, 1994.

3. W. R. Runyan, *Silicon Semiconductor Technology*, McGraw-Hill, New York, 1965.

4. W. G. Pfann, *Zone Melting*, 2nd Ed., Wiley, New York, 1966.

5. E. W. Hass and M. S. Schnoller, "Phosphorus Doping of Silicon by Means of Neutron Irradiation," *IEEE Trans. Electron Devices*, **ED-23,** 803 (1976).

6. M. Hansen, *Constitution of Binary Alloys*, McGraw-Hill, New York, 1958.

7. S. K. Ghandhi, *VLSI Fabrication Principles*, Wiley, New York, 1983.

8. J. R. Arthur, "Vapor Pressures and Phase Equilibria in the GaAs System," *J. Phys. Chem. Solids*, **28,** 2257 (1967).

9. B. El-Kareh, *Fundamentals of Semiconductor Processing Technology*, Kluwer Academic, Boston, 1995.

10. C. A. Wert and R. M. Thomson, *Physics of Solids*, McGraw-Hill, New York, 1964.

11. F. A. Trumbore, "Solid Solubilities of Impurity Elements in Germanium and Silicon," *Bell Syst. Tech. J.*, **39,** 205 (1960); R. Hull, *Properties of Crystalline Silicon*, INSPEC, London, 1999.

12. Y. Matsushita, "Trend of Silicon Substrate Technologies for 0.25 μm Devices," *Proc. VLSI Technol. Workshop*, Honolulu (1996).

13. *The International Technology Roadmap for Semiconductors*, Semiconductor Industry Association, San Jose, CA, 2001.

14. W. F. Beadle, J. C. C. Tsai, and R. D. Plummer, Eds., *Quick Reference Manual for Engineers*, Wiley, New York, 1985.

▶ PROBLEMS

SECTION 2.1: SILICON CRYSTAL GROWTH FROM THE MELT

1. Plot the doping distribution of arsenic at distances of 10, 20, 30, 40, and 45 cm from the seed in a silicon ingot 50 cm long that has been pulled from a melt with an initial doping concentration of 10^{17} cm^{-3}.

2. In silicon, the lattice constant is 5.43 Å. Assume a hard-sphere model. (a) Calculate the radius of a silicon atom. (b) Determine the density of silicon atoms in atoms/cm^3. (c) Use the Avogadro constant to find the density of silicon.

3. Assuming that a 10-kg pure silicon charge is used, what is the amount of boron that must be added to get the boron-doped silicon to have a resistivity of 0.01 Ω-cm when one-half of the ingot is grown?

4. A silicon wafer 1 mm thick having a diameter of 200 mm contains 5.41 mg of boron uniformly distributed in substitutional sites. Find (a) the boron concentration in atoms/cm^3 and (b) the average distance between boron atoms.

5. The seed crystal used in the Czochralski process is usually necked down to a small diameter (5.5 mm) as a means to initiate dislocation-free growth. If the critical-yield strength of silicon is 2×10^6 g/cm^2, calculate the maximum length of a silicon ingot 200 mm in diameter that can be supported by such a seed.

6. Plot the curve of the C_s/C_0 value for $k_0= 0.05$ in the Czochralski technique.

7. A Czochralski-grown crystal is doped with boron. Why is the boron concentration larger at the tail end of the crystal than at the seed end?

8. Why is the impurity concentration larger in the center of the wafer than at its perimeter?

SECTION 2.2: SILICON FLOAT-ZONE PROCESS

9. We use the float-zone process to purify a silicon ingot that contains a uniform gallium concentration of 5×10^{16} cm^{-3}. One pass is made with a molten zone 2 cm long. Over what distance is the resulting gallium concentration below 5×10^{15} cm^{-3}?

10. From Eq. 18 find the C_s/C_0 value at $x/L = 1$ and 2 with $k_e = 0.3$.

11. If $p+$-n abrupt-junction diodes are fabricated using the silicon materials shown in Figure 2.9, find the percentage change of breakdown voltages for the conventionally doped silicon and the neutron-irradiated silicon.

SECTION 2.3: GAAS CRYSTAL GROWTH TECHNIQUES

12. From Figure 2.10, if C_m is 20%, what fraction of the liquid will remain at T_b?

13. From Figure 2.11, explain why the GaAs liquid always becomes gallium rich.

SECTION 2.4: MATERIAL CHARACTERIZATION

14. The equilibrium density of vacancy n_s is given by $N \exp(-E_s/kT)$, where N is the density of semiconductor atoms and E_s is the energy of formation. Calculate n_s in silicon at 27°C, 900°C, and 1200°C. Assume $E_s = 2.3$ eV.

15. Assume the energy of formation (E_f) of a Frenkel-type defect to be 1.1 eV and estimate the defect density at 27°C and 900°C. The equilibrium density of Frenkel-type defects is given by

$$n_f = \sqrt{NN'}\, e^{-E_f/2kT}$$

where N is the atomic density of silicon (cm^{-3}), and N' is the density of available interstitial sites (cm^{-3}) and is represented by $N' = 1 \times 10^{27}\, e^{-3.8(eV)/kT}$ cm^{-3}.

16. How many chips of area 400 mm^2 can be placed on a wafer 300 mm in diameter? Explain your assumptions regarding the chip shape and unused wafer perimeter.

CHAPTER 3

Silicon Oxidation

Many different kinds of thin films are used to fabricate discrete devices and integrated circuits, including thermal oxides, dielectric layers, polycrystalline silicon, and metal films. Figure 3.1 shows a schematic view of a conventional silicon n-channel MOSFET that uses all four groups of films. The first important thin film from the thermal oxide group is the gate oxide layer, under which a conducting channel can be formed between the source and the drain. A related layer is the field oxide, which provides isolation from other devices. Both gate and field oxides generally are grown by a thermal oxidation process because only thermal oxidation can provide the highest-quality oxides having the lowest interface trap densities.

This chapter covers the following topics:

- The thermal oxidation process used to form silicon dioxide (SiO_2)
- Impurity redistribution during oxidation
- Material properties and thickness measurement techniques for SiO_2 films

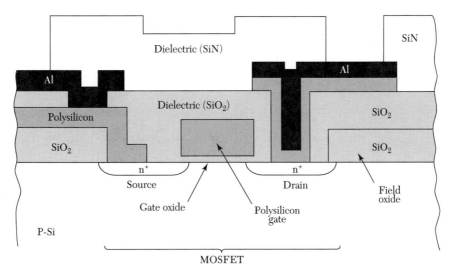

Figure 3.1 Schematic cross section of a metal-oxide-semiconductor field-effect transistor (MOSFET).

▶ 3.1 THERMAL OXIDATION PROCESS

Semiconductors can be oxidized by various methods. These include thermal oxidation, electrochemical anodization, and plasma-enhanced chemical vapor deposition (PECVD, see Chapter 8). Among these methods, thermal oxidation is by far the most important for silicon devices. It is a key process in modern silicon integrated circuit technology. For gallium arsenide, however, thermal oxidation results in generally nonstoichiometric films. The oxides provide poor electrical insulation and semiconductor surface protection; hence, these oxides are rarely used in gallium arsenide technology. Consequently, this chapter concentrates on the thermal oxidation of silicon.

The basic thermal oxidation apparatus is shown in Figure 3.2.[1] The reactor consists of a resistance-heated furnace, a cylindrical fused-quartz tube containing the silicon wafers held vertically in a slotted quartz boat, and a source of either pure dry oxygen or pure water vapor. The loading end of the furnace tube protrudes into a vertical flow hood where a filtered flow of air is maintained. Flow is directed as shown by the arrow in Figure 3.2. The hood reduces dust and particulate matter in the air surrounding the wafers and minimizes contamination during wafer loading. The Oxidation temperature is generally in the range of 900°C to 1200°C, and the typical gas flow rate is about 1 L/min. The oxidation system uses microprocessors to regulate the gas flow sequence, to control the automatic insertion and removal of silicon wafers, to ramp the temperature up (i.e., to increase the furnace temperature linearly) from a low temperature to the oxidation temperature so that the wafers will not warp due to sudden temperature change, to maintain the oxidation temperature to within ±1°C, and to ramp the temperature down when oxidation is completed.

3.1.1 Kinetics of Growth

The following chemical reactions describe the thermal oxidation of silicon in oxygen (dry oxidation) or water vapor (wet oxidation):

$$Si \text{ (solid)} + O_2 \text{ (gas)} \rightarrow SiO_2 \text{ (solid)} \tag{1}$$

$$Si \text{ (solid)} + 2H_2O \text{ (gas)} \rightarrow SiO_2 \text{ (solid)} + 2H_2 \text{ (gas)} \tag{2}$$

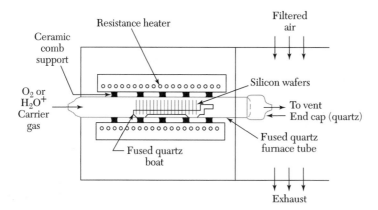

Figure 3.2 Schematic cross section of a resistance-heated oxidation furnace.

The silicon–silicon dioxide interface moves into the silicon during the oxidation process. This creates a fresh interface region, with surface contamination on the original silicon ending up on the oxide surface. The densities and molecular weights of silicon and silicon dioxide are used in the following example to show that growing an oxide of thickness x consumes a layer of silicon $0.44x$ thick (Fig. 3.3).

EXAMPLE 1

If a silicon oxide layer of thickness x is grown by thermal oxidation, what is the thickness of silicon being consumed? The molecular weight of Si is 28.9 g/mol, and the density of Si is 2.33 g/cm^3. The corresponding values for SiO$_2$ are 60.08 g/mol and 2.21 g/cm^3.

SOLUTION The volume of 1 mol of silicon is

$$\frac{\text{Molecular weight of Si}}{\text{Density of Si}} = \frac{28.9 \text{ g/mol}}{2.33 \text{ g/cm}^3} = 12.06 \text{ cm}^3/\text{mol}$$

The volume of 1 mol of silicon dioxide is

$$\frac{\text{Molecular weight of SiO}_2}{\text{Density of SiO}_2} = \frac{60.08 \text{ g/mol}}{2.21 \text{ g/cm}^3} = 27.18 \text{ cm}^3/\text{mol}$$

Since 1 mol of silicon is converted to 1 mol of silicon dioxide,

$$\frac{\text{Thickness of Si} \times \text{area}}{\text{Thickness of SiO}_2 \times \text{area}} = \frac{\text{Volume of 1 mol of Si}}{\text{Volume of 1 mol of SiO}_2}$$

$$\frac{\text{Thickness of Si}}{\text{Thickness of SiO}_2} = \frac{12.06}{27.18} = 0.44$$

Thickness of silicon = 0.44 (thickness of SiO$_2$). For example, to grow a silicon dioxide layer of 100 nm, a layer of 44 nm of silicon is consumed. ◀

The basic structural unit of thermally grown silicon dioxide is a silicon atom surrounded tetrahedrally by four oxygen atoms, as illustrated in Figure 3.4a.[1] The silicon-to-oxygen internuclear distance is 1.6 Å, and the oxygen-to-oxygen internuclear distance is 2.27 Å. These tetrahedra are joined at their corners by oxygen bridges in a variety of ways to form the various phases or structures of silicon dioxide (also called *silica*). Silica has several crystalline structures (e.g., quartz) and an amorphous structure. When silicon is thermally oxidized, the silicon dioxide structure is amorphous. Typically, amorphous silica has a density of 2.21 g/cm^3, compared with 2.65 g/cm^3 for quartz.

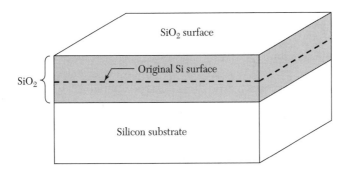

Figure 3.3   Growth of silicon dioxide by thermal oxidation.

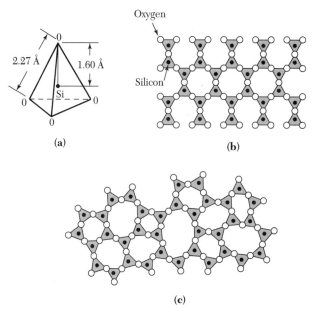

Figure 3.4 (*a*) Basic structural unit of silicon dioxide. (*b*) Two-dimensional representation of a quartz crystal lattice. (*c*) Two-dimensional representation of the amorphous structure of silicon dioxide.[1]

The basic difference between the crystalline and amorphous structures is that the former is a periodic structure, extending over many molecules, whereas the latter has no periodic structure at all. Figure 3.4*b* is a two-dimensional schematic diagram of a quartz crystalline structure made up of rings with six silicon atoms. Figure 3.4*c* is a two-dimensional schematic diagram of an amorphous structure shown for comparison. In the amorphous structure, there is still a tendency to form characteristic rings with six silicon atoms. Note that the amorphous structure in Figure 3.4*c* is quite open because only 43% of the space is occupied by silicon dioxide molecules. The relatively open structure accounts for the lower density and allows a variety of impurities (such as sodium) to enter and diffuse readily through the silicon dioxide layer.

The kinetics of thermal oxidation of silicon can be studied based on a simple model illustrated in Figure 3.5.[2] A silicon slice contacts the oxidizing species (oxygen or water vapor), resulting in a surface concentration of C_0 molecules/cm³ for these species. The magnitude of C_0 equals the equilibrium bulk concentration of the species at the oxidation temperature. The equilibrium concentration generally is proportional to the partial pressure of the oxidant adjacent to the oxide surface. At 1000°C and at a pressure of 1 atm, the concentration C_0 is 5.2×10^{16} molecules/cm³ for dry oxygen and 3×10^{19} molecules/cm³ for water vapor.

The oxidizing species diffuses through the silicon dioxide layer, resulting in a concentration C_s at the surface of silicon. The flux F_1 can be written as

$$F_1 = D\frac{dC}{dx} \cong \frac{D(C_0 - C_s)}{x} \tag{3}$$

where D is the diffusion coefficient of the oxidizing species, and x is the thickness of the oxide layer already present.

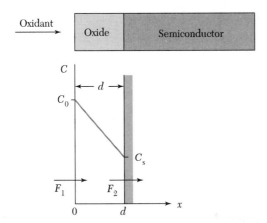

Figure 3.5 Basic model for the thermal oxidation of silicon.[2]

At the silicon surface, the oxidizing species reacts chemically with silicon. Assuming the rate of reaction is proportional to the concentration of the species at the silicon surface, the flux F_2 is given by

$$F_2 = \kappa C_s \tag{4}$$

where κ is the surface reaction rate constant for oxidation. At the steady state, $F_1 = F_2 = F$. Combining Eqs. 3 and 4 gives

$$F = \frac{DC_0}{x + (D/\kappa)} \tag{5}$$

The reaction of the oxidizing species with silicon forms silicon dioxide. Let C_1 be the number of molecules of the oxidizing species in a unit volume of the oxide. There are 2.2×10^{22} silicon dioxide molecules/cm^3 in the oxide, and we add one oxygen molecule (O_2) to each silicon dioxide molecule, whereas we add two water molecules (H_2O) to each SiO_2 molecule. Therefore, C_1 for oxidation in dry oxygen is 2.2×10^{22} cm^{-3}, and for oxidation in water vapor it is twice this number (4.4×10^{22} cm^{-3}). Thus, the growth rate of the oxide layer thickness is given by

$$\frac{dx}{dt} = \frac{F}{C_1} = \frac{DC_0/C_1}{x + (D/\kappa)} \tag{6}$$

We can solve this differential equation subject to the initial condition, $x(0) = d_0$, where d_0 is the initial oxide thickness; d_0 can also be regarded as the thickness of oxide layer grown in an earlier oxidation step. Solving Eq. 6 yields the general relationship for the oxidation of silicon,

$$x^2 + \frac{2D}{\kappa}x = \frac{2DC_0}{C_1}(t + \tau) \tag{7}$$

where $\tau \equiv (d_0^2 + 2Dd_0/\kappa)C_1/2DC_0$, which represents a time coordinate shift to account for the initial oxide layer d_0.

The oxide thickness after an oxidizing time t is given by

$$x = \frac{D}{\kappa}\left[\sqrt{1 + \frac{2C_0\kappa^2(t + \tau)}{DC_1}} - 1\right] \tag{8}$$

For small values of t, Eq. 8 reduces to

$$x \cong \frac{C_0 \kappa}{C_1}(t + \tau) \tag{9}$$

and for larger values of t, it reduces to

$$x \cong \sqrt{\frac{2DC_0}{C_1}(t + \tau)} \tag{10}$$

During the early stages of oxide growth, when surface reaction is the rate-limiting factor, the oxide thickness varies linearly with time. As the oxide layer becomes thicker, the oxidant must diffuse through the oxide layer to react at the silicon–silicon dioxide interface, and the reaction becomes diffusion limited. The oxide growth then becomes proportional to the square root of the oxidizing time, which results in a parabolic growth rate.

Equation 7 is often written in a more compact form:

$$x^2 = Ax = B(t + \tau) \tag{11}$$

where $A = 2D/\kappa$, $B = 2DC_0/C_1$, and $B/A = \kappa C_0/C_1$. Using this form, Eqs. 9 and 10 can be written as

$$x = \frac{B}{A}(t + \tau) \tag{12}$$

for the linear region and as

$$x^2 = B(t + \tau) \tag{13}$$

for the parabolic region. For this reason, the term B/A is referred to as the *linear rate constant,* and B is the *parabolic rate constant*. Experimentally measured results agree with the predictions of this model over a wide range of oxidation conditions. For wet oxidation, the initial oxide thickness d_0 is very small, or $\tau \cong 0$. However, for dry oxidation, the extrapolated value of d_0 at $t = 0$ is about 25 nm. Thus, the use of Eq. 11 for dry oxidation on bare silicon requires a value for τ that can be generated using this initial thickness. Table 3.1 lists the values of the rate constants for wet oxidation of silicon, and Table 3.2 lists the values for dry oxidation.

The temperature dependence of the linear rate constant B/A is shown in Figure 3.6 for both dry and wet oxidation and for (111)- and (100)-oriented silicon wafers.[2] The linear rate constant varies as $\exp(-E_a/kT)$, where the activation energy E_a is about 2 eV for both dry and wet oxidation. This closely agrees with the energy required to break silicon–silicon bonds, 1.83 eV/molecule. Under a given oxidation condition, the linear rate constant depends on crystal orientation. This is because the rate constant is related

TABLE 3.1   Rate Constants for Wet Oxidation of Silicon

Oxidation Temperature (°C)	A (μm)	Parabolic Rate Constant B (μm²/h)	Linear Rate Constant B/A (μm/h)	τ (h)
1200	0.05	0.720	14.40	0
1100	0.11	0.510	4.64	0
1000	0.226	0.287	1.27	0
920	0.50	0.203	0.406	0

TABLE 3.2 Rate Constants for Dry Oxidation of Silicon

Oxidation Temperature (°C)	A (μm)	Parabolic Rate Constant B (μm²/h)	Linear Rate Constant B/A (μm/h)	τ (h)
1200	0.040	0.045	1.12	0.027
1100	0.090	0.027	0.30	0.076
1000	0.165	0.0117	0.071	0.37
920	0.235	0.0049	0.0208	1.40
800	0.370	0.0011	0.0030	9.0
700	…	…	0.00026	81.0

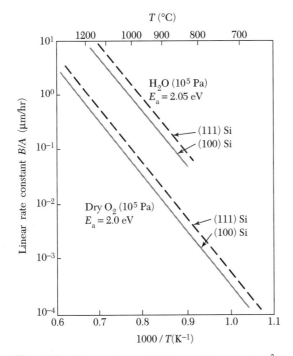

Figure 3.6 Linear rate constant versus temperature.[2]

to the rate of incorporation of oxygen atoms into the silicon. The rate depends on the surface bond structure of silicon atoms, making it orientation dependent. Because the density of available bonds on the (111) plane is higher than that on the (100) plane, the linear rate constant for (111) silicon is larger.

Figure 3.7 shows the temperature dependence of the parabolic rate constant B, which can also be described by $\exp(-E_a/kT)$. The activation energy E_a is 1.24 eV for dry oxidation. The comparable activation energy for oxygen diffusion in fused silica is 1.18 eV. The corresponding value for wet oxidation, 0.71 eV, compares favorably with the value of 0.79 eV for the activation energy of diffusion of water in fused silica. The parabolic rate constant is independent of crystal orientation. This independence is expected because it is a measure of the diffusion process of the oxidizing species through a random network layer of amorphous silica.

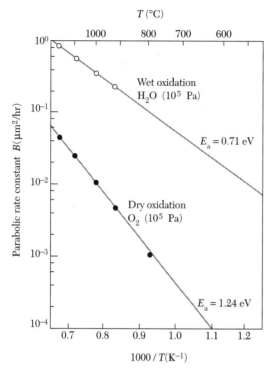

Figure 3.7 Parabolic rate constant versus temperature.[2]

Although oxides grown in dry oxygen have the best electrical properties, considerably more time is required to grow the same oxide thickness at a given temperature in dry oxygen than in water vapor. For relatively thin oxides such as the gate oxide in a MOSFET (typically ≤ 20 nm), dry oxidation is used. However, for thicker oxides such as field oxides (≥ 20 nm) in MOS integrated circuits, and for bipolar devices, oxidation in water vapor (or steam) is used to provide both adequate isolation and passivation.

Figure 3.8 shows the experimental results of silicon dioxide thickness as a function of reaction time and temperature for two substrate orientations.[3] Under a given oxidation condition, the oxide thickness grown on a (111) substrate is larger than that grown on a (100) substrate because of the larger linear rate constant of the (111) orientation. Note that for a given temperature and time, the oxide film obtained using wet oxidation is about 5 to 10 times thicker than that using dry oxidation.

EXAMPLE 2

A silicon sample is oxidized in dry O_2 at 1200°C for 1 hour. (a) What is the thickness of the oxide grown? (b) How much additional time is required to grow 0.1 μm more oxide in wet O_2 at 1200°C?

SOLUTION (a) From Table 3.2, the values of the rate constants for dry O_2 at 1200°C are

$$A = 0.04 \ \mu m \qquad B = 0.045 \ \mu m^2/h$$

and $\tau = 0.027$ h. Using these parameters in Eq. 11, we obtain an oxide thickness of

$$x = 0.196 \ \mu m$$

(b) From Table 3.1, the values of the rate constants for wet O_2 at 1200°C are

$$A = 0.05 \ \mu m \qquad B = 0.72 \ \mu m^2/h$$

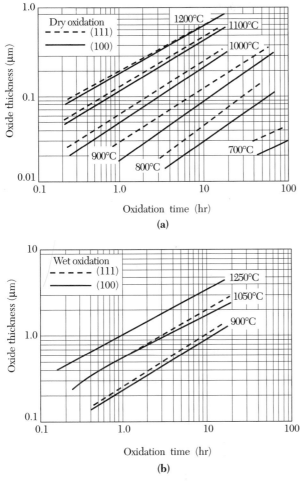

Figure 3.8 Experimental results of silicon dioxide thickness as a function of reaction time and temperature for two substrate orientations. (*a*) Growth in dry oxygen. (*b*) Growth in steam.[3]

Since $d_0 = 0.196$ μm from the first step, we have

$$\tau \equiv \left(d_0^2 + 2Dd_0/\kappa\right)C_1/2DC_0 = \frac{d_0^2 + Ad_0}{B} = 0.067 \text{ h}$$

The final desired thickness is $x = d_0 + 0.1$ μm $= 0.296$ μm. Using these parameters in Eq. 11, we obtain an additional oxidation time of

$$t = 0.76 \quad \text{h} = 4.53 \text{ min}$$

◀

3.1.2 Thin Oxide Growth

Relatively slow growth rates must be used to reproducibly grow thin oxide films of precise thickness. Various approaches to achieve such slow growth rates have been reported, including growth in dry O_2 at atmospheric pressure and lower temperatures (800°C to 900°C); growth at pressures lower than atmospheric pressure; growth in reduced partial pressures of O_2 by using a diluent inert gas, such as N_2, Ar, or He, together with the gas containing the oxidizing species; and the use of composite oxide films, with the gate oxide films consisting of a layer of thermally grown SiO_2 and an overlayer of SiO_2 grown by chemical vapor deposition (CVD). However, the mainstream approach for gate oxides 10 to 15 nm thick is to grow the oxide film at atmospheric pressure and lower temperatures (800°C to 900°C). With this approach, processing using modern *vertical* oxidation furnaces can grow reproducible, high-quality 10-nm oxides to within 0.1 nm across the wafer.

It was noted earlier that for dry oxidation, there is an apparently rapid oxidation that gives rise to an initial oxide thickness d_0 of about 20 nm. Therefore, the simple model presented in Section 3.1.1 is not valid for dry oxidation with an oxide thickness less than or equal to 20 nm. For ULSI, the ability to grow thin (~5–20 nm), uniform, high-quality reproducible gate oxides has become increasingly important. This section briefly considers the growth mechanisms of such thin oxides.

In the early stage of growth in dry oxidation, there is a large compressive stress in the oxide layer that reduces the oxygen diffusion coefficient in the oxide. As the oxide becomes thicker, the stress is reduced because of the viscous flow of silica, and the diffusion coefficient will approach its stress-free value. Therefore, for thin oxides, the value of D/κ may be sufficiently small that we can neglect the term Ax in Eq. 11 and obtain

$$x^2 - d_0^2 = Bt \tag{14}$$

where d_0 is equal to $\sqrt{2DC_0\tau/C_1}$, which is the initial oxide thickness when time is extrapolated to zero, and B is the parabolic rate constant defined previously. We therefore expect the initial growth in dry oxidation to follow a parabolic form.

▶ 3.2 IMPURITY REDISTRIBUTION DURING OXIDATION

Dopant impurities near the silicon surface will be redistributed during thermal oxidation. The redistribution depends on several factors. When two solid phases are brought together, an impurity in one solid will redistribute between the two solids until it reaches equilibrium. This is similar to the previous discussion in Chapter 2 on impurity redistribution in crystal growth from the melt. The ratio of the equilibrium concentration of the impurity in the silicon to that in the silicon dioxide is called the *segregation coefficient* and is defined as

$$k = \frac{\text{Equilibrium concentration of impurity in silicon}}{\text{Equilibrium concentration of impurity in SiO}_2} \tag{15}$$

A second factor that influences impurity distribution is that the impurity may diffuse rapidly through the silicon dioxide and escape to the gaseous ambient. If the diffusivity of the impurity in silicon dioxide is large, this factor will be important. A third factor in the redistribution process is that the oxide is growing, and thus the boundary between the silicon and the oxide is advancing into the silicon as a function of time. The relative rate of this advance compared with the diffusion rate of the impurity through the oxide is important in determining the extent of the redistribution. Note that even if the segregation coefficient of an impurity equals unity, some redistribution of the impurity in the silicon will still take place. As indicated in Figure 3.3, the oxide layer will be about twice as thick as the silicon layer it replaced. Therefore, the same amount of impurity will now be distributed in a larger volume, resulting in depletion of the impurity from the silicon.

Four possible redistribution processes are illustrated in Figure 3.9.[4] These processes can be classified into two groups. In one group, the oxide takes up the impurity (Figs. 3.9a and b for $k < 1$), and in the other the oxide rejects the impurity (Fig. 3.9c and d for $k > 1$). In each case, what happens depends on how rapidly the impurity can diffuse through the oxide. In group 1, the silicon surface is depleted of impurities; an example is boron, with k approximately equal to 0.3. Rapid diffusion of the impurity through the silicon dioxide

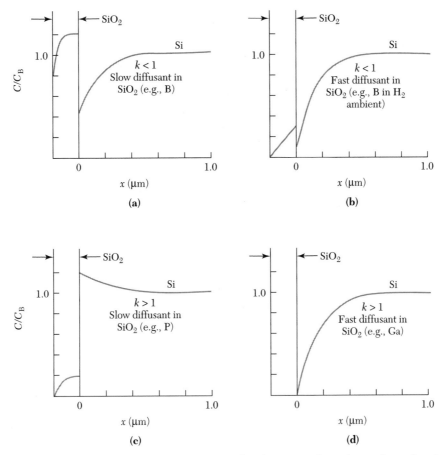

Figure 3.9 Four different cases of impurity redistribution in silicon due to thermal oxidation.[4]

increases the amount of depletion; an example is boron-doped silicon heated in a hydrogen ambient, because hydrogen in silicon dioxide enhances the diffusivity of boron. In group 2, k is greater than unity, so the oxide rejects the impurity. If diffusion of the impurity through the silicon dioxide is relatively slow, the impurity piles up near the silicon surface; an example is phosphorus, with k approximately equal to 10. When diffusion through the silicon dioxide is rapid, so much impurity may escape from the solid to the gaseous ambient that the overall effect will be a depletion of the impurity; an example is gallium, with k approximately equal to 20.

The redistributed dopant impurities in silicon dioxide are seldom electrically active. However, redistribution in silicon has an important effect on processing and device performance. For example, nonuniform dopant distribution will modify the interpretation of the measurements of interface trap properties, and the change of the surface concentration will modify the threshold voltage and device contact resistance.

► 3.3 MASKING PROPERTIES OF SILICON DIOXIDE

A silicon dioxide layer can also provide a selective mask against the diffusion of dopants at elevated temperatures, a very useful property in IC fabrication. Predeposition of dopants (see Chapter 6), whether it be by ion implantation, chemical diffusion, or spin-on techniques, typically results in a dopant source at or near the surface of the oxide. During a subsequent high-temperature drive-in step, diffusion in oxide-masked regions must be slow enough with respect to diffusion in the silicon to prevent dopants from diffusing through the oxide mask to the silicon surface. The required thickness may be determined experimentally by measuring the oxide thickness necessary to prevent the inversion of a lightly doped silicon substrate of opposite conductivity at a particular temperature and time. Typically, oxides used for masking common impurities are 0.5 to 1.0 μm thick.

The values of diffusion constants for various dopants in SiO_2 depend on the concentration, properties, and structure of the oxide. Table 3.3 lists diffusion constants for various common dopants, and Figure 3.10 gives the oxide thickness required to mask boron and phosphorus as a function of diffusion time and temperature. Note that SiO_2 is much more effective for masking boron than phosphorus. Nevertheless, the diffusivities of P, Sb, As, and B in SiO_2 are all orders of magnitude less than their corresponding values in silicon, so they are all compatible with oxide masking. This is not true, however, for Ga or Al. Silicon nitride is used as an alternative masking material for these elements.

TABLE 3.3   Diffusion Constants in SiO$_2$

Dopants	Diffusion Constants at 1100°C (cm²/s)
B	3.4×10^{-17} to 2.0×10^{-14}
Ga	5.3×10^{-11}
P	2.9×10^{-16} to 2.0×10^{-13}
As	1.2×10^{-16} to 3.5×10^{-15}
Sb	9.9×10^{-17}

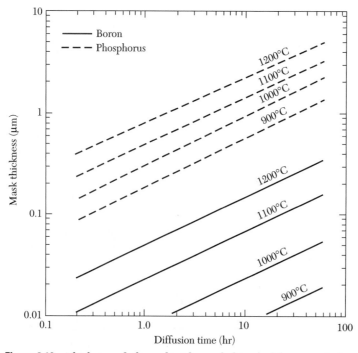

Figure 3.10   Thickness of silicon dioxide needed to mask boron and phosphorus diffusions as a function of diffusion time and temperature.

▶ **3.4 OXIDE QUALITY**

Oxides used for masking are usually grown by wet oxidation. A typical growth cycle consists of a sequence of dry-wet-dry oxidations. Most of the growth in such a sequence occurs in the wet phase, since the SiO_2 growth rate is much higher when water is used as the oxidant. Dry oxidation, however, results in a higher-quality oxide that is denser and has a higher breakdown voltage (5–10 MV/cm). It is for these reasons that the thin gate oxides in MOS devices are usually formed using dry oxidation.

MOS devices are also affected by charges in the oxide and traps at the SiO_2–Si interface. The basic classification of these traps and charges is shown in Figure 3.11. They are the interface trapped charge, fixed oxide charge, oxide trapped charge, and mobile ionic charge.[5]

Interface trapped charges (Q_{it}) are due to the SiO_2–Si interface properties and are dependent on the chemical composition of this interface. The traps are located at the SiO_2–Si interface, with energy states in the silicon forbidden bandgap. The interface trap density (i.e., number of interface traps per unit area and per eV) is orientation dependent. In silicon with a <100> crystal orientation, the interface trap density is about an order of magnitude smaller than that in the <111> orientation. Present-day MOS devices with thermally grown silicon dioxide on silicon have most of the interface trapped charges passivated by low-temperature (450°C) hydrogen annealing (see Chapter 7). The value of Q_{it} for <100>-oriented silicon can be as low as 10^{10} cm^{-2}, which amounts to about one interface trapped charge per 10^5 surface atoms. For <111>-oriented silicon, Q_{it} is about 10^{11} cm^{-2}.

The fixed charge (Q_f) is located within approximately 3 nm of the SiO_2–Si interface. This charge is fixed and very difficult to charge or discharge. Generally, Q_f is positive and depends on oxidation and annealing conditions and on silicon orientation. It has been suggested that when the oxidation is stopped, some ionic silicon is left near the interface. These ions, along with uncompleted silicon bonds (e.g., Si–Si or Si–O bonds) at the surface, may result in the positive interface charge. Q_f can be regarded as a charge sheet

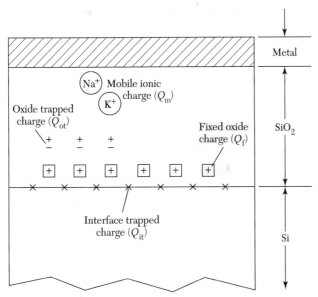

Figure 3.11 Terminology for the charges associated with thermally oxidized silicon.[3]

located at the SiO_2–Si interface. Typical fixed oxide charge densities for a carefully treated SiO_2–Si interface system are about 10^{10} cm^{-2} for a <100> surface and about 5×10^{10} cm^{-2} for a <111> surface. Because of the lower values of Q_{it} and Q_f, the <100> orientation is preferred for silicon MOSFETs.

Oxide trapped charges (Q_{ot}) are associated with defects in the silicon dioxide. These charges can be created, for example, by x-ray radiation or high-energy electron bombardment. The traps are distributed inside the oxide layer. Most process-related Q_{ot} can be removed by low-temperature annealing.

Mobile ionic charges (Q_m), which result from contamination from sodium or other alkali ions, are mobile within the oxide under raised temperatures (e.g., >100°C) and high electric field operations. Trace contamination by alkali metal ions may cause stability problems in semiconductor devices operated under high-bias and high-temperature conditions. Under these conditions mobile ionic charges can move back and forth through the oxide layer and cause threshold voltage shifts. Therefore, special attention must be paid to the elimination of mobile ions in device fabrication. For example, the effects of sodium contamination can be reduced by adding chlorine during oxidation. Chlorine immobilizes the sodium ions. A small amount (6% or less) of anhydrous HCl in the oxidizing gas can accomplish this, but the presence of chlorine during dry oxidation increases both the linear and parabolic rate constants, leading to a higher growth rate.

▶ 3.5 OXIDE THICKNESS CHARACTERIZATION

Perhaps the simplest method for determining the thickness of an oxide is to compare the color of the wafer with a reference color chart, such as the one in Table 3.4.[6] When an oxide-coated wafer is illuminated with white light perpendicular to the surface, the light penetrates the oxide and is reflected by the underlying silicon wafer. Constructive interference leads to enhancement of a certain wavelength of reflected light, and the color of the wafer corresponds to that wavelength. For example, a wafer with a 500-nm silicon dioxide layer will appear blue green.

Clearly, color chart comparisons are subjective and are therefore not the most accurate mechanism for determining oxide thickness. A more accurate measurement can be obtained using techniques such as profilometry or ellipsometry.

Profilometry is a very common method of film thickness measurement. In this technique, a step feature in the grown or deposited film is first created, either by masking during deposition or by etching afterward. The profilometer then drags a fine stylus across the film surface (see Fig. 3.12)[7]. When the stylus encounters a step, a signal variation indicates the step height. This information is then displayed on a chart recorder or CRT screen. Films of thicknesses of less than 100 nm to greater than 5 μm can be measured with this instrument.

Ellipsometry is another widely used measurement technique that is based on the polarization changes that occur when light is reflected from or transmitted through a medium. Changes in polarization are a function of the optical properties of the material (i.e., its complex refractive indices), its thickness, and the wavelength and angle of incidence of the light beam relative to the surface normal. These differences in polarization are measured by an ellipsometer, and the oxide thickness can then be calculated.

▶ 3.6 OXIDATION SIMULATION

As trends toward miniaturization continue and the scale of integration of ICs increases, accurate knowledge of one-, two-, and three-dimensional structural information regarding

TABLE 3.4   Color Chart for Thermally Grown SiO$_2$ Films Observed Perpendicularly under Daylight Fluorescent Lighting

Film Thickness (μm)	Color and Comments	Film Thickness (μm)	Color and Comments
0.05	Tan	0.68	"Bluish" (not blue but borderline between violet and blue green; appears more like a mixture between violet red and blue green and looks grayish)
0.07	Brown		
0.10	Dark violet to red violet		
0.12	Royal blue		
0.15	Light blue to metallic blue		
0.17	Metallic to very light yellow green	0.72	Blue green to green (quite broad)
		0.77	"Yellowish"
0.20	Light gold or yellow; slightly metallic	0.80	Orange (rather broad for orange)
		0.82	Salmon
0.22	Gold with slight yellow orange	0.85	Dull, light red violet
		0.86	Violet
0.25	Orange to melon	0.87	Blue violet
0.27	Red violet	0.89	Blue
0.30	Blue to violet blue	0.92	Blue green
0.31	Blue	0.95	Dull yellow green
0.32	Blue to blue green	0.97	Yellow to "yellowish"
0.34	Light green	0.99	Orange
0.35	Green to yellow green	1.00	Carnation pink
0.36	Yellow green	1.02	Violet red
0.37	Green yellow	1.05	Red violet
0.39	Yellow	1.06	Violet
0.41	Light orange	1.07	Blue violet
0.42	Carnation pink	1.10	Green
0.44	Violet red	1.11	Yellow green
0.46	Red violet	1.12	Green
0.47	Violet	1.18	Violet
0.48	Blue violet	1.19	Red violet
0.49	Blue	1.21	Violet red
0.50	Blue green	1.24	Carnation pink to salmon
0.52	Green (broad)	1.25	Orange
0.54	Yellow green	1.28	"Yellowish"
0.56	Green yellow	1.32	Sky blue to green blue
0.57	Yellow to "yellowish" (not yellow but is in the position where yellow is to be expected; at times appears to be light creamy gray or metallic)	1.40	Orange
		1.45	Violet
		1.46	Blue violet
		1.50	Blue
0.58	Light orange or yellow to pink	1.54	Dull yellow green
0.60	Carnation pink		
0.63	Violet red		

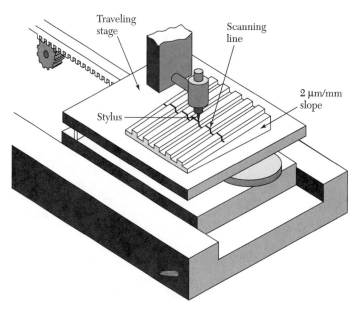

Figure 3.12 Schematic drawing of a surface profilometer.[7]

the features being fabricated becomes more and more critical. However, actual experimental examination and characterization of such structures is both time-consuming and expensive. Therefore, computer simulation is an important tool for investigating ULSI fabrication processes. Sophisticated simulation programs are needed to solve the generalized differential equations that model various processes.

Perhaps the most widely used example of such process simulation software is the Stanford University Process Engineering Modeling (SUPREM) program. SUPREM can accurately simulate multidimensional oxide growth, accounting for the moving Si–SiO$_2$ boundary, impurity segregation during growth, and other physical phenomena. In addition, SUPREM can predict the results of the various deposition, diffusion, epitaxial growth, and ion implantation processes discussed in subsequent chapters.

SUPREM performs oxidation simulations based on the kinetic growth model described in Section 3.1.1. The package incorporates Arrhenius functions to describe the linear and parabolic rate coefficients for wet and dry oxidation, as well as a rudimentary model for chlorinated oxidation. Oxidation is simulated using the command DIFFUSION, with either WETO2 or DRYO2 included as parameters indicating wet or dry oxidation, respectively.[8] SUPREM also requires the specification of process conditions such as times, temperature profiles, and so forth. In the thin oxide regime, SUPREM uses an empirical model of the form[9]

$$\frac{dx}{dt} = \frac{B}{2x + A} + Ce^{-x/L} \tag{16}$$

where B and A are the oxidation rate coefficients, and C and L are empirical constants.

To run SUPREM, an input deck must be provided. This file contains a series of statements and comments. A description of a few commonly used statements is given in Appendix I. The deck begins with a TITLE statement, which is merely a comment repeated on each page of the program output. The next command, INITIALIZE, is a control statement that sets the substrate type, orientation, and doping. This command can also be used to specify the thickness of the region to be simulated and establish a grid. After the substrate and materials are established, a series of statements is used to specify the sequence of process steps as they occur. Finally, the output of the simulation can be printed or plotted using PRINT or PLOT statements, respectively. Simulation ends with a STOP statement. Several COMMENT statements will typically appear throughout the deck. The user is encouraged to use these statements to better facilitate documentation of the process flow. These ideas are illustrated in Example 3.

EXAMPLE 3

Suppose we want to perform a dry-wet-dry oxidation sequence on a <100> silicon wafer at 1100°C for 5 minutes in dry O_2, 2 hours in wet O_2, and finally, for 5 more minutes in dry O_2. If the silicon substrate is doped with phosphorus at a level of 10^{16} cm^{-3}, use SUPREM to determine the final oxidation thickness and the phosphorus doping profile in the oxide and silicon layers.

SOLUTION   The SUPREM input listing is as follows:

```
TITLE         Oxidation Example
COMMENT       Initialize silicon substrate
INITIALIZE    <100> Silicon Phosphor Concentration=1e16
COMMENT       Ramp furnace up to 1100 C over 10 minutes in N2
DIFFUSION     Time=10 Temperature=900 Nitrogen T.rate=20
COMMENT       Oxidize the wafers for 5 minutes at 1100 C in dry O2
DIFFUSION     Time=5 Temperature=1100 DryO2
COMMENT       Oxidize the wafers for 120 minutes at 1100 C in wet O2
DIFFUSION     Time=120 Temperature=1100 WetO2
COMMENT       Oxidize the wafers for 5 minutes at 1000 C in dry O2
DIFFUSION     Time=5 Temperature=1100 DryO2
COMMENT       Ramp furnace down to 900 C over 10 minutes in N2
DIFFUSION     Time=10 Temperature=1100 Nitrogen T.rate=-20
PRINT         Layers Chemical Concentration Phosphor
PLOT          Active Net Cmin=1e14
STOP          End oxidation example
```

We assume the furnace has an idle temperature of 900°C when the wafer is inserted, so we use a 10-minute ramp up at 20°C/minute to 1100°C at the beginning of processing and a 10-minute ramp down at −20°C/minute back to 900°C at the end. The ramp up and ramp down are performed in a nitrogen ambient.

After oxidation is complete, we print and plot the phosphorus concentration as a function of depth into the silicon substrate. The results are shown in Figure 3.13, which indicates a final oxide thickness of 0.909 μm and depicts the phosphorus incorporation in the oxide layer.   ◀

▶ 3.7 SUMMARY

Silicon dioxide is a high-quality insulator that can be thermally grown on silicon wafers. It can also serve as a barrier layer during impurity diffusion or implantation, and it is a key component of MOS devices and circuits. These factors have contributed significantly to silicon's current status as the dominant semiconductor material in use today.

This chapter described the mechanism of thermal oxidation of silicon and presented a kinetic model of oxide growth. This model accurately predicts oxide growth rate for a wide range of process conditions. The chapter also discussed dopant redistribution and the masking properties of oxides. Oxide characterization methods and oxide quality were

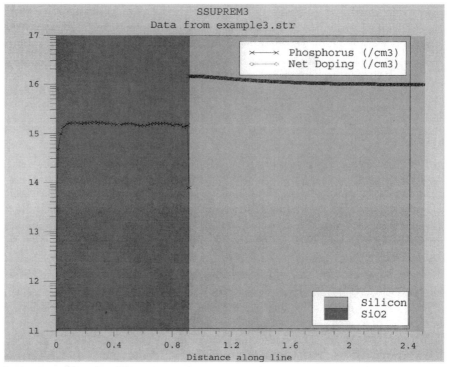

Figure 3.13 Plot of phosphorus concentration as a function of depth into the silicon substrate, using SUPREM software.

described as well. Finally, the process simulation software package SUPREM was introduced. The use of SUPREM, however, is not limited to oxidation, and it will be revisited in subsequent chapters.

▶ REFERENCES

1. E. H. Nicollian and J. R. Brews, *MOS Physics and Technology*, Wiley, New York, 1982.

2. B. E. Deal and A. S. Grove, "General Relationship for the Thermal Oxidation of Silicon," *J. Appl. Phys.*, **36,** 3770 (1965).

3. J. D. Meindl, et al., "Silicon Epitaxy and Oxidation," in F. Van de Wiele, W. L. Engl, and P. O. Jespers, Eds., *Process and Device Modeling for Integrated Circuit Design*, Noorhoff, Leyden, 1977.

4. A. S. Grove, *Physics and Technology of Semiconductor Devices*, Wiley, New York, 1967.

5. B. E. Deal, "Standardized Terminology for Oxide Charge Associated with Thermally Oxidized Silicon," *IEEE Trans. Electron Devices*, **ED-27,** 606 (1980).

6. W. Pliskin and E. Conrad, "Nondestructive Determination of Thickness and Refractive Index of Transparent Films," *IBM J. Res. Develop.*, **8,** 43–51 (1964).

7. S. Wolf and R. Tauber, *Silicon Processing for the VLSI Era*, Lattice Press, Sunset Beach, CA, 2000.

8. *SSUPREM3 User's Manual*, Silvaco International, Santa Clara, CA, 1995.

9. H. Massoud, C. Ho, and J. Plummer, in J. Plummer, Ed., *Computer Aided Design of Integrated Circuit Fabrication Processes for VLSI Devices*, Stanford University Technical Report, Stanford, CA, 1982.

▶ PROBLEMS

Asterisks denote difficult problems.

1. A *p*-type <100>-oriented silicon wafer with a resistivity of 10 Ω-cm is placed in a wet oxidation system to grow a field oxide of 0.45 μm at 1050°C. Determine the time required to grow the oxide.

°2. After the first oxidation as given in Problem 1, a window is opened in the oxide to grow a gate oxide at 1000°C for 20 minutes in dry oxidation. Find the thicknesses of the gate oxide and the total field oxide.

3. Show that Eq. 11 reduces to $x^2 = Bt$ for long times and to $x = B/A(t + \tau)$ for short times.

4. Determine the diffusion coefficient D for dry oxidation of <100>-oriented silicon samples at 980°C and 1 atm.

5. Define *segregation coefficient*.

6. Assume that the Cu concentration in the SiO_2 layer is 5×10^{13} atoms/cm^3 after vapor phase deposition and is measured with atomic absorption spectrometry. The Cu concentration in the Si layer is 3×10^{11} atoms/cm^3 after HF/H_2O_2 dissolution. Calculate the segregation coefficient of Cu in SiO_2/Si layers.

°7. A bare and undoped <100> silicon sample is oxidized for 1 hour at 1100°C in dry O_2. It is then covered and has the oxide removed over half the wafer. Next, it is re-oxidized in wet O_2 at 1000°C for 30 minutes. Use SUPREM to determine the thickness in the two regions. How high are the step on the surface and the step in the substrate?

4

Photolithography

Photolithography is the process of transferring patterns of geometric shapes on a mask to a thin layer of photosensitive material (called *photoresist*) covering the surface of a semiconductor wafer.[1] These patterns define the various regions in an integrated circuit, such as the implantation regions, the contact windows, and the bonding-pad areas. The resist patterns defined by the lithographic process are not permanent elements of the final device, but only replicas of circuit features. To produce circuit features, these resist patterns must be transferred once more into the underlying layers comprising the device. Pattern transfer is accomplished by an etching process that selectively removes unmasked portions of a layer (see Chapter 5).[2] A brief description of pattern transfer was given in Section 1.4.2. The present chapter covers the following topics:

- The importance of a clean room for lithography
- The most widely used lithographic method—optical lithography—and its resolution enhancement techniques
- Advantages and limitations of other lithographic methods

▶ 4.1 OPTICAL LITHOGRAPHY

The vast majority of lithographic equipment for IC fabrication is optical equipment using ultraviolet light (wavelength or $\lambda \cong 0.2$–$0.4\ \mu m$). This section considers the exposure tools, masks, resists, and resolution enhancement techniques used for optical lithography. It also considers the pattern transfer process, which serves as a basis for other lithographic systems. The section first briefly discusses the clean room, because all lithographic processes must be performed in an ultraclean environment.

4.1.1 The Clean Room

An IC fabrication facility requires a clean processing room, especially in the area used for photolithography. The need for such a clean room arises because dust particles in the air can settle on semiconductor wafers and lithographic masks and can cause defects in the devices, which result in circuit failure. For example, a dust particle on a semiconductor surface can disrupt the single-crystal growth of an epitaxial film, causing the formation of dislocations. A dust particle incorporated into the gate oxide can result in enhanced conductivity and cause device failure due to low breakdown voltage. The situation is even more critical in the lithographic area. When dust particles adhere to the surface of a photomask, they behave as opaque patterns on the mask, and these patterns will be transferred to the underlying layer along with the circuit patterns on the mask. Figure 4.1 shows three dust particles on a photomask.[3] Particle 1 may result in the formation of

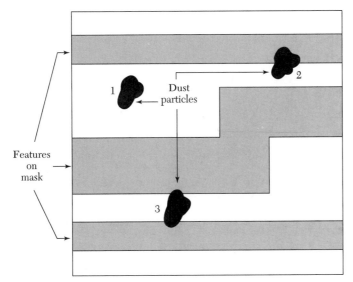

Figure 4.1 Various ways in which dust particles can interfere with photomask patterns.[3]

a pinhole in the underlying layer. Particle 2 is located near a pattern edge and may cause a constriction of current flow in a metal runner. Particle 3 can lead to a short circuit between the two conducting regions and render the circuit useless.

In a clean room, the total number of dust particles per unit volume must be tightly controlled, along with the temperature and humidity. Figure 4.2 shows the particle-size distribution curves for various classes of clean rooms. Two systems are used to define the classes of clean room.[4] In the English system, the numerical designation of the class is taken from the maximum allowable number of particles that are 0.5 μm and larger per cubic foot of air. In the metric system, the class is taken from the logarithm (base 10) of the maximum allowable number of particles that are 0.5 μm and larger per cubic meter. For example, a class 100 clean room (English system) has a dust count of 100 particles/ft^3 with particle diameters of 0.5 μm and larger, whereas a class M 3.5 clean room (metric system) has a dust count of $10^{3.5}$, or about 3500 particles/m^3 with particle diameters of 0.5 μm or larger. Since 100 particles/ft^3 = 3500 particles/m^3, a class 100 in the English system corresponds to a class M 3.5 in the metric system.

Because the number of dust particles increases as particle size decreases, a more stringent control of the clean room environment is required when the minimum feature lengths of ICs are reduced to the deep submicron range. For most IC fabrication areas, a class 100 clean room is required; that is, the dust count must be about four orders of magnitude lower than that of ordinary room air. However, for the lithography area, a class 10 clean room or one with a lower dust count is required.

EXAMPLE 1

If we expose a 200-mm wafer for 1 minute to an air stream under a laminar-flow condition at 30 m/min, how many dust particles will land on the wafer in a class 10 clean room?

SOLUTION For a class 10 clean room, there are 350 particles (0.5 μm and larger) per cubic meter. The air volume that goes over the wafer in 1 minute is

$$(30 \text{ m / min}) \times \pi \left(\frac{0.2 \text{ m}}{2} \right)^2 \times 1 \text{ minute} = 0.942 \text{ m}^3$$

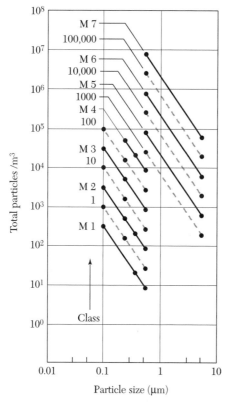

Figure 4.2 Particle-size distribution curve for English (- - -) and metric (—) classes of clean rooms.[4]

The number of dust particles (0.5 μm and larger) contained in the air volume is $350 \times 0.942 = 330$ particles.

Therefore, if there are 400 IC chips on the wafer, the particle count amounts to one particle on each of 82% of the chips. Fortunately, only a fraction of the particles that land adhere to the wafer surface, and of those only a fraction are at a circuit location critical enough to cause a failure. However, the calculation indicates the importance of the clean room. ◀

4.1.2 Exposure Tools

The pattern transfer process is accomplished by using a lithographic exposure tool. The performance of an exposure tool is determined by three parameters: resolution, registration, and throughput. *Resolution* is the minimum feature dimension that can be transferred with high fidelity to a resist film on a semiconductor wafer. *Registration* is a measure of how accurately patterns on successive masks can be aligned (or overlaid) with respect to previously defined patterns on the wafer. *Throughput* is the number of wafers that can be exposed per hour for a given mask level.

There are basically two optical exposure methods: shadow printing and projection printing.[5,6] Shadow printing may have the mask and wafer in direct contact with one another (as in *contact printing*) or in close proximity (as in *proximity printing*). Figure 4.3a shows a basic setup for contact printing, in which a resist-coated wafer is brought into physical contact with a mask, and the resist is exposed by a nearly collimated beam of ultraviolet light through the back of the mask for a fixed time. The intimate contact between resists and mask provides a resolution of approximately 1 μm. However, contact printing suffers a major drawback caused by dust particles. A dust particle or a speck of silicon dust on the wafer can be imbedded into the mask when the mask makes contact with the wafer. The imbedded particle causes permanent damage to the mask and results in defects in the wafer with each succeeding exposure.

To minimize mask damage, the proximity exposure method is used. Figure 4.3*b* shows the basic setup. It is similar to the contact printing method, except that there is a small gap (10–50 μm) between the wafer and the mask during exposure. The small gap results in optical diffraction at feature edges on the photomask; that is, when light passes by the edges of an opaque mask feature, fringes are formed and some light penetrates into the shadow region. As a result, the resolution is degraded to the 2 to 5-μm range.

In shadow printing, the minimum linewidth [or critical dimension (CD)] that can be printed is roughly

$$CD \cong \sqrt{\lambda g} \tag{1}$$

where λ is the wavelength of the exposure radiation and g is the gap between the mask and the wafer and includes the thickness of the resist. For $\lambda = 0.4$ μm and $g = 50$ μm, the CD is 4.5 μm. If we reduce λ to 0.25 μm (a wavelength range of 0.2 to 0.3 μm is in the deep UV spectral region) and g to 15 μm, the CD becomes 2 μm. Thus, there is an advantage in reducing both λ and g. However, for a given distance g, any dust particle with a diameter larger than g potentially can cause mask damage.

To avoid the mask damage problem associated with shadow printing, projection-printing exposure tools have been developed to project an image of the mask patterns onto a resist-coated wafer many centimeters away from the mask. To increase resolution, only a small portion of the mask is exposed at a time. The small image area is scanned or stepped over the wafer to cover the entire wafer surface. Figure 4.4*a* shows a 1:1 wafer scan projection system.[6,7] A narrow, arc-shaped image field approximately 1 mm in width serially transfers the slit image of the mask onto the wafer. The image size on the wafer is the same as that on the mask.

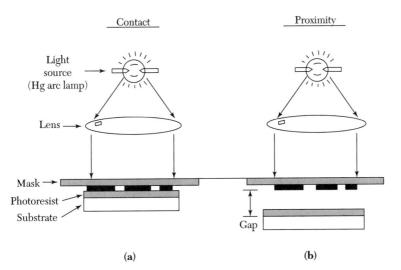

Figure 4.3 Schematic of optical shadow printing techniques.[1] (*a*) Contact printing. (*b*) Proximity printing.

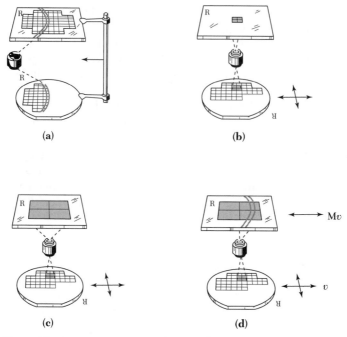

Figure 4.4   Image partitioning techniques for projection printing. (*a*) Annual-field wafer scan. (*b*) 1:1 step-and-repeat. (*c*) M:1 reduction step-and-repeat. (*d*) M:1 reduction step-and-scan.[6, 7]

The small image field can also be stepped over the surface of the wafer by two-dimensional translations of the wafer only, while the mark remains stationary. After the exposure of one chip site, the wafer is moved to the next chip site and the process is repeated. Figures 4.4*b* and 4.4*c* show the partitioning of the wafer image by *step-and-repeat projection* with a ratio of 1:1 or at a demagnification ratio of *M*:1 (e.g., 10:1 for a 10 times reduction on the wafer), respectively. The demagnification ratio is an important factor in our ability to produce both the lens and the mask from which we wish to print. The 1:1 optical systems are easier to design and fabricate than 10:1 or 5:1 reduction systems, but it is much more difficult to produce defect-free masks at 1:1 than it is at a 10:1 or a 5:1 demagnification ratio.

Reduction projection lithography can also print larger wafers without redesigning the stepper lens, as long as the field size (i.e., the exposure area onto the wafer) of the lens is large enough to contain one or more IC chips. When the chip size exceeds the field size of the lens, further partitioning of the image on the reticle is necessary. In Figure 4.4*d*, the image field on the reticle can be a narrow arc shape for *M*:1 step-and-scan projection lithography. The step-and-scan system yields two-dimensional translations of the wafer with speed *v*, and one-dimensional translation of the mask with a speed *M* times that of the wafer speed.

The resolution of a projection system is given by

$$l_{\mathrm{m}} = k_1 \frac{\lambda}{\mathrm{NA}} \tag{2}$$

where λ is again the exposure wavelength, k_1 is a process-dependent factor, and NA is the numerical aperture, which is given by

$$NA = \bar{n} \sin\theta \qquad (3)$$

where $\bar{n}$ is the index of refraction in the image medium (usually air, where $\bar{n} = 1$), and θ is the half-angle of the cone of light converging to a point image at the wafer, as shown in Figure 4.5.[5] Also shown in the figure is the depth of focus (DOF), which can be expressed as

$$DOF = \frac{\pm l_m/2}{\tan\theta} \approx \frac{\pm l_m/2}{\sin\theta} = k_2 \frac{\lambda}{(NA)^2} \qquad (4)$$

where k_2 is another process-dependent factor.

Equation 2 indicates that resolution can be improved (i.e., smaller l_m) by either reducing the wavelength or increasing NA or both. However, Eq. 4 indicates that the DOF degrades much more rapidly by increasing NA than by decreasing λ. This explains the trend toward shorter-wavelength sources in optical lithography.

The high-pressure mercury-arc lamp is widely used in exposure tools because of its high intensity and reliability. The mercury-arc spectrum is composed of several peaks, as shown in Figure 4.6. The terms *G-line, H-line,* and *I-line* refer to the peaks at 436 nm, 405 nm, and 365 nm, respectively. I-line lithography with 5:1 step-and-repeat projection can offer a resolution of 0.3 μm with resolution enhancement techniques (see Section 4.1.6). Advanced exposure tools such as the 248-nm lithographic system using a KrF excimer laser, the 193-nm lithographic system using an ArF excimer laser, and the 157-nm lithographic system using a F_2 excimer laser have been developed for mass production with a resolution of 0.18 μm (180 nm), 0.10 μm (100 nm), and 0.07 μm (70 nm), respectively.

4.1.3 Masks

Masks used for IC manufacturing are usually reduction reticles. The first step in mask making is to use a computer-aided design (CAD) system in which designers can completely describe the circuit patterns electrically. The digital data produced by the CAD system then drives a pattern generator, which is an electron beam lithographic system (see Section 4.2.1) that transfers the patterns directly to electron-sensitized mask. The mask consists of a fused-silica substrate covered with a chromium layer. The circuit pattern is first transferred to the electron-sensitized layer (electron resist), which is transferred once more

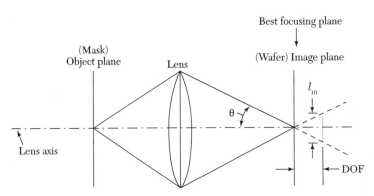

Figure 4.5 Simple image system.[5]

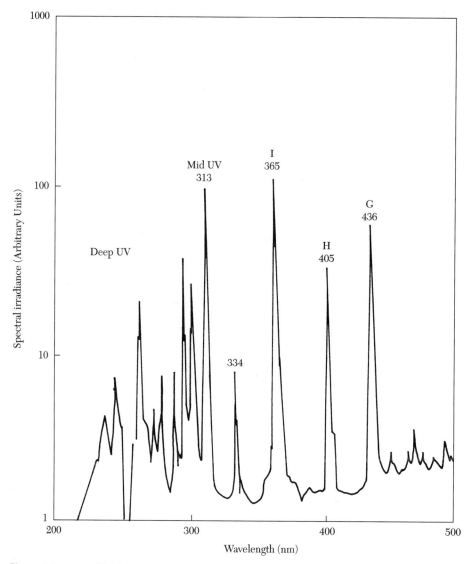

Figure 4.6 Typical high-pressure mercury-arc lamp spectrum.

into the underlying chromium layer for the finished mask. The details of pattern transfer are considered in Section 4.1.5.

The patterns on a mask represent one level of an IC design. The composite layout is broken into mask levels that correspond to the IC process sequence, such as the isolation region on one level, the gate region on another, and so on. Typically, 15 to 20 different mask levels are required for a complete IC process cycle.

The standard-size mask substrate is a fused-silica plate 15 × 15 cm square, 0.6 cm thick. The size is needed to accommodate the lens field sizes for 4:1 or 5:1 optical exposure tools, whereas the thickness is required to minimize pattern placement errors due to substrate distortion. The fused-silica plate is needed for its low coefficient of thermal expansion, its high transmission at shorter wavelengths, and its mechanical strength. Figure 4.7 shows a

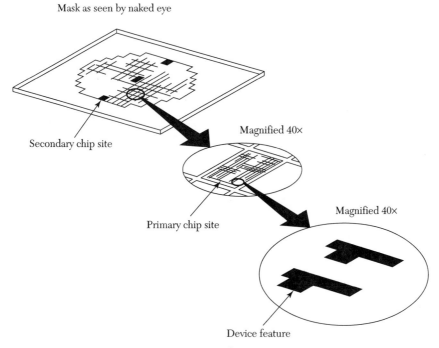

Mask as seen by naked eye

Secondary chip site

Magnified 40×

Primary chip site

Magnified 40×

Device feature

Figure 4.7 An integrated circuit photomask.[1]

mask on which patterns of geometric shapes have been formed. A few secondary chip sites, used for process evaluation, are also included in the mask.

One of the major concerns about masks is defect density. Mask defects can be introduced during the manufacture of the mask or during subsequent lithographic processes. Even a small mask-defect density has a profound effect on the final IC yield. *Yield* is defined as the ratio of good chips per wafer to the total number of chips per wafer (see Chapter 10). As a first-order approximation, the yield Y for a given masking level can be expressed as

$$Y \cong e^{-D_0 A_c} \tag{5}$$

where D_0 is the average number of "fatal" defects per unit area, and A_c is the defect-sensitive area (or "critical area") of the IC chip. If D_0 remains the same for all mask levels (e.g., $N = 10$ levels), then the final yield becomes

$$Y \cong e^{-ND_0 A_c} \tag{6}$$

Figure 4.8 shows the mask limit yield for a 10-level lithographic process as a function of chip size for various values of defect densities. For example, for $D_0 = 0.25$ defect/cm^2, the yield is 10% for a chip size of 90 mm^2, and it drops to about 1% for a chip size of 180 mm^2. Therefore, inspection and cleaning of masks are important to achieve high yields on large chips. Of course, an ultraclean processing area is mandatory for lithographic processing.

4.1.4 Photoresist

Photoresist is a radiation-sensitive compound that can be classified as positive or negative, depending on how it responds to radiation. For *positive resists*, the exposed regions become more soluble and are thus more easily removed in the development process. The

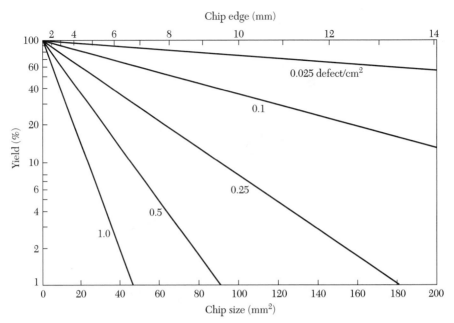

Figure 4.8 Yield for a 10-mask lithographic process with various defect densities per level.

net result is that the patterns formed (also called images) in the positive resist are the same as those on the mask. For *negative resists*, the exposed regions become less soluble, and the patterns formed in the negative resist are the reverse of the mask patterns.

Positive photoresists consist of three components: a photosensitive compound, a base resin, and an organic solvent. Prior to exposure, the photosensitive compound is insoluble in the developer solution. After exposure, the photosensitive compound absorbs radiation in the exposed pattern areas, changes its chemical structure, and becomes soluble in the developer solution. After development, the exposed areas are removed.

Negative photoresists are polymers combined with a photosensitive compound. After exposure, the photosensitive compound absorbs the optical energy and converts it into chemical energy to initiate a polymer cross-linking reaction. This reaction causes cross linking of the polymer molecules. The cross-linked polymer has a higher molecular weight and becomes insoluble in the developer solution. After development, the unexposed areas are removed. One major drawback of a negative photoresist is that in the development process, the whole resist mass swells by absorbing developer solvent. This swelling action limits the resolution of negative photoresists.

Figure 4.9*a* shows a typical exposure response curve and image cross section for a positive resist.[1] The response curve describes the percentage of resist remaining after exposure and development versus the exposure energy. Note that the resist has a finite solubility in its developer, even without exposure to radiation. As the exposure energy increases, the solubility gradually increases until at a threshold energy E_T, the resist becomes completely soluble. The sensitivity of a positive resist is defined as the energy required to produce complete solubility in the exposed region. Thus, E_T corresponds to the sensitivity. In addition to E_T, a parameter γ, the contrast ratio, is defined to characterize the resist:

$$\gamma \equiv \left[\ln\left(\frac{E_T}{E_1}\right)\right]^{-1} \qquad (7)$$

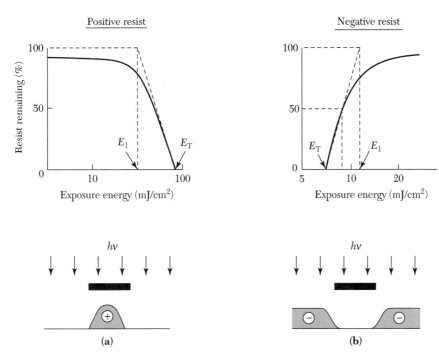

Figure 4.9 Exposure response curve and cross section of the resist image after development.[1] (*a*) Positive photoresist. (*b*) Negative photoresist.

where E_1 is the energy obtained by drawing the tangent at E_T to reach 100% resist thickness, as shown in Figure 4.9*a*. A larger γ implies a higher solubility of the resist with an incremental increase of exposure energy and results in sharper images.

The image cross section in Figure 4.9*a* illustrates the relationship between the edges of a photomask image and the corresponding edges of the resist images after development. The edges of the resist image are generally not at the vertically projected positions of the mask edges because of *diffraction.* The edge of the resist image corresponds to the position where the total absorbed optical energy equals the threshold energy E_T.

Figure 4.9*b* shows the exposure response curve and image cross section for a negative resist. The negative resist remains completely soluble in the developer solution for exposure energies lower than E_T. Above E_T, more of the resist film remains after development. At exposure energies twice the threshold energy, the resist film becomes essentially insoluble in the developer. The sensitivity of a negative resist is defined as the energy required to retain 50% of the original resist film thickness in the exposed region. The parameter γ is defined similarly to γ in Eq. 7, except that E_1 and E_T are interchanged. The image cross section for the negative resist (Fig. 4.9*b*) is also influenced by the diffraction effect.

EXAMPLE 2

Find the parameter γ for the photoresists shown in Figure 4.9.

SOLUTION For the positive resist, $E_T = 90$ mJ/cm^2 and $E_1 = 45$ mJ/cm^2, so

$$\gamma \equiv \left[\ln\left(\frac{E_T}{E_1} \right) \right]^{-1} = \left[\ln\left(\frac{90}{45} \right) \right]^{-1} = 1.4$$

For the negative resist, $E_T = 7$ mJ/cm^2 and $E_1 = 12$ mJ/cm^2, so

$$\gamma \equiv \left[\ln\left(\frac{E_1}{E_T} \right) \right]^{-1} = \left[\ln\left(\frac{12}{7} \right) \right]^{-1} = 1.9$$

For deep UV lithography (e.g., 248 and 193 nm), we cannot use conventional photoresists because these resists require a high-dose exposure in deep UV, which will cause lens damage and lower throughput. *Chemical-amplified resist* (CAR) has been developed for the deep UV process. CAR consists of a photo-acid generator, a resin polymer, and a solvent. CAR is very sensitive to deep UV radiation, and the exposed and unexposed regions differ greatly in their solubility in the developer solution. ◀

4.1.5 Pattern Transfer

Figure 4.10 illustrates the steps to transfer IC patterns from a mask to a silicon wafer that has an insulating SiO$_2$ layer formed on its surface.[8] The wafer is placed in a clean room, which typically is illuminated with yellow light, since photoresists are not sensitive to wavelengths greater than 0.5 μm. To ensure satisfactory adhesion of the resist, the surface must be changed from hydrophilic to hydrophobic. This change can be made by the application of an adhesion promoter, which can provide a chemically compatible surface for the resist. The most common adhesion promoter for silicon ICs is hexamethyl-disilazane (HMDS). After the application of this adhesion layer, the wafer is held on a vacuum spindle, and 2 to 3 cm^3 of liquidous resist is applied to the center of the wafer. The wafer is then rapidly accelerated to a constant rotational speed, which is maintained for about 30 seconds. Spin speed is generally in the range of 1000 to 10,000 rpm to coat a uniform film about 0.5 to 1 μm thick, as shown in Figure 4.10a. The thickness of photoresist is correlated with its viscosity.

After the spinning step, the wafer is "soft baked" (typically at 90–120°C for 60–120 seconds) to remove the solvent from the photoresist film and to increase resist adhesion to the wafer. The wafer is aligned with respect to the mask in an optical lithographic system, and the resist is exposed to UV light, as shown in Figure 4.10b. If a positive photoresist is used, the exposed resist is dissolved in the developer, as shown on the left side of Figure 4.10c. Photoresist development is usually done by flooding the wafer with the developer solution. The wafer is then rinsed and dried. After development, "post baking" at approximately 100°C to 180°C may be required to increase the adhesion of the resist to the substrate. The wafer is then put in an ambient that etches the exposed insulation layer but does not attack the resist, as shown in Figure 4.10d. Finally, the resist is stripped (e.g., using solvents or plasma oxidation), leaving behind an insulator image (or pattern) that is the same as the opaque image on the mask (left side of Fig. 4.10e).

For negative photoresist, the procedures described are also applicable, except that the unexposed areas are removed. The final insulator image (right side of Figure 4.10e) is the reverse of the opaque image on the mask.

The insulator image can be used as a mask for subsequent processing. For example, *ion implantation* (see Chapter 7) can be done to dope the exposed semiconductor region, but not the area covered by the insulator. The dopant pattern is a duplicate of the design pattern on the photomask for a negative photoresist or is its complementary pattern for a positive photoresist. The complete circuit is fabricated by aligning the next mask in the sequence to the previous pattern and repeating the lithographic transfer process.

A related pattern transfer process is the *liftoff* technique, shown in Figure 4.11. A positive resist is used to form the resist pattern on the substrate (Figs. 4.11a and 4.11b).

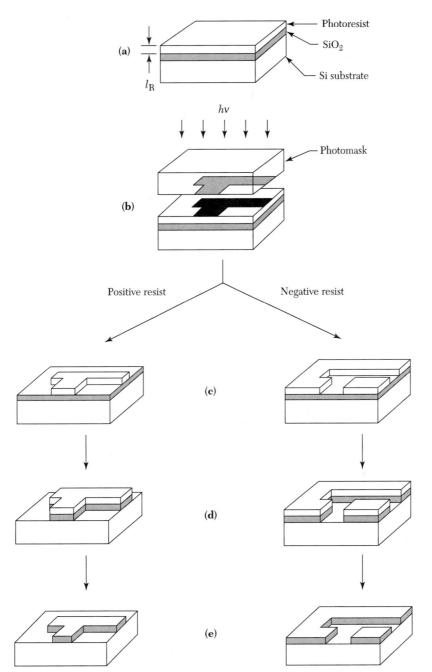

Figure 4.10 Details of the optical lithographic pattern transfer process.[8]

The film (e.g., aluminum) is deposited over the resist and the substrate (Fig. 4.11c). The film thickness must be smaller than that of the resist. Those portions of the film on the resist are removed by selectively dissolving the resist layer in an appropriate liquid etchant so that the overlying film is lifted off and removed (Fig. 4.11d). The liftoff technique is capable of high resolution and is used extensively for discrete devices such as high-power MESFETs. However, it is not as widely applicable for ultralarge-scale integration, in which dry etching is the preferred technique.

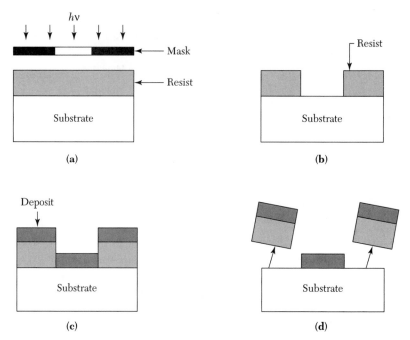

Figure 4.11 The liftoff process for pattern transfer.

4.1.6   Resolution Enhancement Techniques

Optical lithography has been continuously challenged to provide better resolution, greater DOF, and wider exposure latitude in IC processing. These challenges have been met by reducing the wavelength of the exposure tools and developing new resists. In addition, many resolution enhancement techniques have been developed to extend the capability of optical lithography to even smaller feature lengths.

An important resolution enhancement technique is the *phase-shifting mask* (PSM). The basic concept is shown in Figure 4.12.[9] For a conventional mask (Fig. 4.12a), the electric field has the same phase at every aperture (clear area). Diffraction and the limited resolution of the optical system spread the electric field at the wafer, as shown by the dotted lines. Interference between waves diffracted by the adjacent apertures enhances the field between them. Because the intensity (I) is proportional to the square of the electric field, it becomes difficult to separate the two images that are projected close to one another. The phase-shift layer that covers adjacent apertures reverses the sign of the electric field, as shown in Figure 4.12b. Because the intensity at the mask is unchanged, the electric field of the images at the wafer can be cancelled. Therefore, images that are projected close to one another can be separated. A 180° phase change can be obtained by using a transparent layer of thickness $d = \lambda/2(\bar{n} - 1)$, where $\bar{n}$ is the refractive index and λ is the wavelength that covers one aperture, as shown in Figure 4.12b.

Another resolution enhancement technique is *optical proximity correction* (OPC), which uses modified shapes of adjacent subresolution geometry to improve imaging capability. For example, a square contact hole with dimensions near the resolution limit will print nearly as a circle. Modifying the contact-hole pattern with additional geometry at the corners will help to print a more accurate square hole.

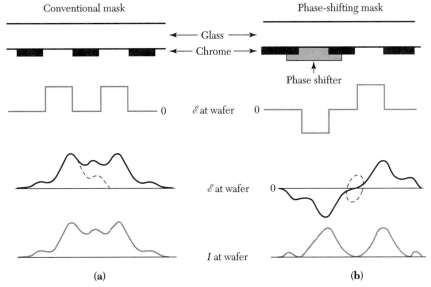

Figure 4.12 The principle of phase-shift technology. (*a*) Conventional technology. (*b*) Phase-shift technology.[9]

► 4.2 NEXT-GENERATION LITHOGRAPHIC METHODS

Why is optical lithography so widely used and what makes it such a promising method? The reasons are that it has high throughput, good resolution, low cost, and ease in operation.

However, due to deep-submicron IC process requirements, optical lithography has some limitations that have not yet been solved. Although we can use PSM or OPC to extend its useful span, the complexity of mask production and mask inspection cannot be easily resolved. In addition, the cost of masks is very high. Therefore, we need to find alternatives to optical lithography to process deep-submicron or nanometer ICs.

Various types of next-generation lithographic methods for IC fabrication are discussed in this section. Electron beam lithography, extreme UV lithography, x-ray lithography, and ion beam lithography are considered, as are the differences among these methods.

4.2.1 Electron Beam Lithography

Electron beam (or *e-beam*) lithography is primarily used to produce photomasks. Relatively few tools are dedicated to direct exposure of the resist by a focused electron beam without a mask. Figure 4.13 shows a schematic of an electron beam lithography system.[10] The electron gun is a device that can generate a beam of electrons with a suitable current density. A tungsten thermionic emission cathode or single-crystal lanthanum hexaboride (LaB_6) is used for the electron gun. Condenser lenses are used to focus the electron beam to a spot size 10 to 25 nm in diameter. Beam-blanking plates, which turn the electron beam on and off, and beam deflection coils are computer controlled and operated at MHz or higher rates to direct the focused electron beam to any location in the scan field on the substrate. Because the scan field (typically 1 cm) is much smaller than the substrate diameter, a precision mechanical stage is used to position the substrate to be patterned.

The advantages of electron beam lithography include the generation of submicron resist geometries, highly automated and precisely controlled operation, greater depth

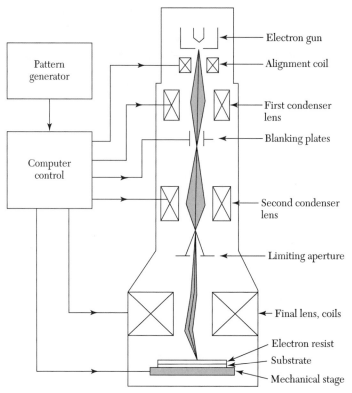

Figure 4.13 Schematic of an electron beam lithography machine.[10]

of focus than that available from optical lithography, and direct patterning on a semi-conductor wafer without using a mask. The disadvantage is that electron beam lithography machines have low throughput—approximately 10 wafers per hour at less than 0.25 μm resolution. This throughput is adequate for the production of photomasks, for situations that require small numbers of custom circuits, and for design verification. However, for maskless direct writing, the machine must have the highest possible throughput, and therefore the largest beam diameter possible consistent with the minimum device dimensions.

There are basically two ways to scan the focused electron beam: raster scan and vector scan.[11] In a *raster scan* system, resist patterns are written by a vertically oriented beam that moves through a regular mode, as shown in Figure 4.14a. The beam scans sequentially over every possible location on the mask and is blanked (turned off) where no exposure is required. All patterns on the area to be written must be subdivided into individual addresses, and a given pattern must have a minimum incremental interval that is evenly divisible by the beam address size.

In a *vector scan* system, shown in Figure 4.14b, the beam is directed only to the requested pattern features and jumps from feature to feature, rather than scanning the whole chip, as in raster scan. For many chips, the average exposed region is only 20% of the chip area, so time is saved using a vector scan system.

Figure 4.14c shows several types of electron beams employed for e-beam lithography: the Gaussian spot beam (round beam), the variable-shaped beam, and cell projection. In the variable-shaped beam system, the patterning beam has a rectangular cross section of variable size and aspect ratio. It offers the advantage of exposing several address units simultaneously. Therefore, the vector scan method using a variable-shaped beam

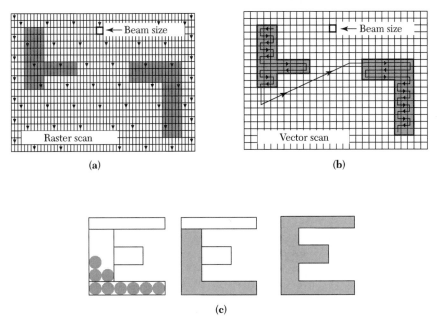

Figure 4.14 (*a*) Raster scan writing scheme. (*b*) Vector scan writing schemes. (*c*) Shapes of electron beam: round, variable, and cell projection.[12]

has higher throughput than the conventional Gaussian spot beam. It is also possible to pattern a complex geometric shape in one exposure with an electron beam system; this is called *cell projection*, as shown in the far right of Figure 4.14*c*. The cell projection technique[12] is particularly suitable for highly repetitive designs, as in MOS memory cells, since several memory cell patterns can be exposed at once. Cell projection has not yet achieved the throughput of optical exposure tools.

Electron Resist

Electron resists are polymers. The behavior of an electron beam resist is similar to that of a photoresist; that is, a chemical or physical change is induced in the resist by irradiation. This change allows the resist to be patterned. For a positive electron resist, the polymer–electron interaction causes chemical bonds to be broken (chain scission) to form shorter molecular fragments, as shown in Figure 4.15*a*.[13] As a result, the molecular weight is reduced in the irradiated area, which can be dissolved subsequently in a developer solution that attacks the low-molecular-weight material. Common positive electron resists include poly-methyl methacrylate (PMMA) and poly-butene-1 sulfone (PBS). Positive electron resists can achieve resolution of 0.1 μm or better.

For a negative electron resist, the irradiation causes radiation-induced polymer linking, as shown in Figure 4.15*b*. The cross linking creates a complex three-dimensional structure with a molecular weight higher than that of the nonirradiated polymer. The nonirradiated resist can be dissolved in a developer solution that does not attack the high-molecular-weight material. Poly-glycidyl methacrylate-co-ethyl-acrylate (COP) is a common negative electron resist. COP, like most negative photoresists, also swells during development, so resolution is limited to about 1 μm.

The Proximity Effect

In optical lithography, the resolution is limited by diffraction of light. In electron beam lithography, the resolution is not limited by diffraction (because the wavelengths associated

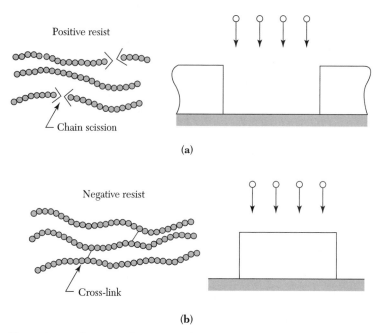

Figure 4.15   Schematic of positive and negative resists used in electron beam lithography.[13]

with electrons of a few keV and higher energies are less than 0.1 nm) but by electron scattering. When electrons penetrate the resist film and underlying substrate, they undergo collisions. These collisions lead to energy losses and path changes. Thus, the incident electrons spread out as they travel through the material until either all of their energy is lost or they leave the material because of backscattering.

Figure 4.16*a* shows computed electron trajectories of 100 electrons with initial energy of 20 keV incident at the origin of a 0.4-µm PMMA film on a thick silicon substrate.[14] The electron beam is incident along the *z*-axis, and all trajectories have been projected onto the *xz* plane. This figure shows qualitatively that the electrons are distributed in an oblong pear-shaped volume with a diameter on the same order of magnitude as the electron penetration depth (~3.5 µm). Also, many electrons undergo backscattering collisions and travel backward from the silicon substrate into the PMMA resist film and leave the material.

Figure 4.16*b* shows the normalized distributions of the forward-scattering and backscattering electrons at the resist-substrate interface. Because of backscattering, electrons effectively can irradiate several micrometers away from the center of the exposure beam. Since the dose of a resist is given by the sum of the irradiations from all surrounding areas, electron beam irradiation at one location will affect the irradiation in neighboring locations. This phenomenon is called the *proximity effect*. The proximity effect places a limit on the minimum spacings between pattern features. To correct for the proximity effect, patterns are divided into smaller segments. The incident electron dose in each segment is adjusted so that the integrated dose from all its neighboring segments is the correct exposure dose. This approach further decreases the throughput of the electron beam system because of the additional computer time required to expose the subdivided resist patterns.

4.2.2   Extreme Ultraviolet Lithography

Extreme ultraviolet (EUV) lithography is a promising next-generation lithographic technology to extend the minimum linewidths to 30 nm without throughput losses.[15] Figure 4.17 shows

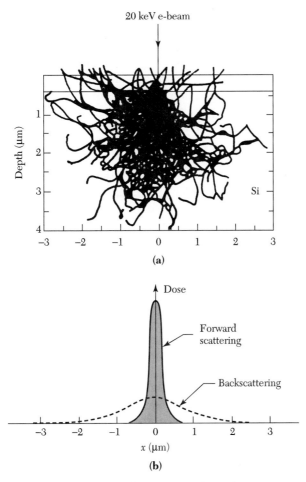

(a)

(b)

Figure 4.16 (*a*) Simulated trajectories of 100 electrons in PMMA for a 20-keV electron beam.[14] (*b*) Dose distribution for forward scattering and backscattering at the resist–substrate interface.

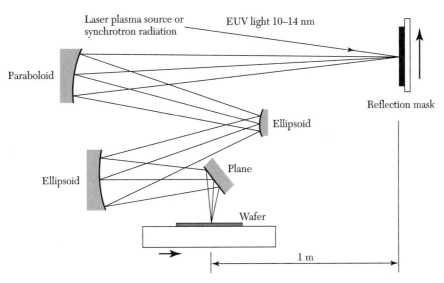

Figure 4.17 Schematic representation of an extreme ultraviolet (EUV) lithography system.[15]

a schematic diagram of an EUV lithography system. A laser-produced plasma or synchrotron radiation can serve as the source of EUV having a wavelength of 10 to 14 nm. The EUV radiation is reflected by a mask that is produced by patterning an absorber material deposited on a multilayer coated flat silicon or glass-plate mask blank. EUV radiation is reflected from the nonpatterned regions (i.e., nonabsorbing regions) of the mask through a 4× reduction camera and imaged into a thin layer of resist on the wafer.

Since the EUV radiation beam is narrow, the mask must be scanned by the beam to illuminate the entire pattern field that describes the circuit mask layer. Also, for a 4×4 mirror (i.e., the one-paraboloid, two-ellipsoid, and one-plane mirror) reduction camera, the wafer must be scanned at one-fourth the mask speed in a direction opposite to the mask movement to reproduce the image field on all chip sites on the wafer surface. A precision system is required to perform the chip-site alignment and to control the wafer and mask stage movements and the exposure dose during the scanning process.

EUV lithography is capable of printing 50-nm features with PMMA resist using 13-nm radiation. However, the production of EUV exposure tools has a number of challenges. Since EUV is strongly absorbed in all materials, the lithography process must be performed in a vacuum. The camera must use reflective lens elements, and the mirrors must be coated with multilayer coatings that produce distributed quarter-wave Bragg reflectors. In addition, the mask blank must also be multilayer coated to maximize its reflectivity at λ of 10 to 14 nm.

4.2.3 X-Ray Lithography

X-ray lithography[16] (XRL) is a potential candidate to succeed optical lithography for the fabrication of integrated circuits at 100 nm. The synchrotron storage ring is the choice of x-ray source for high-volume manufacturing. It can provide a large amount of collimated flux and can easily accommodate 10 to 20 exposure tools.

XRL uses a shadow printing method similar to optical proximity printing. Figure 4.18 shows a schematic XRL system. The x-ray wavelength is about 1 nm, and the printing is through a 1× mask in close proximity (10–40 μm) to the wafer. Since x-ray absorption

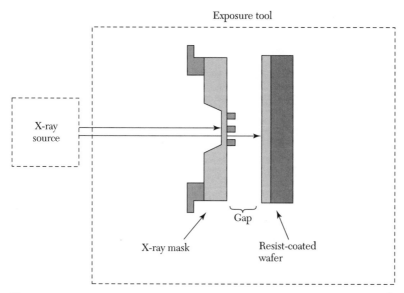

Figure 4.18 Schematic representation of a proximity x-ray lithography system.[17]

depends on the atomic number of the material and most materials have low transparency at $\lambda \cong 1$ nm, the mask substrate must be a thin membrane (1–2 μm thick) made of low-atomic-number material, such as silicon carbide or silicon. The pattern itself is defined in a thin (~0.5 μm), relatively high-atomic-number material, such as tantalum, tungsten, gold, or one of their alloys, which is supported by the thin membrane.

Masks are the most difficult and critical element of an XRL system, and the construction of an x-ray mask is much more complicated than that of a photomask. To avoid absorption of the x-rays between the source and mask, the exposure generally takes place in a helium environment. The x-rays are produced in vacuum, which is separated from the helium by a thin vacuum window (usually of beryllium). The mask substrate will absorb 25% to 35% of the incident flux and must therefore be cooled. An x-ray resist 1 μm thick will absorb about 10% of the incident flux. There are no reflections from the substrate to create standing waves, so antireflection coatings are unnecessary.

We can use electron beam resists as x-ray resists because when an x-ray is absorbed by an atom, the atom goes to an excited state with the emission of an electron. The excited atom returns to its ground state by emitting an x-ray having a different wavelength than the incident x-ray. This x-ray is absorbed by another atom, and the process repeats. Since all the processes result in the emission of electrons, a resist film under x-ray irradiation is equivalent to one being irradiated by a large number of secondary electrons from any of the other processes. Once the resist film is irradiated, chain cross linking or chain scission will occur, depending on the type of resist.

4.2.4 Ion Beam Lithography

Ion beam lithography can achieve higher resolution than optical, x-ray, or electron beam lithographic techniques because ions have a higher mass and therefore scatter less than electrons. The most important application is the repair of masks for optical lithography, a task for which commercial systems are available.

Figure 4.19 shows the computer-simulated trajectories of 50 H$^+$ ions implanted at 60 keV into PMMA and various substrates.[17] Note that the spread of the ion beam at a depth of 0.4 μm is only 0.1 μm in all cases (compare with Fig. 4.16a for electrons). Backscattering is completely absent for the silicon substrate, and there is only a small

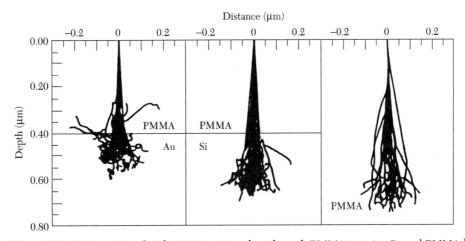

Figure 4.19 Trajectories of 60-keV H$^+$ ions traveling through PMMA into Au, Si, and PMMA.[17]

amount of backscattering for the gold substrate. However, ion beam lithography may suffer from random (or stochastic) space-charge effects, causing broadening of the ion beam.

There are two types of ion beam lithography systems: a scanning focused-beam system and a mask-beam system. The former system is similar to the electron beam machine (Fig. 4.13), in which the ion source can be Ga^+ or H^+. The latter system is similar to an optical 5× reduction projection step-and-repeat system, which projects 100-keV light ions such as H_2^+ through a stencil mask.

4.2.5 Comparison of Various Lithographic Methods

The lithographic methods discussed earlier all have 100 nm or better resolution. A comparison of various lithographic technologies is shown in Table 4.1. However, each method has its own limitations: the diffraction effect in optical lithography, the proximity effect in electron beam lithography, mask fabrication complexities in x-ray lithography, difficulty in mask blank production for EUV lithography, and stochastic space charge in ion beam lithography.

For IC fabrication, many mask levels are involved. However, it is not necessary to use the same lithographic method for all levels. A mix-and-match approach can take advantage of the unique features of each lithographic process to improve resolution and to maximize throughput. For example, a 4:1 EUV method can be used for the most critical mask levels, whereas a 4:1 or 5:1 optical system can be used for the rest.

According to the Semiconductor Industry Association's *International Technology Roadmap for Semiconductors*, IC manufacturing technology will reach the 50 nm generation around 2010.[18] With each new technology generation, lithography has become

TABLE 4.1   Comparison of Various Lithographic Technologies

	Optical 248/193 nm	SCALPEL	EUV	X-ray	Ion Beam
Exposure Tool					
Source	Laser	Filament	Laser plasma	Synchrotron	Multicusp
Diffraction limited	Yes	No	Yes	Yes	No
Optics	Refractive	Refractive	Refractive	No optics	Full-field refractive
Step and scan	Yes	Yes	Yes	Yes	Stepper
Throughput of 200-mm wafers/hr	40	30–35	20–30	30	30
Mask					
Demagnification	4×	4×	4×	1×	4×
Optical proximity correction	Yes	No	Yes	Yes	No
Radiation path	Transmission	Transmission	Reflection	Transmission	Stencil
Resist					
Single or multilayer	Single	Single	Surface imaging	Single	Single
Chemical-amplified resist	Yes	Yes	No	Yes	No

SCALPEL, Scattering with angular limitation projection electron beam lithography; EUV, extreme ultraviolet.

an even more important key driver for the semiconductor industry because of the requirements of smaller feature size and tighter overlay tolerance. In addition, lithography tool costs have become higher relative to the total equipment costs for an IC manufacturing facility. Currently, the technology development of next-generation lithography is conducted by multinational research projects or by industrial partners.

▶ 4.3 PHOTOLITHOGRAPHY SIMULATION

As is the case with oxidation (see Chapter 3), computer simulation is also an important tool for studying the photolithography process. The SUPREM package, unfortunately, is not capable of photolithographic simulation. However, another popular tool, PROLITH, does provide this capability.

PROLITH is a Windows-based program that uses a positive/negative photoresist optical lithography model originally developed by Chris Mack.[19] PROLITH simulates the complete one- and two-dimensional optical lithography process from aerial image formation through resist exposure and development. The output of the program is an accurate prediction of the final resist profile, which is presented in a wide variety of images, plots, graphs, and calculations. In particular, PROLITH is able to simulate the following:

- Formation of an image of a mask feature by an optical projection system
- Exposure of photoresist by this image
- Diffusion of the image
- Development of the exposed photoresist

PROLITH accepts lithography information in the form of data files and input parameters and uses this information to simulate standard and advanced lithography processes. To run PROLITH, the user simply clicks on the PROLITH icon from the Windows Start menu. After a successful license search, the Imaging Tool parameters window appears (see Fig. 4.20). As the user makes choices from the View menu, PROLITH displays windows in which parameters may be entered in order to view simulation results. These can be observed from the Graphs menu. These ideas are illustrated in Example 3.

EXAMPLE 3

Use PROLITH to view the resist profile for the cylindrical mask feature in Figure 4.20 after exposure and development. Assume the following process conditions:

Photoresist type = SPR 500
Pre-bake temperature = 95°C
Pre-bake time = 60 seconds
Numerical aperture of the lens = 0.5
Exposure wavelength = 365 nm
Exposure energy = 150 mJ/cm^2
Post-exposure bake temperature = 110°C
Post-exposure bake time = 60 seconds
Development time = 60 seconds
Developer = MFT 245/501

SOLUTION All of the given values can be entered from the Parameters menu or by clicking on the appropriate icons on the toolbar. The resulting resist profile is shown in Figure 4.21. ◀

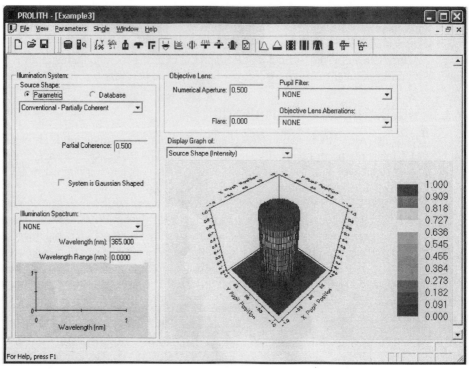

Figure 4.20 The Imaging Tools windows in PROLITH.

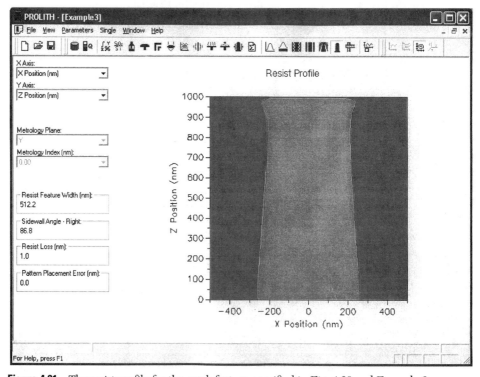

Figure 4.21 The resist profile for the mask feature specified in Fig. 4.20 and Example 3.

▶ 4.4 SUMMARY

The continued growth of the semiconductor industry is a direct result of the capability to transfer smaller and smaller circuit patterns onto semiconductor wafers. Currently, the vast majority of lithographic equipment is optical systems. This chapter considered various exposure tools, masks, photoresists, and the clean room for optical lithography. The primary factor limiting resolution in optical lithography is diffraction. However, because of advancements in excimer lasers, photoresist chemistry, and resolution enhancement techniques such as PSM and OPC, optical lithography will remain the mainstream technology, at least to the 100 nm generation.

Electron beam lithography is the technology of choice for mask making and nanofabrication, in which new device concepts are explored. Other lithographic processing technologies are EUV, x-ray, and ion beam lithography. Although all these have 100 nm or better resolution, each process has its own limitations: proximity effect in electron beam lithography, mask blank production difficulties in EUV lithography, mask fabrication complexity in x-ray lithography, and stochastic space charge in ion beam lithography.

At the present time, no obvious successor to optical lithography can be identified unambiguously. However, a mix-and-match approach can take advantage of the unique features of each lithography process to improve resolution and to maximize throughput.

▶ REFERENCES

1. For a more detailed discussion on lithography, see (a) K. Nakamura, "Lithography," in C. Y. Chang and S. M. Sze, Eds., *ULSI Technology*, McGraw-Hill, New York, 1996. (b) P. Rai-Choudhurg, *Handbook of Microlithography, Micromachining, and Microfabrication*, Vol. 1, SPIE, Washington, DC, 1997. (c) D. A. McGillis, "Lithography," in S. M. Sze, Ed., *VLSI Technology*, McGraw-Hill, New York, 1983.

2. For a more detailed discussion on etching, see Y. J. T. Liu, "Etching," in C. Y. Chang and S. M. Sze, Eds., *ULSI Technology*, McGraw-Hill, New York, 1996.

3. J. M. Duffalo and J. R. Monkowski, "Particulate Contamination and Device Performance," *Solid State Technol.* **27**, 3, 109 (1984).

4. H. P. Tseng and R. Jansen, "Cleanroom Technology," in C. Y. Chang and S. M. Sze, Eds., *ULSI Technology*, McGraw-Hill, New York, 1996.

5. M. C. King, "Principles of Optical Lithography," in N. G. Einspruch, Ed., *VLSI Electronics*, Vol. 1, Academic, New York, 1981.

6. J. H. Bruning, "A Tutorial on Optical Lithography," in D. A. Doane, et al., Eds., *Semiconductor Technology*, Electrochemical Soc., Penningston, 1982.

7. R. K. Watts and J. H. Bruning, "A Review of Fine-Line Lithographic Techniques: Present and Future," *Solid State Technol.*, **24**, 5, 99 (1981).

8. W. C. Till and J. T. Luxon, *Integrated Circuits, Materials, Devices, and Fabrication*, Prentice-Hall, Englewood Cliffs, NJ, 1982.

9. M. D. Levenson, N. S. Viswanathan, and R. A. Simpson, "Improving Resolution in Photolithography with a Phase-Shift Mask," *IEEE Trans. Electron Devices*, **ED-29**, 18–28 (1982).

10. D. P. Kern, et al., "Practical Aspects of Microfabrication in the 100-nm Region," *Solid State Technol.*, **27**, 2, 127 (1984).

11. J. A. Reynolds, "An Overview of e-Beam Mask-Making," *Solid State Technol.*, **22**, 8, 87 (1979).

12. Y. Someda, et al., "Electron-Beam Cell Projection Lithography: Its Accuracy and Its Throughput," *J. Vac. Sci. Technol.*, **B12** (6), 3399 (1994).

13. W. L. Brown, T. Venkatesan, and A. Wagner, "Ion Beam Lithography," *Solid State Technol.*, **24**, 8, 60 (1981).

14. D. S. Kyser and N. W. Viswanathan, "Monte Carlo Simulation of Spacially Distributed Beams in Electron-Beam Lithography," *J. Vac. Sci. Technol.*, **12**, 1305 (1975).

15. Charles Gwyn, et al., "Extreme Ultraviolet Lithography," White Paper, Sematech, Next Generation Lithography Workshop, Colorado Springs, Dec. 7–10, 1998.

16. J. P. Silverman, "Proximity X-Ray Lithography," White Paper, Sematech, Next Generation Lithography Workshop, Colorado Springs, Dec. 7–10, 1998.

17. L. Karapiperis, et al., "Ion Beam Exposure Profiles in PMMA-Computer Simulation," *J. Vac. Sci. Technol.*, **19**, 1259 (1981).

18. *The International Technology Roadmap for Semiconductors*, Semiconductor Ind. Assoc., San Jose, CA, 2001.

19. *PROLITH/2 User's Manual*, FINLE Technologies, Austin, TX, 1998.

▶ PROBLEMS

SECTION 4.1: OPTICAL LITHOGRAPHY

1. For a class 100 clean room, find the number of dust particles per cubic meter with particle sizes (a) between 0.5 and 1 μm, (b) between 1 and 2 μm, and (c) above 2 μm.

2. Find the final yield for a nine-mask-level process in which the average fatal defect density per cm^2 is 0.1 for four levels, 0.25 for four levels, and 1.0 for one level. The chip area is 50 mm^2.

3. An optical lithographic system has an exposure power of 0.3 mW/cm^2. The required exposure energy for a positive photoresist is 140 mJ/cm^2, and for a negative photoresist is 9 mJ/cm^2. Assuming negligible times for loading and unloading wafers, compare the wafer throughput for positive photoresist and negative photoresist.

4. (a) For an ArF excimer laser 193-nm optical lithographic system with NA = 0.65, k_1 = 0.60, and k_2 = 0.50, what are the theoretical resolution and depth of focus for this tool? (b) What can we do in practice to adjust NA, k_1, and k_2 parameters to improve resolution? (c) What parameter does the phase-shift mask (PSM) technique change to improve resolution?

5. The plots in Figure 4.9 are called *response curves* in microlithography. (a) What are the advantages and disadvantages of using resists with high γ values? (b) Conventional resists cannot be used for 248-nm or 193-nm lithography. Why not?

SECTION 4.2: NEXT-GENERATION LITHOGRAPHIC METHODS

6. (a) Explain why a shaped beam promises higher throughput than a Gaussian beam in e-beam lithography. (b) How can alignment be performed for e-beam lithography? Why is alignment in x-ray lithography so difficult? (c) What are the potential advantages of x-ray lithography over e-beam lithography?

7. Why has the operating mode of optical lithographic systems evolved from proximity printing to 1:1 projection printing and finally to 5:1 projection step-and-repeat? (b) Is it possible to build a step-and-scan x-ray lithographic system? Why or why not?

SECTION 4.3: PHOTOLITHOGRAPHY SIMULATION

8. Repeat Example 3 with the following revised process conditions:

   Pre-bake temperature = 100°C

   Pre-bake time = 5 minutes

   Exposure energy = 50 mJ/cm^2

   Post-exposure bake temperature = 120°C

   Post-exposure bake time = 15 minutes

   Development time = 60 seconds

   Developer = MF 319

   Explain any differences in the resist profile.

CHAPTER 5

Etching

As discussed in the previous chapter, lithography is the process of transferring patterns to photoresist covering the surface of a semiconductor wafer. To produce circuit features, these resist patterns must be transferred into the underlying layers comprising the device. The pattern transfer is accomplished by an etching process that selectively removes unmasked portions of a layer.[1] A brief description of etching was given in Section 1.4.2. The present chapter covers the following topics:

- Mechanisms for wet chemical etching of semiconductors, insulators, and metal films
- Plasma-assisted etching (also called dry etching) for high-fidelity pattern transfer

5.1 WET CHEMICAL ETCHING

Wet chemical etching is used extensively in semiconductor processing. Starting from semiconductor wafers sliced from an ingot (Chapter 2), chemical etchants are used for lapping and polishing to give an optically flat, damage-free surface. Prior to thermal oxidation (Chapter 3) or epitaxial growth (Chapter 8), semiconductor wafers are chemically cleaned to remove contamination that results from handling and storing. Wet chemical etching is especially suitable for blanket etches (i.e., over the whole wafer surface) of polysilicon, oxide, nitride, metals, and III-V compounds.

The mechanisms for wet chemical etching involve three essential steps, as illustrated in Figure 5.1: The reactants are transported by diffusion to the reacting surface, chemical reactions occur at the surface, and the products from the surface are removed by diffusion. Both agitation and the temperature of the etchant solution influence the etch rate, which is the amount of film removed by etching per unit time. In IC processing, most wet chemical etches proceed by immersing the wafers in a chemical solution or by spraying the wafers with the etchant solution. For immersion etching, the wafer is immersed in the etch solution, and mechanical agitation is usually required to ensure etch uniformity and a consistent etch rate. Spray etching has gradually replaced immersion etching because it greatly increases the etch rate and uniformity by constantly supplying fresh etchant to the wafer surface.

For semiconductor production lines, highly uniform etch rates are important. Etch rates must be uniform across a wafer, from wafer to wafer, from run to run, and for any variations in feature sizes and pattern densities. Etch rate uniformity is given by the following equation:

$$\text{Etch rate uniformity } (\%) = \frac{\left(\text{Maximum etch rate} - \text{Minimum etch rate}\right)}{\left(\text{Maximum etch rate} + \text{Minimum etch rate}\right)} \times 100\% \quad (1)$$

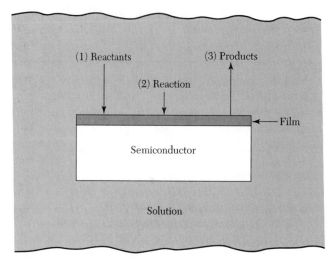

Figure 5.1 Basic mechanisms in wet chemical etching.

EXAMPLE 1

Calculate the Al average etch rate and etch rate uniformity on a 200-mm diameter silicon wafer, assuming the etch rates at the center, left, right, top, and bottom of the wafer are 750, 812, 765, 743, and 798 nm/min, respectively.

SOLUTION

$$\text{Al average etch rate} = (750 + 812 + 765 + 743 + 798) \div 5 = 773.6 \text{ nm/min}$$
$$\text{Etch rate uniformity} = (812 - 743) \div (812 + 743) \times 100\% = 4.4\%$$
◀

5.1.1 Silicon Etching

For semiconductor materials, wet chemical etching usually proceeds by oxidation, followed by the dissolution of the oxide by a chemical reaction. For silicon, the most commonly used etchants are mixtures of nitric acid (HNO_3) and hydrofluoric acid (HF) in water or acetic acid (CH_3COOH). Nitric acid oxidizes silicon to form a SiO_2 layer.[2] The oxidation reaction is

$$Si + 4HNO_3 \rightarrow SiO_2 + 2H_2O + 4NO_2 \qquad (2)$$

Hydrofluoric acid is used to dissolve the SiO_2 layer. The reaction is

$$SiO_2 + 6HF \rightarrow H_2SiF_6 + 2H_2O \qquad (3)$$

Water can be used as a diluent for this etchant. However, acetic acid is preferred because it reduces the dissolution of the nitric acid.

Some etchants dissolve a given crystal plane of single-crystal silicon much faster than another plane; this results in orientation-dependent etching.[3] For a silicon lattice, the (111) plane has more available bonds per unit area than the (110) and (100) planes; therefore, the etch rate is expected to be slower for the (111) plane. A commonly used orientation-dependent etch for silicon consists of a mixture of KOH in water and isopropyl alcohol.

For example, a solution with 19 wt % KOH in deionized (DI) water at about 80°C removes the (100) plane at a much higher rate than the (110) and (111) planes. The ratio of the etch rates for the (100), (110), and (111) planes is 100:16:1.

Orientation-dependent etching of <100>-oriented silicon through a patterned silicon dioxide mask creates precise V-shaped grooves,[4] the edges being the (111) planes at an angle of 54.7° from the (111) surface, as shown at the left of Figure 5.2a. If the window in the mask is sufficiently large or if the etching time is short, a U-shaped groove will be formed, as shown at the right of Figure 5.2a. The width of the bottom surface is given by

$$W_b = W_0 - 2l \cot 54.7°$$

or

$$W_b = W_0 - \sqrt{2}l \tag{4}$$

where W_0 is the width of the window on the wafer surface and l is the etched depth. If $<\bar{1}10>$-oriented silicon is used, essentially straight-walled grooves with sides of (111) planes can be formed, as shown in Figure 5.2b. We can use the large orientation dependence in the etch rates to fabricate device structures with submicron feature lengths.

5.1.2 Silicon Dioxide Etching

The wet etching of silicon dioxide is commonly achieved in a dilute solution of HF with or without the addition of ammonium fluoride (NH_4F). Adding NH_4F creates what is

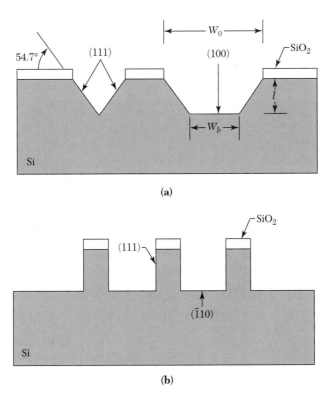

Figure 5.2 Orientation-dependent etching.[4] (*a*) Through window patterns on <100>-oriented silicon. (*b*) Through window patterns on <110>-oriented silicon.

referred to as a *buffered HF solution* (BHF), also called *buffered oxide etch* (BOE). The addition of NH_4F to HF controls the pH value and replenishes the depletion of the fluoride ions, thus maintaining stable etching performance. The overall reaction for SiO_2 etching is the same as that in Eq. 3. The etch rate of SiO_2 etching depends on etchant solution, etchant concentration, agitation, and temperature. In addition, density, porosity, microstructure, and the presence of impurities in the oxide influence the etch rate. For example, a high concentration of phosphorus in the oxide results in a rapid increase in the etch rate, and a loosely structured oxide formed by chemical vapor deposition (CVD) or sputtering exhibits a faster etch rate than thermally grown oxide.

Silicon dioxide can also be etched in vapor-phase HF. Vapor-phase HF oxide etch technology has a potential for submicron feature etching because the process can be well controlled.

5.1.3 Silicon Nitride and Polysilicon Etching

Silicon nitride films are etchable at room temperature in concentrated HF or buffered HF and in a boiling H_3PO_4 solution. Selective etching of nitride to oxide is done with 85% H_3PO_4 at 180°C because this solution attacks silicon dioxide very slowly. The etch rate is typically 10 nm/min for silicon nitride, but less than 1 nm/min for silicon dioxide. However, photoresist adhesion problems are encountered when etching nitride with boiling H_3PO_4 solution. Better patterning can be achieved by depositing a thin oxide layer on top of the nitride film before resist coating. The resist pattern is transferred to the oxide layer, which then acts as a mask for subsequent nitride etching.

Etching polysilicon is similar to etching single-crystal silicon. However, the etch rate is considerably faster because of the grain boundaries. The etch solution is usually modified to ensure that it does not attack the underlying gate oxide. Dopant concentrations and temperature may affect the etch rate of polysilicon.

5.1.4 Aluminum Etching

Aluminum and aluminum alloy films are generally etched in heated solutions of phosphoric acid, nitric acid, acetic acid, and DI water. The typical etchant is a solution of 73% H_3PO_4, 4% HNO_3, 3.5% CH_3COOH, and 19.5% DI water at 30°C to 80°C. The wet etching of aluminum proceeds as follows: HNO_3 oxidizes aluminum, and H_3PO_4 then dissolves the oxidized aluminum. The etch rate depends on etchant concentration, temperature, agitation of the wafers, and impurities or alloys in the aluminum film. For example, the etch rate is reduced when copper is added to the aluminum.

Wet etching of insulating and metal films is usually done with similar chemicals that dissolve these materials in bulk form and involve their conversion into soluble salts or complexes. Generally, film materials will be etched more rapidly than their bulk counterparts. Also, the etch rates are higher for films that have a poor microstructure, built-in stress, or departure from stoichiometry, or that have been irradiated. Some useful etchants for insulating and metal films are listed in Table 5.1.

5.1.5 Gallium Arsenide Etching

A wide variety of etches have been investigated for gallium arsenide; however, few of them are truly isotropic.[5] This is because the surface activities of the (111) Ga and (111) As faces are very different. Most etches give a polished surface on the arsenic face, but the gallium face tends to show crystallographic defects and etches more slowly. The most commonly used etchants are the H_2SO_4–H_2O_2–H_2O and H_3PO_4–H_2O_2–H_2O systems. For

TABLE 5.1 **Etchants for Insulators and Conductors**

Material	Etchant Composition	Etch Rate (nm/min)
SiO_2	28 ml HF 170 ml HF ⎱ Buffered HF 113 g NH_4F ⎰	100
	15 ml HF 10 ml HNO_3 ⎱ P etch 300 ml H_2O ⎰	12
Si_3N_4	Buffered HF	0.5
Al	H_3PO_4	10
	4 ml HNO_3 3.5 ml CH_3COOH 73 ml H_3PO_4 19.5 ml H_2O	30
Au	4 g KI	1000
Mo	1 g I_2 40 ml H_2O 5 ml H_3PO_4 2 ml HNO_3 4 ml CH_3COOH 150 ml H_2O	500
Pt	1 ml HNO_3 7 ml HCl 8 ml H_2O	50
W	34 g KH_2PO_4 13.4 g KOH 33 g $K_3Fe(CN)_6$ H_2O to make 1 liter	160

an etchant with an 8:1:1 volume ratio of H_2SO_4:H_2O_2:H_2O, the etch rate is 0.8 μm/min for the <111> Ga face and 1.5 μm/min for all other faces. For an etchant with a 3:1:50 volume ratio of H_3PO_4:H_2O_2:H_2O, the etch rate is 0.4 μm/min for the <111> Ga face and 0.8 μm for all other faces.

▶ 5.2 DRY ETCHING

In pattern transfer operations, a resist pattern is defined by a lithographic process to serve as a mask for etching of its underlying layer (Fig. 5.3a).[6] Most of the layer materials (e.g., SiO_2, Si_3N_4, and deposited metals) are amorphous or polycrystalline thin films. If they are etched in a wet chemical etchant, the etch rate is generally isotropic (i.e., the lateral and vertical etch rates are the same), as illustrated in Figure 5.3b. If h_f is the thickness of the layer material and l the lateral distance etched underneath the resist mask, we can define the degree of anisotropy (A_f) by

$$A_f \equiv 1 - \frac{l}{h_f} = 1 - \frac{R_l t}{R_v t} = 1 - \frac{R_l}{R_v} \tag{5}$$

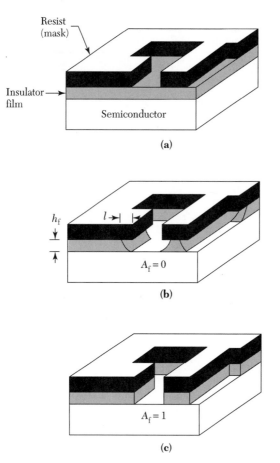

Figure 5.3 Comparison of wet chemical etching and dry etching for pattern transfer.[6]

where t is time and R_l and R_v are the lateral and vertical etch rates, respectively. For isotropic etching, $R_l = R_v$ and $A_f = 0$.

The major disadvantage of wet chemical etching in pattern transfer is the undercutting of the layer underneath the mask, resulting in a loss of resolution in the etched pattern. In practice, for isotropic etching, the film thickness should be about one-third or less of the resolution required. If patterns are required with resolutions much smaller than the film thickness, anisotropic etching (i.e., $1 \geq A_f > 0$) must be used. In practice, the value of A_f is chosen to be close to unity. Figure 5.3c shows the limiting case where $A_f = 1$, corresponding to $l = 0$ (or $R_l = 0$).

To achieve a high-fidelity transfer of the resist patterns required for ultralarge-scale integration processing, dry etching methods have been developed. Dry etching is synonymous with *plasma-assisted etching*, which denotes several techniques that use plasma in the form of low-pressure discharges. Dry etch methods include plasma etching, reactive ion etching (RIE), sputter etching, magnetically enhanced RIE (MERIE), reactive ion beam etching, and high-density plasma (HDP) etching.

5.2.1 Plasma Fundamentals

A plasma is a fully or partially ionized gas composed of equal numbers of positive and negative charges and a different number of unionized molecules. A plasma is produced when an electric field of sufficient magnitude is applied to a gas, causing the gas to break down and become ionized. The plasma is initiated by free electrons that are released by some means, such as field emission from a negatively biased electrode. The free electrons gain kinetic energy from the electric field. In the course of their travel through the gas, the electrons collide with gas molecules and lose their energy. The energy transferred in the collision causes the gas molecules to be ionized (i.e., to free electrons). The free electrons gain kinetic energy from the field, and the process continues. Therefore, when the applied voltage is larger than the breakdown potential, a sustained plasma is formed throughout the reaction chamber.

The electron concentrations in the plasma for dry etchings are relatively low, typically on the order of 10^9 to 10^{12} cm^{-3}. At a pressure of 1 Torr, the concentrations of gas molecules are 10^4 to 10^7 times higher than the electron concentrations. This results in an average gas temperature in the range of 50°C to 100°C. Therefore, plasma-assisted dry etching is a low-temperature process.

EXAMPLE 2

The electron densities in RIE and HDP systems range from 10^9 to 10^{10} cm^{-3} and 10^{11} to 10^{12} cm^{-3}, respectively. Assuming the RIE chamber pressure is 200 mTorr and HDP chamber pressure is 5 mTorr, calculate the ionization efficiency in RIE reactors and HDP reactors at room temperature. The ionization efficiency is the ratio of the electron density to the density of molecules.

SOLUTION

$$PV = nRT$$

where P is the pressure in atm (1 atm = 760,000 mTorr), V is the volume in liters, n is the number of moles, R is the gas constant (0.082 liter-atm/mol-K), and T is the absolute temperature in K, respectively.

For the RIE system,

$$n/V = P/RT = (200/760,000)/(0.082 \times 300) = 1.06 \times 10^{-5} \text{ (mol/liter)}$$
$$= 1.06 \times 10^{-5} \times 6.02 \times 10^{23} \div 1000$$
$$= 6.38 \times 10^{15} \text{ (cm}^{-3})$$
$$\text{Ionization efficiency} = (10^9 \sim 10^{10})/(6.38 \times 10^{15})$$
$$= 1.56 \times 10^{-7} \sim 1.56 \times 10^{-6}$$

For the HDP system,

$$n/V = P/RT = (5/760,000)/(0.082 \times 300) = 2.66 \times 10^{-7} \text{ (mol/liter)}$$
$$= 2.66 \times 10^{-7} \times 6.02 \times 10^{23} \div 1000$$
$$= 1.6 \times 10^{14} \text{ (cm}^{-3})$$
$$\text{Ionization efficiency} = (10^{11} \sim 10^{12})/(1.6 \times 10^{14})$$
$$= 6.25 \times 10^{-4} \sim 6.25 \times 10^{-3}$$

Therefore, HDP has much higher ionization efficiency than RIE. ◄

5.2.2 Etch Mechanism, Plasma Diagnostics, and End-Point Control

Plasma etching is a process in which a solid film is removed by a chemical reaction with ground-state or excited-state neutral species. Plasma etching is often enhanced or induced by energetic ions generated in a gaseous discharge. The basic etch mechanism, plasma diagnostics, and end-point control are introduced briefly in this section.

Etch Mechanism

Plasma etching proceeds in five steps, as illustrated in Figure 5.4. First, the etchant species is generated in the plasma. The reactant is then transported by diffusion through a stagnant gas layer to the surface. Next, the reactant is adsorbed on the surface. A chemical reaction (along with physical effects such as ion bombardment) follows to form volatile compounds. Finally, the compounds are desorbed from the surface, diffused into the bulk gas, and pumped out by the vacuum system.[7]

Plasma etching is based on the generation of plasma in a gas at low pressure. Two basic types of methods are used: physical methods and chemical methods. The former includes sputter etching, and the latter includes pure chemical etching. In physical etching, positive ions bombard the surface at high speed; small amounts of negative ions formed in the plasma cannot reach the wafer surface and therefore play no direct role in plasma etching. In chemical etching, neutral reactive species generated by the plasma interact with the material surface to form volatile products. Chemical and physical etch mechanisms have different characteristics. Chemical etching exhibits a high etch rate and good selectivity (i.e., the ratio of etch rates for different materials) and produces low ion bombardment–induced damage, but yields isotropic profiles. Physical etching can yield anisotropic profiles, but is associated with low etch selectivity and high bombardment–induced damage. Combinations of chemical and physical etching give anisotropic etch profiles, reasonably good selectivity, and moderate bombardment–induced damage. An example is the RIE process, which uses a physical method to assist chemical etching or creates reactive ions to participate in chemical etching.

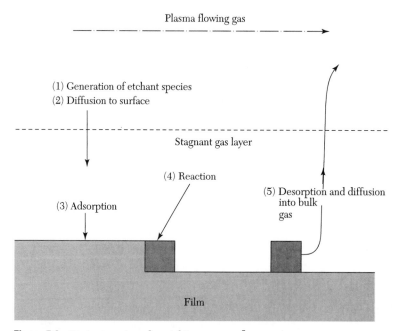

Figure 5.4 Basic steps in a dry etching process.[7]

Plasma Diagnostics

Most processing plasmas emit radiation in the range from infrared to ultraviolet. A simple analytical technique is to measure the intensity of these emissions versus wavelength with the aid of optical emission spectroscopy (OES). Using observed spectral peaks, it is usually possible to determine the presence of neutral and ionic species by correlating these emissions with previously determined spectral series. Relative concentrations of the species can be obtained by correlating changes in intensity with the plasma parameters. The emission signal derived from the primary etchant or by-product begins to rise or fall at the end of the etch cycle.

End-Point Control

Dry etching differs from wet chemical etching in that dry etching has less etch selectivity to the underlying layer. Therefore, the plasma reactor must be equipped with a monitor that indicates when the etching process is to be terminated (i.e., an *end-point detection system*). Laser interferometry of the wafer surface is used to continuously monitor etch rates and to determine the end point. During etching, the intensity of laser light reflected off a thin film surface oscillates. This oscillation occurs because of the phase interference between the light reflected from the outer and inner interfaces of the etching layer. This layer must therefore be optically transparent or semitransparent to observe the oscillation. Figure 5.5 shows a typical signal from a silicide/polycrystalline Si gate etch. The period of the oscillation is related to the change in film thickness by

$$\Delta d = \lambda / 2\overline{n} \qquad (6)$$

where Δd is the change in film thickness for one period of reflected light, λ is the wavelength of the laser light, and $\overline{n}$ is the refractive index of the layer being etched. For example, Δd for polysilicon is 80 nm, measured by using a helium-neon laser source for which $\lambda = 632.8$ nm.

5.2.3 Reactive Plasma Etching Techniques and Equipment

Plasma reactor technology in the IC industry has changed dramatically since the first application of plasma processing to photoresist stripping. A reactor for plasma etching contains a vacuum chamber, pump system, power supply generators, pressure sensors, gas flow control units, and end-point detector. Table 5.2 shows the similarities and differences

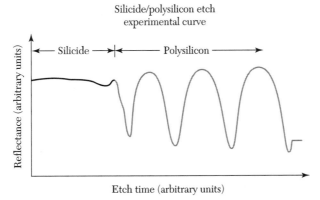

Figure 5.5 The relative reflectance of the etching surface of a composite silicide/poly-Si layer. The end point of the etch is indicated by the cessation of the reflectance oscillation.

TABLE 5.2   Etch Mechanisms and Pressure Ranges of Plasma Reactors

Etch Tool Configuration	Etch Mechanism	Pressure Range (Torr)
Barrel etching	Chemical	0.1–10
Downstream plasma etching	Chemical	0.1–10
Reactive ion etching (RIE)	Chemical and physical	0.01–1
Magnetic enhanced RIE	Chemical and physical	0.01–1
Magnetic confinement triode RIE	Chemical and physical	0.001–0.1
Electron cyclotron resonance plasma etch	Chemical and physical	0.001–0.1
Inductively coupled plasma or transformer-coupled plasma	Chemical and physical	0.001–0.1
Surface wave coupled plasma or helicon plasma etching	Chemical and physical	0.001–0.1

in the types of etching equipment that are commercially available. A comparison of pressure operating ranges and ion energies for different types of reactors is shown in Figure 5.6. Each etch tool is designed empirically and uses a particular combination of pressure, electrode configuration and type, and source frequency to control the two primary etch mechanisms—chemical and physical. Higher etch rates and tool automation are required for most etchers used in manufacturing.

Reactive Ion Etching

RIE has been extensively used in the microelectronics industry. In a parallel-plate diode system, a radio frequency capacitively coupled bottom electrode holds the wafer. This allows the grounded electrode to have a significantly larger area because it is, in fact, the chamber itself. The larger grounded area combined with the lower operating pressure (<500 mTorr) causes the wafers to be subjected to a heavy bombardment of energetic ions from the plasma as a result of the large negative self-bias at the wafer surface.

The etch selectivity of this system is relatively low compared with traditional barrel-etch systems because of strong physical sputtering. However, selectivity can be improved

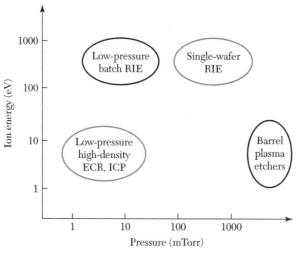

Figure 5.6   Comparison of ion energy and operating pressure ranges for different types of plasma reactors.

by choosing the proper etch chemistry, for example, by polymerizing the silicon surface with fluorocarbon polymers to obtain selectivity of SiO$_2$ over silicon. Alternatively, a triode-configuration RIE etch, as shown in Figure 5.7, can separate plasma generation from ion transport. Ion energy is controlled through a separate bias on the wafer electrode, thereby minimizing the loss of selectivity and the ion bombardment–induced damage observed in most traditional RIE systems.

Electron Cyclotron Resonance Plasma Etching

Most parallel-plate plasma etchers, except triode RIE, do not provide the ability to control plasma parameters such as electron energy, plasma density, and reactant density independently. As a result, ion bombardment–induced damage becomes a serious problem. The *electron cyclotron resonance* (ECR) reactor combines microwave power with a static magnetic field to force electrons to circulate around the magnetic field lines at an angular frequency. When this frequency equals the applied microwave frequency, a resonance coupling occurs between the electron energy and the applied electric field that results in a high degree of dissociation and ionization (10^{-2} for ECR compared with 10^{-6} for RIE). Figure 5.8 shows an ECR reaction chamber configuration. Microwave power is coupled through a microwave window into the ECR source region. The magnetic field is supplied from the magnetic coils. ECR plasma systems can also be used in thin film deposition. High efficiency in exciting the reactants in ECR plasmas allows the deposition of films at room temperature without the need for thermal activation.

Other High-Density Plasma Etchers

As feature sizes for ULSI continue to decrease, the limits of the conventional RIE system are being approached. In addition to the ECR system, other types of high-density plasma sources, such as the inductively coupled plasma (ICP) source, the transformer-coupled plasma (TCP) source, and surface wave–coupled plasma (SWP) sources, have been developed. These etchers have high plasma density (10^{11}–10^{12} cm^{-3}) and low processing pressure (<20 mTorr). In addition, they allow the wafer platen to be powered independently of the source, providing significant decoupling between the ion energy (wafer bias) and the ion flux (plasma density, primarily driven by source power). The primary

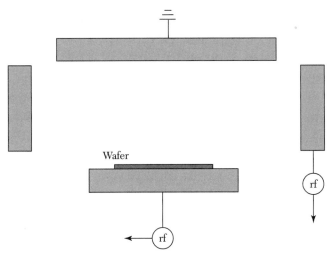

Figure 5.7 Schematic of a triode reactive ion etch reactor. The ion energy is separately controlled by a bias voltage on the bottom electrode. rf, radio frequency.

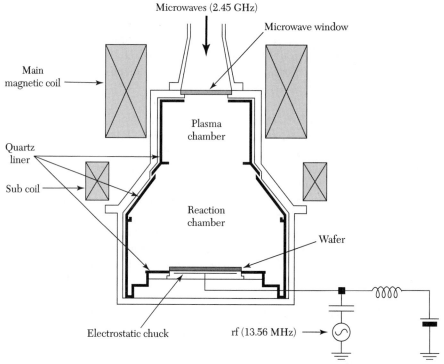

Figure 5.8 Schematic of an electron cyclotron resonance reactor.

processing advantages of HDP sources are better critical dimension (CD) control, higher etching rates, and better selectivity.

In addition, HDP sources provide low substrate damage (because of independent biasing of the substrate and the side electrode potentials) and high anisotropy (because of low pressure, yet high active species density). However, because of their complexity and higher cost, these systems may not be used for less critical applications, such as spacer etching or planarization.[8] Figure 5.9 shows a TCP plasma reactor. A high-density, low pressure plasma is generated by a flat spiral coil that is separated from the plasma by a

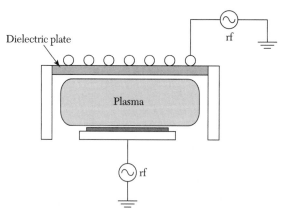

Figure 5.9 Schematic of a transformer-coupled plasma reactor.

dielectric plate on the top of the reactor. The wafer is located away from the coil, so it is not affected by the electromagnetic field generated by the coil. There is little plasma density loss because plasma is generated only a few mean free paths away from the wafer surface. Therefore, a high-density plasma and high etch rates are achieved.

Clustered Plasma Processing

Semiconductor wafers are processed in clean rooms to minimize exposure to ambient particulate contamination. As device dimensions shrink, particulate contamination becomes a more serious problem. To minimize particulate contamination, clustered plasma tools use a wafer handler to pass wafers from one process chamber to another in a vacuum environment. Clustered plasma processing tools can also increase throughput. Figure 5.10 shows the multilayer metal interconnect (TiW/AlCu/TiW) etching process with clustered tools in an AlCu etch chamber, a TiW etch chamber, and a strip passivation chamber. Clustered tools provide an economic advantage by having high chip yield because the wafer is exposed to less ambient contamination and is handled less.

5.2.4 Reactive Plasma Etching Applications

Plasma etching has rapidly evolved from simple batch resist stripping to large and single-wafer processing. Etching systems continue to be improved, from the conventional RIE tool to the high-density plasma tool for pattern transfer of deep-submicron devices. Aside from the etching tool, etch chemistry also plays a critical role in the performance of the etch process. Table 5.3 lists some etch chemistries for different etch processes. Developing an etch process usually means optimizing etch rate, selectivity, profile control, critical dimension, damage, and so forth by adjusting a large number of process parameters.

Silicon Trench Etching

As device feature size decreases, a corresponding decrease is needed in the wafer surface area occupied by the isolation between circuit elements and the storage capacitor

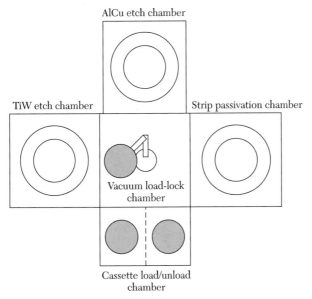

Figure 5.10 Cluster reactive ion etch tool for multilayer metal (TiW/AlCu/TiW) interconnect etching.[2]

TABLE 5.3 Etch Chemistries of Different Etch Processes

Material Being Etched	Etching Chemistry
Deep Si trench	$HBr/NF_3/O_2/SF_6$
Shallow Si trench	$HBr/Cl_2/O_2$
Poly Si	$HBr/Cl_2/O_2$, HBr/O_2, BCl_3/Cl_2, SF_6
Al	BCl_3/Cl_2, $SiCl_4/Cl_2$, HBr/Cl_2
AlSiCu	$BCl_3/Cl_2/N_2$
W	SF_6 only, NF_3/Cl_2
TiW	SF_6 only
WSi_2, $TiSi_2$, $CoSi_2$	CCl_2F_2/NF_3, CF_4/Cl_2, $Cl_2/N_2/C_2F_6$
SiO_2	$CF_4/CHF_3/Ar$, C_2F_6, C_3F_8, C_4F_8/CO, C_5F_8, CH_2F_2
Si_3N_4	CHF_3/O_2, CH_2F_2, CH_2CHF_2

of a DRAM cell. This surface area can be reduced by etching trenches into the silicon substrate and filling them with suitable dielectric or conductive materials. Deep trenches, usually with a depth greater than 5 μm, are used mainly for forming storage capacitors. Shallow trenches, usually with a depth less than 1 μm, are often used for isolation.

Chlorine-based and bromine-based chemistries have a high silicon etch rate and high etch selectivity to the silicon dioxide mask. The combination of $HBr + NF_3 + SF_6 + O_2$ gas mixtures is used to form a trench capacitor with a depth of approximately 7 μm. This combination is also used for shallow trench isolation etching. *Aspect ratio–dependent etching* (i.e., variation in etch rate with aspect ratio) is often observed in deep silicon trench etching, caused by limited ion and neutral transport within the trench. Figure 5.11 shows the dependence of average silicon trench etch rate on aspect ratio. Trenches with large aspect ratios are etched more slowly than trenches with small aspect ratios.

Polysilicon and Polycide Gate Etching
Polysilicon or polycide (i.e., low-resistance metal silicides over polysilicon) is usually used as a gate material for MOS devices. Anisotropic etching and high etch selectivity to the

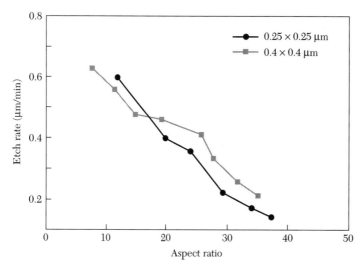

Figure 5.11 Dependence of average silicon trench etch rate on aspect ratio.[2]

gate oxide are the most important requirements for gate etching. For example, the selectivity required in 1G DRAM is more than 150 (i.e., the ratio of etch rates for polycide and gate oxide is 150:1). Achieving high selectivity and etch anisotropy at the same time is difficult for most ion-enhanced etching processes. Therefore, multistep processing is used, in which different etch steps in the process are optimized for etch anisotropy and selectivity. On the other hand, the trend in plasma technology for anisotropic etching and high selectivity is to utilize a low-pressure, high-density plasma using a relatively low power. Most chlorine-based and bromine-based chemistries can be used for gate etching to achieve the required etch anisotropy and selectivity.

Dielectric Etching

The patterning of dielectrics, especially silicon dioxide and silicon nitride, is a key process in the manufacture of modern semiconductor devices. Because of their higher bonding energies, dielectric etching requires aggressive ion-enhanced, fluorine-based plasma chemistry. Vertical profiles are achieved by sidewall passivation, typically by introducing a carbon-containing fluorine species to the plasma (e.g., CF_4, CHF_3, C_4F_8). High ion-bombardment energies are required to remove this polymer layer from the oxide, as well as to mix the reactive species into the oxide surface to form SiF_X products.

A low-pressure, high-density plasma is advantageous for aspect ratio–dependent etching. However, the HDP generates high-temperature electrons and subsequently generates a high degree of dissociation of ions and radicals. It generates far more active radicals and ions than RIE or MERIE plasmas. In particular, a high F concentration worsens the selectivity to silicon. Various methods have been tried to enhance the selectivities in the high-density plasma. First, a parent gas with a high C/F ratio, such as C_2F_6, C_4F_8, or C_5F_8, was attempted. Other methods to scavenge F radicals have also been developed.[9]

Interconnect Metal Etching

Etching of a metallization layer is a very important step in IC fabrication. Aluminum, copper, and tungsten are the most popular materials used for interconnection. These materials usually require anisotropic etching. The reaction of aluminum with fluorine results in nonvolatile AlF_3, which has a vapor pressure of only 1 Torr at 1240°C. Chlorine-based chemistry (e.g., a Cl_2/BCl_3 mixture) has been widely used for aluminum etching. Chlorine has a very high chemical etch rate with aluminum and tends to produce an undercut during etching. Carbon-containing gas (e.g., CHF_3) or N_2 is added to form sidewall passivation during aluminum etching to obtain anisotropic etching.

Exposure to the ambient is another problem in aluminum etching. Residual chlorine on the aluminum sidewall and the photoresist tends to react with atmospheric water to form HCl, which corrodes aluminum. An *in situ* exposure of the wafer to a CF_4 discharge to exchange Cl with F and then to an oxygen discharge to remove the resist, followed by immediate immersion in deionized water, can eliminate aluminum corrosion. Figure 5.12 shows 0.35-μm TiN/Al/Ti lines and spaces on a wafer maintained at ambient for 72 hours. No corrosion is present even after prolonged exposure to the ambient.

Copper has drawn much attention as a new metallization material in ULSI circuits because of its low resistivity (~1.7 Ω-cm) and superior resistance to electromigration compared with Al or Al alloys. However, because of the low volatility of copper halides, plasma etching at room temperature is difficult. Process temperatures higher than 200°C are required to etch copper films. Therefore, the *damascene* process is used to form Cu interconnection without dry etching. Damascene processing involves the creation of interconnect lines by first etching a trench or canal in a planar dielectric layer and then filling that trench with metal, such as aluminum or copper. In dual damascene processing (Fig. 5.13), a second level is involved where a series of holes (i.e., contacts or vias) are etched and filled

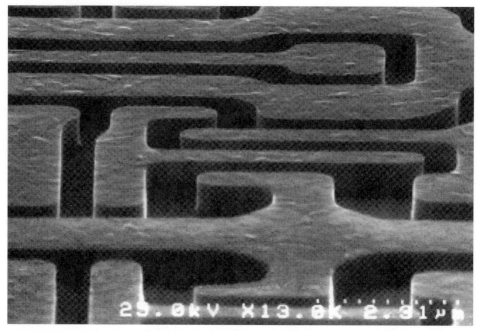

Figure 5.12 0.35-μm TiN/Al/TiN lines and spaces on a wafer maintained at ambient for 72 hours after a microwave strip are not corroded.

in addition to the trench. After filling, the metal and dielectric are planarized by *chemical mechanical polishing* (CMP, see Chapter 8). The advantage of damascene processing is that it eliminates the need for metal etch. This is an important concern as the industry moves from aluminum to copper interconnections.

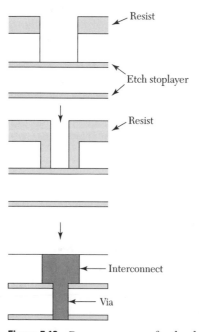

Figure 5.13 Process sequence for the dual damascene process.

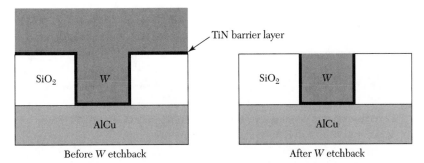

Figure 5.14   Formation of tungsten plug in a contact hole by depositing blanket low-pressure chemical vapor deposition W and then using reaction ion etching etchback.

Low-pressure CVD (LPCVD) tungsten (W) has been widely used for filling contact holes and first-level metallization because of its excellent deposition conformability. Both fluorine- and chlorine-based chemistries etch W and form volatile etch products. An important tungsten etch process is the blanket W etchback to form a W plug. The blanket LPCVD W is deposited on top of a TiN barrier layer, as shown in Figure 5.14. A two-step process is usually used. First, 90% of the W is etched at a high etch rate, and then the etch rate is reduced to remove the remaining W with an etchant that has a high W-to-TiN selectivity.

▶ 5.3 ETCH SIMULATION

SUPREM may be used to simulate the etching process. However, etch simulation using SUPREM is rudimentary at best. Simulation results may be achieved using the ETCH command, which allows the user to etch all or part of any given layer at the top of the current structure. If the material at the top of the structure is not the material specified, then no etching takes place. If the amount to be etched is not specified, then the entire layer is removed.

EXAMPLE 3

Suppose we want to simulate the etching of 0.3 μm of the oxide grown after the dry-wet-dry sequence in Example 3 of Chapter 3.

SOLUTION   The SUPREM input listing is as follows:

```
TITLE          Etching Example
COMMENT        Initialize silicon substrate
INITIALIZE     <100> Silicon Phosphor Concentration=1e16
COMMENT        Ramp furnace up to 1100 C over 10 minutes in N2
DIFFUSION      Time=10 Temperature=900 Nitrogen T.rate=20
COMMENT        Oxidize the wafers for 5 minutes at 1100 C in dry O2
DIFFUSION      Time=5 Temperature=1100 DryO2
COMMENT        Oxidize the wafers for 120 minutes at 1100 C in wet O2
DIFFUSION      Time=120 Temperature=1100 WetO2
COMMENT        Oxidize the wafers for 5 minutes at 1000 C in dry O2
DIFFUSION      Time=5 Temperature=1100 DryO2
COMMENT        Ramp furnace down to 900 C over 10 minutes in N2
DIFFUSION      Time=10 Temperature=1100 Nitrogen T.rate=-20
```

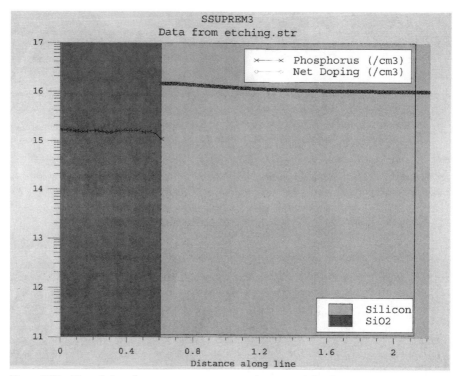

Figure 5.15   Plot of phosphorus concentration as a function of depth into the silicon substrate, using SUPREM.

```
ETCH              Oxide Thickness = 0.3
PRINT             Layers Chemical Concentration Phosphor
PLOT              Active Net Cmin=1e14
STOP              End etching example
```

After oxidation is complete, we print and plot the phosphorus concentration as a function of depth into the silicon substrate. The results are shown in Figure 5.15, which indicates a final oxide thickness of 0.609 μm and depicts the phosphorus incorporation in the oxide layer.   ◀

▶ 5.4 SUMMARY

The two major processes to transfer patterns in IC fabrication are photolithography and etching. Wet chemical etching is used extensively in semiconductor processing. It is particularly suitable for blanket etching. This chapter discussed wet chemical etching processes for silicon and gallium arsenide, insulators, and metal interconnections. Wet chemical etching was used for pattern transfer. However, undercutting of the layer underneath the mask resulted in loss of resolution in the etched pattern.

Dry etching methods are used to achieve high-fidelity pattern transfer. Dry etching is synonymous with plasma-assisted etching. This chapter considered plasma fundamentals and various dry etching systems, which have grown from relatively simple, parallel-plate configurations to complex chambers with multiple frequency generators and a variety of process-control sensors.

The challenges for future etching technology are high etch selectivity, better dimensioned control, low aspect ratio–dependent etching, and low plasma-induced damage. Low-pressure, high-density plasma reactors are necessary to meet these requirements.

As processing evolves from 200-mm to 300-mm and even larger diameter wafers, continued improvements are required for etch uniformity across the wafer. New gas chemistries must be developed to provide the improved selectivity necessary for advanced integration schemes.

▶ REFERENCES

1. For a more detailed discussion on etching, see Y. J. T. Liu, "Etching," in C. Y. Chang and S. M. Sze, Eds., *ULSI Technology*, McGraw-Hill, New York, 1996.

2. H. Robbins and B. Schwartz, "Chemical Etching of Silicon II, the System HF, HNO_3, H_2O and $HC_2H_3O_2$," *J. Electrochem. Soc.*, **107,** 108 (1960).

3. K. E. Bean, "Anisotropic Etching in Silicon," *IEEE Trans. Electron Devices*, **ED-25,** 1185 (1978).

4. D. P. Kern, et al., "Practical Aspects of Microfabrication in the 100-nm Region," *Solid State Technol.*, **27,** 2, 127 (1984).

5. S. Iida and K. Ito, "Selective Etching of Gallium Arsenide Crystal in H_2SO_4-H_2O_2-H_2O System," *J. Electrochem. Soc.*, **118,** 768 (1971).

6. E. C. Douglas, "Advanced Process Technology for VLSI Circuits," *Solid State Technol.*, **24,** 5, 65 (1981).

7. J. A. Mucha and D. W. Hess, "Plasma Etching," in L. F. Thompson and C. G. Willson, Eds., *Microcircuit Processing: Lithography and Dry Etching*, American Chemical Society, Washington, DC, 1984.

8. M. Armacost, et al., "Plasma-Etching Processes for ULSI Semiconductor Circuits," *IBM J. Res. Dev.*, **43,** 39 (1999).

9. C. O. Jung, et al., "Advanced Plasma Technology in Microelectronics," *Thin Solid Films*, **341,** 112 (1999).

▶ PROBLEMS

Asterisks denote difficult problems.

SECTION 5.1: WET CHEMICAL ETCHING

1. If the mask and the substrate cannot be etched by a particular etchant, sketch the edge profile of an isotropically etched feature in a film of thickness h_f for (a) etching just to completion, (b) 100% overetch, and (c) 200% overetch.

2. A <100>-oriented silicon crystal is etched in a KOH solution through a 1.5 μm × 1.5 μm window defined in silicon dioxide. The etch rate normal to (100) planes is 0.6 μm/min. The etch rate ratios are 100:16:1 for the (100):(110):(111) planes. Show the etched profile after 20 seconds, 40 seconds, and 60 seconds.

3. Repeat the previous problem with a <$\bar{1}$10>-oriented silicon etched with a thin SiO_2 mask in KOH solution. Show the etched pattern profiles on <$\bar{1}$10> Si.

4. A <100>-oriented silicon wafer 150 mm in diameter is 625 μm thick. The wafer has 1000 μm × 1000 μm ICs on it. The IC chips are to be separated by orientation-dependent etching. Describe two methods for doing this and calculate the fraction of the surface area that is lost in these processes.

SECTION 5.2: DRY ETCHING

°5. The average distance traveled by particles between collisions is called the mean free path (λ), $\lambda \cong 5 \times 10^{-3}/P$ (cm), where P is pressure in Torr. In typical plasmas of interest, the chamber pressure ranges from 1 Pa to 150 Pa. What are the corresponding density of gas molecules (cm^{-3}) and the mean free path?

6. Fluorine (F) atoms etch Si at a rate given by

$$\text{Etch rate (nm/min)} = 2.86 \times 10^{-13} \, n_F \times T^{1/2} \exp(-E_a/RT)$$

where n_F is the concentration of F atoms (cm^{-3}), T the temperature (K), and E_a and R the activation energy (2.48 kcal/mol) and gas constant (1.987 cal-K), respectively. If n_F is 3×10^{15}, calculate the etch rate of Si at room temperature.

7. SiO$_2$ etched by F atoms could also be expressed by

$$\text{Etch rate (nm/min)} = 0.614 \times 10^{-13} \, n_F \times T^{1/2} \exp(-E_a/RT)$$

where n_F is 3×10^{15} (cm^{-3}) and E_a is 3.76 kcal/mol. Calculate the etch rate of SiO$_2$ and etch selectivity of SiO$_2$ over Si at room temperature.

8. A multiple-step etch process is required for etching a polysilicon gate with thin gate oxide. How do you design an etch process that has no micromasking, has an anisotropic etch profile, and is selective to thin gate oxide?

9. Find the etch selectivity required to etch a 400-nm polysilicon layer without removing more than 1 nm of its underlying gate oxide, assuming that the polysilicon is etched with a process having a 10% etch-rate uniformity.

10. A 1-μm Al film is deposited over a flat field oxide region and patterned with photoresist. The metal is then etched with a mixture of BCl$_3$/Cl$_2$ gases at a temperature of 70°C in a Helicon etcher. The selectivity of Al over photoresist is maintained at 3. Assuming a 30% overetch, what is the minimum photoresist thickness required to ensure that the top metal surface is not attacked?

11. In an ECR plasma, a static magnetic field B forces electrons to circulate around the magnetic field lines at an angular frequency, ω_e, that is given by

$$\omega_e = qB/m_e$$

where q is the electronic charge and m_e the electron mass. If the microwave frequency is 2.45 GHz, what is the required magnetic field?

12. What are the major distinctions between traditional reactive ion etching and high-density plasma etching (ECR, ICP, etc.)?

13. Describe how to eliminate the corrosion issues in Al lines after etching with chlorine-based plasma.

6

Diffusion

Impurity doping is the introduction of controlled amounts of impurity dopants into semiconductors. The practical use of impurity doping mainly has been to change the electrical properties of the semiconductors. Diffusion and ion implantation are the two key methods of impurity doping. Both diffusion and ion implantation are used for fabricating discrete devices and integrated circuits because these processes generally complement each other.[1,2] For example, diffusion is used to form a deep junction (e.g., a twin well in CMOS), whereas ion implantation, which is discussed in Chapter 7, is used to form a shallow junction (e.g., a source/drain junction of a MOSFET).

Until the early 1970s, impurity doping was done mainly by diffusion at elevated temperatures, as shown in Figure 6.1. In this method the dopant atoms are placed on or near the surface of the wafer by deposition from the gas phase of the dopant or by using doped oxide sources. The doping concentration decreases monotonically from the surface, and the profile of the dopant distribution is determined mainly by the temperature and diffusion time.

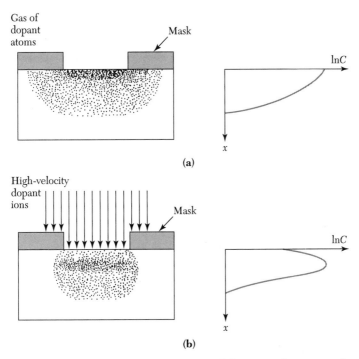

Figure 6.1 Comparison of (*a*) diffusion and (*b*) ion-implantation techniques for the selective introduction of dopants into the semiconductor substrate.

This chapter focuses on the diffusion process. Ion implantation is covered in Chapter 7. Specifically, this chapter covers the following topics:

- The movement of impurity atoms in the crystal lattice under high-temperature and high-concentration-gradient conditions
- Impurity profiles for constant diffusivity and concentration-dependent diffusivity
- The impact of lateral diffusion and impurity redistribution on device characteristics
- The simulation of diffusion using SUPREM

▶ 6.1 BASIC DIFFUSION PROCESS

Diffusion of impurities is typically done by placing semiconductor wafers in a carefully controlled, high-temperature quartz-tube furnace and passing a gas mixture that contains the desired dopant through it. The temperature usually ranges between 800°C and 1200°C for silicon and 600°C and 1000°C for gallium arsenide. The number of dopant atoms that diffuse into the semiconductor is related to the partial pressure of the dopant impurity in the gas mixture.

For diffusion in silicon, boron is the most popular dopant for introducing a p-type impurity, whereas arsenic and phosphorus are used extensively as n-type dopants. These three elements are highly soluble in silicon; they have solubilities above 5×10^{20} cm^{-3} in the diffusion temperature range. These dopants can be introduced in several ways, including solid sources (e.g., BN for boron, As$_2$O$_3$ for arsenic, and P$_2$O$_5$ for phosphorus), liquid sources (BBr$_3$, AsCl$_3$, and POCl$_3$), and gaseous sources (B$_2$H$_6$, AsH$_3$, and PH$_3$). However, liquid sources are most commonly used. A schematic diagram of the furnace and gas flow arrangement for a liquid source is shown in Figure 6.2. This arrangement is similar to that used for thermal oxidation. An example of the chemical reaction for phosphorus diffusion using a liquid source is

$$4POCl_3 + 3O_2 \rightarrow 2P_2O_5 + 6Cl_2 \uparrow \tag{1}$$

The P$_2$O$_5$ forms a glass-on-silicon wafer and is then reduced to phosphorus by silicon,

$$2P_2O_5 + 5Si \rightarrow 4P + 5SiO_2 \tag{2}$$

The phosphorus is released and diffuses into the silicon, and Cl$_2$ is vented.

For diffusion in gallium arsenide, the high vapor pressure of arsenic requires special methods to prevent the loss of arsenic by decomposition or evaporation.[2] These methods include diffusion in sealed ampules with an overpressure of arsenic and diffusion in an open-tube furnace with a doped oxide capping layer (e.g., silicon nitride). Most of the studies on p-type diffusion have been confined to the use of zinc in the forms of Zn–Ga–As

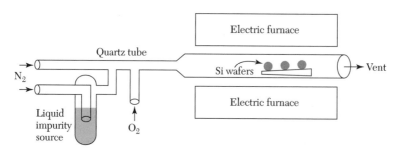

Figure 6.2 Schematic diagram of a typical open-tube diffusion system.

alloys and $ZnAs_2$ for the sealed-ampule approach or $ZnO–SiO_2$ for the open-tube approach. The n-type dopants in gallium arsenide include selenium and tellurium.

6.1.1 Diffusion Equation

Diffusion in a semiconductor can be visualized as the atomic movement of the diffusant (dopant atoms) in the crystal lattice by vacancies or interstitials. Figure 6.3 shows the two basic atomic diffusion models in a solid.[1,3] The open circles represent the host atoms occupying the equilibrium lattice positions. The solid dots represent impurity atoms. At elevated temperatures, the lattice atoms vibrate around the equilibrium lattice sites. There is a finite probability that a host atom will acquire sufficient energy to leave the lattice site and become an interstitial atom, thereby creating a vacancy. When a neighboring impurity atom migrates to the vacancy site, as illustrated in Figure 6.3a, the mechanism is called *vacancy diffusion*. If an interstitial atom moves from one place to another without occupying a lattice site (Fig. 6.3b), the mechanism is *interstitial diffusion*. An atom smaller than the host atom often moves interstitially.

The basic diffusion process of impurity atoms is similar to that of charge carriers (electrons and holes). We define a flux F as the number of dopant atoms passing through a unit area in a unit time and C as the dopant concentration per unit volume. We then have

$$F = -D\frac{\partial C}{\partial x} \tag{3}$$

where the proportionality constant D is the *diffusion coefficient* or *diffusivity*. Note that the basic driving force of the diffusion process is the concentration gradient $\partial C/\partial x$. The flux is proportional to the concentration gradient, and the dopant atoms will move (diffuse) away from a high-concentration region toward a lower-concentration region.

If we substitute Eq. 3 into the one-dimensional continuity equation under the condition that no materials are formed or consumed in the host semiconductor, we obtain

$$\frac{\partial C}{\partial t} = -\frac{\partial F}{\partial x} = \frac{\partial}{\partial x}\left(D\frac{\partial C}{\partial x}\right) \tag{4}$$

When the concentration of the dopant atoms is low, the diffusion coefficient can be considered to be independent of doping concentration, and Eq. 4 becomes

$$\frac{\partial C}{\partial t} = D\frac{\partial^2 C}{\partial x^2} \tag{5}$$

Equation 5 is often referred to as *Fick's diffusion equation* or *Fick's law*.

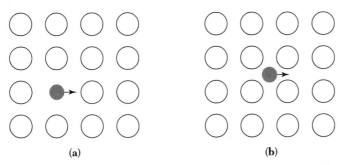

(a)   (b)

Figure 6.3 Atomic diffusion mechanisms for a two-dimensional lattice.[1,3] (a) Vacancy mechanism. (b) Interstitial mechanism.

Figure 6.4 shows the measured diffusion coefficients for low concentrations of various dopant impurities in silicon and gallium arsenide.[4,5] The logarithm of the diffusion coefficient plotted against the reciprocal of the absolute temperature gives a straight line in most of the cases. This implies that over the temperature range, the diffusion coefficients can be expressed as

$$D = D_0 \exp\left(\frac{-E_a}{kT}\right) \qquad (6)$$

where D_0 is the diffusion coefficient in cm²/s extrapolated to infinite temperature, and E_a is the activation energy in eV.

For the interstitial diffusion model, E_a is related to the energies required to move dopant atoms from one interstitial site to another. The values of E_a are found to be between 0.5 and 2 eV in both silicon and gallium arsenide. For the vacancy diffusion model, E_a is related to both the energies of motion and the energies of formation of vacancies. Thus, E_a for vacancy diffusion is larger than that for interstitial diffusion, usually between 3 and 5 eV.

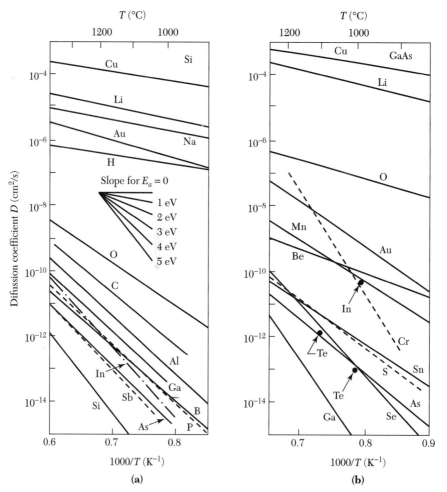

Figure 6.4 Diffusion coefficient (also called diffusivity) as a function of the reciprocal of temperature for (*a*) silicon and (*b*) gallium arsenide.[4, 5]

For fast diffusants, such as Cu in Si and GaAs, shown in the upper portion of Figures 6.4*a* and *b*, the measured activation energies are less than 2 eV, and interstitial atomic movement is the dominant diffusion mechanism. For slow diffusants, such as As in Si and GaAs, shown in the lower portion of Figures 6.4*a* and *b*, E_a is larger than 3 eV, and vacancy diffusion is the dominant mechanism.

6.1.2 Diffusion Profiles

The diffusion profile of the dopant atoms is dependent on the initial and boundary conditions. This subsection considers two important cases, namely, constant surface concentration diffusion and constant total dopant diffusion. In the first case, impurity atoms are transported from a vapor source onto the semiconductor surface and diffused into the semiconductor wafers. The vapor source maintains a constant level of surface concentration during the entire diffusion period. In the second case, a fixed amount of dopant is deposited onto the semiconductor surface and is subsequently diffused into the wafers.

Constant Surface Concentration
The initial condition at $t = 0$ is

$$C(x, 0) \tag{7}$$

which states that the dopant concentration in the host semiconductor is initially zero. The boundary conditions are

$$C(0, t) = C_s \tag{8a}$$

and

$$C(\infty, t) = 0 \tag{8b}$$

where C_s is the surface concentration (at $x = 0$), which is independent of time. The second boundary condition states that at large distances from the surface there are no impurity atoms.

The solution of Fick's diffusion equation that satisfies the initial and boundary conditions is given by[6]

$$C(x,t) = C_s \operatorname{erfc}\left(\frac{x}{2\sqrt{Dt}}\right) \tag{9}$$

where erfc is the complementary error function and $\sqrt{Dt}$ is the diffusion length. The definition of erfc and some properties of the function are summarized in Table 6.1. The diffusion profile for the constant surface concentration condition is shown in Figure 6.5*a*, which plots, on both linear (upper) and logarithmic (lower) scales, the normalized concentration as a function of depth for three values of the diffusion length corresponding to three consecutive diffusion times and a fixed D for a given diffusion temperature. Note that as time progresses, the dopant penetrates deeper into the semiconductor.

The total number of dopant atoms per unit area of the semiconductor is given by

$$Q(t) = \int_0^\infty C(x,t)\, dx \tag{10}$$

Substituting Eq. 9 into Eq. 10 yields

$$Q(t) = \frac{2}{\sqrt{\pi}} C_s \sqrt{Dt} \cong 1.13 C_s \sqrt{Dt} \tag{11}$$

TABLE 6.1   Error Function Algebra

$$\text{erf}(x) \equiv \frac{2}{\sqrt{\pi}} \int_0^x e^{-y^2} dy$$

$$\text{erfc}(x) \equiv 1 - \text{erf}(x)$$

$$\text{erf}(0) = 0$$

$$\text{erf}(\infty) = 1$$

$$\text{erf}(x) \cong \frac{2}{\sqrt{\pi}} x \quad \text{for } x \ll 1$$

$$\text{erfc}(x) \cong \frac{1}{\sqrt{\pi}} \frac{e^{-x^2}}{x} \quad \text{for } x \gg 1$$

$$\frac{d}{dx} \text{erf}(x) = \frac{2}{\sqrt{\pi}} e^{-x^2}$$

$$\frac{d^2}{dx^2} \text{erf}(x) = -\frac{4}{\sqrt{\pi}} x e^{-x^2}$$

$$\int_0^x \text{erfc}(y') dy' = x \, \text{erfc}(x) + \frac{1}{\sqrt{\pi}} \left(1 - e^{-x^2}\right)$$

$$\int_0^\infty \text{erfc}(x) dx = \frac{1}{\sqrt{\pi}}$$

This expression can be interpreted as follows. The quantity $Q(t)$ represents the area under one of the diffusion profiles of the linear plot in Figure 6.5a. These profiles can be approximated by triangles with height C_s and base $2\sqrt{Dt}$. This leads to $Q(t) \cong C_s \sqrt{Dt}$, which is close to the exact result obtained from Eq. 11.

A related quantity is the gradient of the diffusion profile $\partial C/\partial x$. The gradient can be obtained by differentiating Eq. 9:

$$\left.\frac{\partial C}{\partial x}\right|_{x,t} = \frac{C_s}{\sqrt{\pi Dt}} e^{-x^2/4Dt} \tag{12}$$

EXAMPLE 1

For a boron diffusion in silicon at 1000°C, the surface concentration is maintained at 10^{19} cm^{-3} and the diffusion time is 1 hour. Find $Q(t)$ and the gradient at $x = 0$ and at a location where the dopant concentration reaches 10^{15} cm^{-3}.

SOLUTION   The diffusion coefficient of boron at 1000°C, as obtained from Figure 6.4, is about 2×10^{14} cm^2/s, so the diffusion length is

$$\sqrt{Dt} = \sqrt{2 \times 10^{-14} \times 3600} = 8.48 \times 10^{-6} \text{ cm}$$

$$Q(t) = 1.13 C_s \sqrt{Dt} = 1.13 \times 10^{19} \times 8.48 \times 10^{-6} = 9.5 \times 10^{13} \text{ atoms/cm}^2$$

$$\left.\frac{dC}{dx}\right|_{x=0} = -\frac{C_s}{\sqrt{\pi Dt}} = \frac{-10^{19}}{\sqrt{\pi} \times 8.48 \times 10^{-6}} = -6.7 \times 10^{23} \text{ cm}^{-4}$$

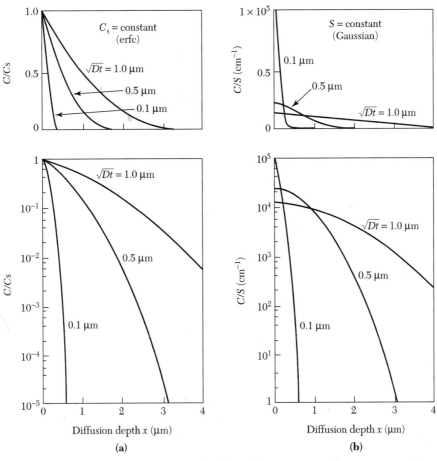

Figure 6.5 Diffusion profiles. (a) Normalized complementary error function versus distance for successive diffusion times. (b) Normalized Gaussian function versus distance.

When $C = 10^{15}$ cm^{-3}, the corresponding distance x_j is given by Eq. 9, or

$$x_j = 2\sqrt{Dt}\ \mathrm{erfc}^{-1}\left(\frac{10^{15}}{10^{19}}\right) = 2\sqrt{Dt}\,(2.75) = 4.66\times10^{-5}\ \mathrm{cm} = 0.466\ \mu\mathrm{m}$$

$$\left.\frac{dC}{dx}\right|_{x=0.466\,\mu m} = -\frac{C_s}{\sqrt{\pi Dt}}e^{-x^2/4Dt} = -3.5\times10^{20}\ \mathrm{cm}^{-4}$$

◀

Constant Total Dopant

For this case, a fixed (or constant) amount of dopant is deposited onto the semiconductor surface in a thin layer, and the dopant subsequently diffuses into the semiconductor. The initial condition is the same as in Eq. 7. The boundary conditions are

$$\int_0^\infty C(x,t) = S \tag{13a}$$

and

$$C(\infty,t) = 0 \tag{13b}$$

where S is the total amount of dopant per unit area.

The solution of the diffusion equation that satisfies the above conditions is

$$C(x,t) = \frac{S}{\sqrt{\pi Dt}} \exp\left(-\frac{x^2}{4Dt}\right) \tag{14}$$

This expression is the Gaussian distribution. Since the dopant will move into the semiconductor as time increases, in order to keep the total dopant S constant, the surface concentration must decrease. This is indeed the case, since the surface concentration is given by Eq. 14 with $x = 0$:

$$C(x,t) = \frac{S}{\sqrt{\pi Dt}} \tag{15}$$

Figure 6.5b shows the dopant profile for a Gaussian distribution in which the normalized concentration (C/S) is plotted as a function of the distance for three increasing diffusion lengths. Note the reduction of the surface concentration as the diffusion time increases. The gradient of the diffusion profile is obtained by differentiating Eq. 14 and is

$$\frac{dC}{dx}\bigg|_{x,t} = -\frac{xS}{2\sqrt{\pi}(Dt)^{3/2}} = -\frac{x}{2Dt}C(x,t) \tag{16}$$

The gradient (or slope) is zero at $x = 0$ and at $x = \infty$, and the maximum gradient occurs at $x = \sqrt{2Dt}$.

In integrated circuit processing, a two-step diffusion process is commonly used, in which a *predeposition* diffused layer is first formed under the constant surface concentration condition. This step is followed by a *drive-in* diffusion (also called *redistribution* diffusion) under a constant total dopant condition. For most practical cases, the diffusion length $\sqrt{Dt}$ for the predeposition diffusion is much smaller than the diffusion length for the drive-in diffusion. Therefore, the predeposition profile can be considered a delta function at the surface, and the extent of the penetration of the predeposition profile can be regarded to be negligibly small compared with that of the final profile that results from the drive-in step.

EXAMPLE 2

Arsenic was predeposited by arsine gas, and the resulting total amount of dopant per unit area was 1×10^{14} atoms/cm^2. How long would it take to drive the arsenic in to a junction depth of 1 μm? Assume a background doping of $C_B = 1 \times 10^{15}$ atoms/cm^3, and a drive-in temperature of 1200°C. For As diffusion, $D_0 = 24$ cm^2/s, and $E_a = 4.08$ eV.

SOLUTION

$$D = D_0\exp\left(\frac{-E_a}{kT}\right) = 24\exp\left(\frac{-4.08}{8.614\times10^{-5}\times1473}\right) = 2.602\times10^{-13} \text{ cm}^2/\text{s}$$

$$x_j^2 = 10^{-8} = 4Dt\ln\left(\frac{S}{C_B\sqrt{Dt}}\right) = 1.04\times10^{-12}t\ln\left(\frac{1.106\times10^5}{\sqrt{t}}\right)$$

$$t \bullet \log t - 10.09t + 8350 = 0$$

The solution to this equation can be determined by the cross point of equation $y = t \bullet \log t$ and $y = 10.09t - 8350$. Therefore, $t = 1190$ seconds, or approximately 20 minutes. ◀

6.1.3 Evaluation of Diffused Layers

The results of a diffusion process can be evaluated by three measurements: the junction depth, the sheet resistance, and the dopant profile of the diffused layer. The junction depth can be delineated by cutting a groove into the semiconductor and etching the surface with a solution (e.g., 100 cm^3 HF and a few drops of HNO$_3$ for silicon) that stains the p-type region darker than the n-type region, as illustrated in Figure 6.6a. If R_0 is the radius of the tool used to form the groove, then the junction depth x_j is given by

$$x_j = \sqrt{R_0^2 - b^2} - \sqrt{R_0^2 - a^2} \tag{17}$$

where a and b are indicated in the figure. In addition, if R_0 is much larger than a and b, then

$$x_j \cong \frac{a^2 - b^2}{2R_0} \tag{18}$$

The junction depth x_j, as illustrated in Figure 6.6b, is the position where the dopant concentration equals the substrate concentration C_B, or

$$C(x_j) = C_B \tag{19}$$

Thus, if the junction depth and C_B are known, the surface concentration C_s and the impurity distribution can be calculated, provided the diffusion profile follows one or the other simple equation derived in Section 6.1.2.

The resistance of a diffused layer can be measured by the *four-point probe* technique shown in Figure 6.7. The probes are equally spaced. A small current I from a constant-current source is passed through the outer two probes, and a voltage V is measured between the inner two probes. For a thin semiconductor sample with thickness W that is much smaller than the sample diameter d, the *resistivity* (ρ) is given by

$$\rho = \frac{V}{I} \cdot W \cdot CF \ \ \Omega\text{-cm} \tag{20}$$

where CF is known as the correction factor. The correction factor depends on the ratio of d/s, where s is the probe spacing. When $d/s > 20$, the correction factor approaches 4.54.

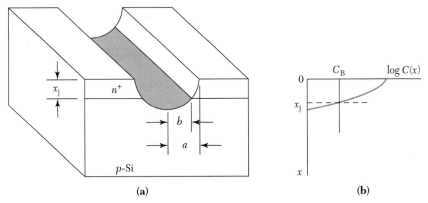

p-Si

(a)

(b)

Figure 6.6 Junction-depth measurement. (a) Grooving and staining. (b) Position in which dopant and substrate concentrations are equal.

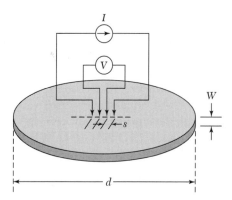

Figure 6.7 Measurement of resistivity using a four-point probe.[3]

The *sheet resistance* (R_s) is related to the junction depth (x_j), the carrier mobility (μ, which is a function of the total impurity concentration), and the impurity distribution $C(x)$ by the following expression:[7]

$$R_s = \frac{1}{q\int_0^{x_j} \mu C(x)\, dx}$$

(21)

For a given diffusion profile, the average resistivity ($\overline{\rho} = R_s x_j$) is uniquely related to the surface concentration (C_s) and the substrate doping concentration for an assumed diffusion profile. Design curves relating C_s and $\overline{\rho}$ have been calculated for simple diffusion profiles, such as the erfc or Gaussian distribution.[8] To use these curves correctly, we must be sure that the diffusion profiles agree with the assumed profiles. For low concentration and deep diffusions, the diffusion profiles generally can be represented by the aforementioned simple functions. However, as discussed in the next section, for high concentration and shallow diffusions, the diffusion profiles cannot be represented by these simple functions.

The diffusion profile can be measured using a capacitance-voltage technique. The majority carrier profile (n), which is equal to the impurity profile if impurities are fully ionized, can be determined by measuring the reverse-bias capacitance of a p–n junction or a Schottky barrier diode as a function of the applied voltage. This is due to the relationship[9]

$$n = \frac{2}{q\varepsilon_s}\left[\frac{-1}{d\left(1/C'^2\right)/dV}\right]$$

(22)

where q is the charge of an electron, ε_s is the permittivity of the semiconductor, C' is capacitance per unit area of the sample, and V is the applied voltage.

A more elaborate method is the secondary ion mass spectroscopic (SIMS) technique, which measures the total impurity profile. In the SIMS technique, an ion beam sputters material off the surface of a semiconductor, and the ion component is detected and mass analyzed. This technique has high sensitivity to many elements, such as boron and arsenic, and is an ideal tool for providing the precision needed for profile measurements in high-concentration or shallow-junction diffusions.[10]

▶ 6.2 EXTRINSIC DIFFUSION

The diffusion profiles described in Section 6.1 are for constant diffusivities. These profiles occur when the doping concentration is lower than the intrinsic carrier concentration (n_i) at the diffusion temperature. For example, at $T = 1000°C$, n_i equals 5×10^{18} cm^{-3} for silicon and 5×10^{17} cm^{-3} for gallium arsenide. The diffusivity at low concentrations is often referred to as the *intrinsic diffusivity*. Doping profiles that have concentrations less than $n_i(T)$ are in the intrinsic diffusion region, as indicated in the left side of Figure 6.8. In this region, the resulting dopant profiles of sequential or simultaneous diffusions of *n*- and *p*-type impurities can be determined by superposition; that is, the diffusions can be treated independently. However, when the impurity concentration, including both the substrate and the dopant, is greater than $n_i(T)$, the semiconductor becomes *extrinsic*, and the diffusivity is considered to be extrinsic. In the extrinsic diffusion region, the diffusivity becomes concentration dependent.[11] In the extrinsic diffusion region, the diffusion profiles are more complicated, and there are interactions and cooperative effects among the sequential or simultaneous diffusions.

6.2.1 Concentration-Dependent Diffusivity

As mentioned previously, when a host atom acquires sufficient energy from the lattice vibration to leave its lattice site, a vacancy is created. Depending on the charges associated with a vacancy, we can have a neutral vacancy V^0, an acceptor vacancy V^-, a double-charged acceptor vacancy V^{2-}, a donor vacancy V^+, and so forth. We expect that the vacancy density of a given charge state (i.e., the number of vacancies per unit volume, C_V) has a temperature dependence similar to that of the carrier density, that is,

$$C_V = C_i \exp\left(\frac{E_F - E_i}{kT}\right) \tag{23}$$

where C_i is the intrinsic vacancy density, E_F is the Fermi level, and E_i is the intrinsic Fermi level.

If the dopant diffusion is dominated by the vacancy mechanism, the diffusion coefficient is expected to be proportional to the vacancy density. At low doping concentrations

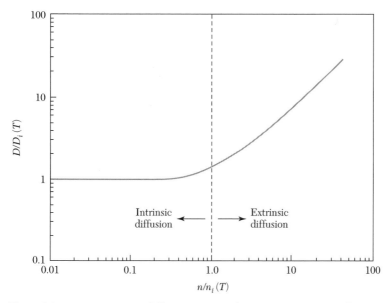

Figure 6.8 Donor impurity diffusivity versus electron concentration showing regions of intrinsic and extrinsic diffusion.[11]

$(n < n_i)$, the Fermi level coincides with the intrinsic Fermi level $(E_F = E_i)$. The vacancy density is equal to C_i and is independent of doping concentration. The diffusion coefficient, which is proportional to C_i, also is independent of doping concentration. At high concentrations $(n > n_i)$, the Fermi level will move toward the conduction band edge (for donor-type vacancies), and the term $[\exp(E_F - E_i)/kT]$ becomes larger than unity. This causes C_V to increase, which in turn causes the diffusion coefficient to increase, as shown in the right side of Figure 6.8.

When the diffusion coefficient varies with dopant concentration, Eq. 4 should be used as the diffusion equation instead of Eq. 5, in which D is independent of C. We consider the case where the diffusion coefficient can be written as

$$D = D_s \left(\frac{C}{C_s} \right)^\gamma \tag{24}$$

where C_s is the surface concentration, D_s is the diffusion coefficient at the surface, and γ is a parameter to describe the concentration dependence. For such a case, we can write Eq. 4 as an ordinary differential equation and solve it numerically.

Figure 6.9 shows the solutions[12] for constant surface concentration diffusion with different values of γ. For $\gamma = 0$, we have the case of constant diffusivity, and the profile is the same as that shown in Figure 6.5a. For $\gamma > 0$, the diffusivity decreases as the concentration decreases, and increasingly steep and boxlike concentration profiles result for increasing γ. Therefore, highly abrupt junctions are formed when diffusions are made into a background of an opposite impurity type. The abruptness of the doping profile results in a junction depth virtually independent of the background concentration. Note that the junction depth (see Fig. 6.9) is given by

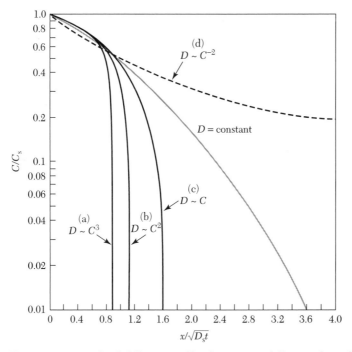

Figure 6.9 Normalized diffusion profiles for extrinsic diffusion where the diffusion coefficient becomes concentration dependent.[12]

$$x_j = 1.6\sqrt{D_s t} \text{ for } D \sim C \ (\gamma = 1)$$

$$x_j = 1.1\sqrt{D_s t} \text{ for } D \sim C^2 \ (\gamma = 2) \tag{25}$$

$$x_j = 0.87\sqrt{D_s t} \text{ for } D \sim C^3 \ (\gamma = 3)$$

In the case of $\gamma = -2$, the diffusivity increases with decreasing concentration, which leads to a concave profile, as opposed to the convex profiles for other cases.

6.2.2 Diffusion Profiles

Diffusion in Silicon

The measured diffusion coefficients of boron and arsenic in silicon have a concentration dependence with $\gamma \cong 1$. Their concentration profiles are abrupt, as depicted in curve c of Figure 6.9. For gold and platinum diffusion in silicon, γ is close to -2, and their concentration profiles have the concave shape shown in curve d of Figure 6.9.

The diffusion of phosphorus in silicon is associated with the doubly charged acceptor vacancy V^{2-}, and the diffusion coefficient at high concentration varies as C^2. We would expect that the diffusion profile of phosphorus resembles that shown in curve b of Figure 6.9. However, because of a dissociation effect, the diffusion profile exhibits anomalous behavior.

Figure 6.10 shows phosphorus diffusion profiles for various surface concentrations after diffusion into silicon for 1 hour at 1000°C.[13] When the surface concentration is low, corresponding to the intrinsic diffusion region, the diffusion profile is given by an erfc (curve a). As the concentration increases, the profile begins to deviate from the simple expression (curves b and c). At very high concentration (curve d), the profile near the surface is indeed similar to that shown in curve b of Figure 6.9. However, at concentration n_e, a kink occurs and is followed by a rapid diffusion in the tail region. The concentration n_e corresponds to a Fermi level 0.11 eV below the conduction band. At this energy level, the coupled impurity–vacancy pair $(P^+ V^{2-})$ dissociates to P^+, V^-, and an electron. Thus, the dissociation generates a large number of singly charged acceptor vacancies V^-, which in turn enhances the diffusion in the tail region of the profile. The diffusivity in the tail region is over 10^{-12} cm²/s, which is about two orders of magnitude larger than the intrinsic diffusivity at 1000°C. Because of its high diffusivity, phosphorus is commonly used to form deep junctions, such as the n-tubs in CMOS.

Zinc Diffusion in Gallium Arsenide

We expect diffusion in gallium arsenide to be more complicated than that in silicon because the diffusion of impurities may involve atomic movements on both the gallium and arsenic sublattices. Vacancies play a dominant role in diffusion processes in gallium arsenide because both p- and n-type impurities must ultimately reside in lattice sites. However, the charge states of the vacancies have not been established.

Zinc is the most extensively studied diffusant in gallium arsenide. Its diffusion coefficient is found to vary as C^2. Therefore, the diffusion profiles are steep, as shown in Figure 6.11,[13] and resemble curve b of Figure 6.9. Note that even for the case of the lowest surface concentration, the diffusion is in the extrinsic diffusion region, because n_i for GaAs at 1000°C is less than 10^{18} cm⁻³. As seen in Figure 6.11, the surface concentration has a profound effect on the junction depth. The diffusivity varies linearly with the partial pressure of the zinc vapor, and the surface concentration is proportional to the square root of the partial pressure. Therefore, from Eq. 25, the junction depth is linearly proportional to the surface concentration.

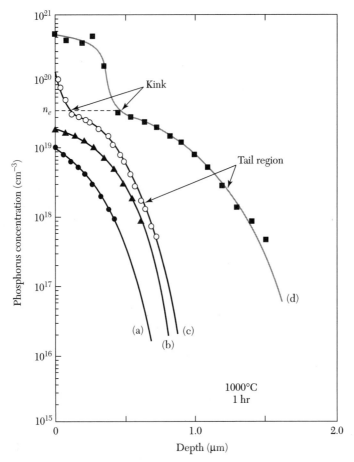

Figure 6.10 Phosphorus diffusion profiles[13] for various surface concentrations after diffusion into silicon for 1 hour at 1000°C.

▶ 6.3 LATERAL DIFFUSION

The one-dimensional diffusion equation discussed previously can describe the diffusion process satisfactorily, except at the edge of the mask window. Here the impurities will diffuse downward and sideways (i.e., laterally). In this case, we must consider a two-dimensional diffusion equation and use a numerical technique to obtain the diffusion profiles under different initial and boundary conditions.

Figure 6.12 shows the contours of constant doping concentration for a constant surface concentration diffusion condition, assuming that the diffusivity is independent of concentration.[14] At the far right of the figure, the variation of the dopant concentration from $0.5\ C_s$ to $10^{-4}\ C_s$ (where C_s is the surface concentration) corresponds to the erfc distribution given by Eq. 9. The contours are in effect a map of the location of the junctions created by diffusing into various background concentrations. For example, at $C/C_s = 10^{-4}$ (i.e., the background doping is 10^4 times lower than the surface concentration), we see from this constant concentration curve that the vertical penetration is about 2.8 μm, whereas the lateral penetration is about 2.3 μm (i.e., the penetration along the diffusion mask–semiconductor interface). Therefore, the lateral penetration is about 80% of the

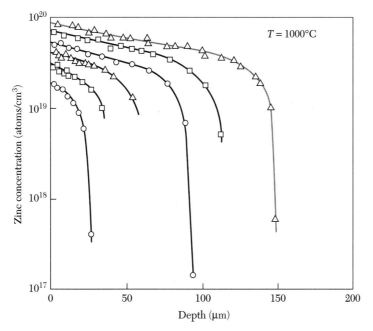

Figure 6.11 Diffusion profiles[13] of zinc in GaAs after annealing at 1000°C for 2.7 hours. The different surface concentrations are obtained by maintaining the Zn source at temperatures in the range 600°C to 800°C.

penetration in the vertical direction for concentrations three or more orders of magnitude below the surface concentration. Similar results are obtained for a constant total dopant diffusion condition. The ratio of lateral to vertical penetration is about 75%. For concentration-dependent diffusivities, the ratio is found to be reduced slightly, to about 65% to 70%.

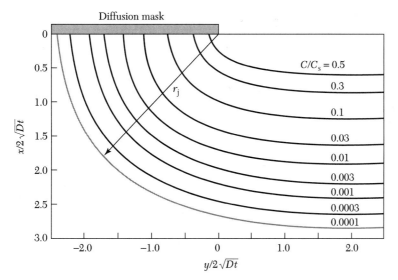

Figure 6.12 Diffusion contours at the edge of an oxide window, where r_j is the radius of curvature.[14]

Because of the lateral diffusion effect, the junction consists of a central plane (or flat) region with approximately cylindrical edges with a radius of curvature r_j, as shown in Figure 6.12. In addition, if the diffusion mask contains sharp corners, the shape of the junction near the corner will be roughly spherical because of lateral diffusion. Since the electric field intensities are higher for cylindrical and spherical junction regions, the avalanche breakdown voltages of such regions can be substantially lower than that of a plane junction having the same background doping.

▶ 6.4 DIFFUSION SIMULATION

The various complications that arise in the computation of diffusion profiles (such as concentration-dependent diffusivity) tend to preclude the use of analytical hand calculations for all but the simplest examples. Fortunately, the SUPREM software package introduced in Chapter 3 also includes complete models for diffusion. SUPREM can simulate one- or two-dimensional diffusion profiles. This is accomplished using the DIFFUSION command. The output of the program is typically the chemical, carrier, and vacancy concentrations as functions of depth into the semiconductor substrate.

All diffusion process simulators, including SUPREM, are based upon three basic equations.[15] The first equation is for the flux (J), which in one dimension is given by

$$J_i = -D_i \frac{dC_i}{dx} + Z_i \mu_i C_i E \tag{26}$$

where Z_i is the charge state, μ_i is the mobility of the impurity, and E is the electric field. The subscript i indicates the SUPREM grid location. The second relationship is the continuity equation, which is given by

$$\frac{dC_i}{dt} + \frac{dJ_i}{dx} = G_i \tag{27}$$

where G_i is the generation/recombination rate of the impurity. The final key relationship is Poisson's equation, which in one dimension is given by

$$\frac{d}{dx}[\varepsilon_s E] = q\left(p - n + N_D^+ - N_A^-\right) \tag{28}$$

where ε_s is the permittivity, n and p are the electron and hole concentrations, and N_D^+ and N_A^- are the concentrations of ionized donors and acceptors, respectively. SUPREM solves Eqs. 26 to 28 simultaneously over a one-dimensional grid specified by the user. The diffusivity values used by SUPREM are based on the vacancy model of Fair.[11] The values of E_a and D_0 for B, Sb, and As are included in a look-up table. Empirical models are used to account for field-aided, oxidation-enhanced, and oxidation-retarded diffusion.

EXAMPLE 3

Suppose we want to simulate the predeposition of boron into an n-type <100> silicon wafer at 850°C for 15 minutes. If the silicon substrate is doped with phosphorus at a level of 10^{16} cm^{-3}, use SUPREM to determine the boron doping profile and the junction depth.

SOLUTION The SUPREM input listing is as follows:

```
TITLE          Predeposition Example
COMMENT        Initialize silicon substrate
INITIALIZE     <100> Silicon Phosphor Concentration=1e16
COMMENT        Diffuse boron
DIFFUSION      Time=15 Temperature=850 Boron Solidsol
PRINT          Layers Chemical Concentration Phosphorus Boron Net
PLOT           Active Net Cmin=1e15
STOP           End predeposition example
```

Note that surface concentration of the boron is set to the solid solubility limit by the `Solidsol` parameter in the **DIFFUSION** command. After predeposition is complete, we print and plot the boron concentration as a function of depth into the silicon substrate. The results are shown in Figure 6.13, which indicates a junction depth of 0.0555 μm.   ◀

▶ 6.5 SUMMARY

Diffusion is a key method of impurity doping. This chapter first considered the basic diffusion equation for constant diffusivity. The complementary error function (erfc) and the Gaussian function were obtained for the constant surface concentration case and constant total dopant cases, respectively. The results of a diffusion process can be evaluated by measurements of the junction depth, the sheet resistance, and the dopant profile.

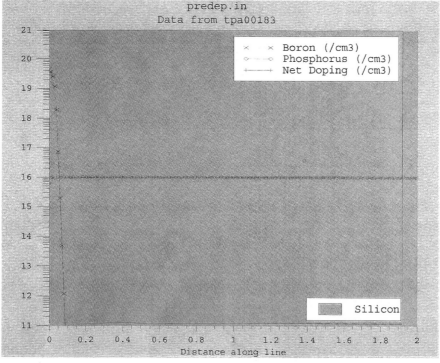

Figure 6.13 Plot of boron concentration as a function of depth into the silicon substrate, using SUPREM.

When the doping concentration is higher than the intrinsic carrier concentration n_i at the diffusion temperature, the diffusivity becomes concentration dependent. This dependence has a profound effect on the resulting doping profile. For example, arsenic and boron diffusivity in silicon vary linearly with the impurity concentration. Their doping profiles are much more abrupt than the erfc profile. Phosphorus diffusivity in silicon varies as the square of concentration. This dependence and a dissociation effect give rise to a phosphorus diffusivity that is 100 times larger than its intrinsic diffusivity.

Lateral diffusion at the edge of a mask and impurity redistribution during oxidation are two processes in which diffusion can have an important impact on device performance. The former can substantially reduce the breakdown voltage, and the latter will influence the threshold voltage as well as the contact resistance.

▶ REFERENCES

1. S. M. Sze, Ed., *VLSI Technology*, 2nd Ed., McGraw-Hill, New York, 1988, Ch. 7, 8.

2. S. K. Ghandhi, *VLSI Fabrication Principles*, 2nd Ed., Wiley, New York, 1994, Ch. 4, 6.

3. W. R. Runyan and K. E. Bean, *Semiconductor Integrated Circuit Processing Technology*, Addison-Wesley, Boston, 1990, Ch. 8.

4. H. C. Casey and G. L. Pearson, "Diffusion in Semiconductors," in J. H. Crawford and L. M. Slifkin, Eds., *Point Defects in Solids*, Vol. 2, Plenum, New York, 1975.

5. J. P. Joly, "Metallic Contamination of Silicon Wafers," *Microelectron. Eng., 40*, 285 (1998).

6. A. S. Grove, *Physics and Technology of Semiconductor Devices*, Wiley, New York, 1967.

7. ASTM Method F374-88, "Test Method for Sheet Resistance of Silicon Epitaxial, Diffused, and Ion-Implanted Layers Using a Collinear Four-Probe Array," V10, 249 (1993).

8. J. C. Irvin, "Evaluation of Diffused Layers in Silicon," *Bell Syst. Tech. J., 41*, 2 (1962).

9. S. M. Sze, *Semiconductor Devices: Physics and Technology*, 2nd Ed., Wiley, New York, 2002, Ch. 7.

10. ASTM Method E1438-91, "Standard Guide for Measuring Width of Interfaces in Sputter Depth Profiling Using SIMS," V10, 578 (1993).

11. R. B. Fair, "Concentration Profiles of Diffused Dopants," in F. F. Y. Wang, Ed., *Impurity Doping Processes in Silicon*, North-Holland, Amsterdam, 1981.

12. L. R. Weisberg and J. Blanc, "Diffusion with Interstitial-Substitutional Equilibrium, Zinc in GaAs," *Phys. Rev., 131*, 1548 (1963).

13. F. A. Cunnell and C. H. Gooch, "Diffusion of Zinc in Gallium Arsenide," *J. Phys. Chem. Solid, 15*, 127 (1960).

14. D. P. Kennedy and R. R. O'Brien, "Analysis of the Impurity Atom Distribution Near the Diffusion Mask for a Planar *p-n* Junction," *IBM J. Res. Dev., 9*, 179 (1965).

15. S. A. Campbell, *The Science and Engineering of Microelectronic Fabrication*, 2nd Ed., Oxford University Press, New York, 2001, Ch. 3.

▶ PROBLEMS

Asterisks denote difficult problems.

SECTION 6.1: BASIC DIFFUSION PROCESS

1. Calculate the junction depth and the total amount of dopant introduced after boron pre-deposition performed at 950°C for 30 minutes in a neutral ambient. Assume the substrate is *n*-type silicon with $N_D = 1.8 \times 10^{16}$ cm^{-3} and the boron surface concentration is $C_s = 1.8 \times 10^{20}$ cm^{-3}.

2. If the sample in Problem 1 is subjected to a neutral drive-in at 1050°C for 60 minutes, calculate the diffusion profile and the junction depth.

3. Assume the measured phosphorus profile can be represented by a Gaussian function with a diffusivity $D = 2.3 \times 10^{-13}$ cm^2/s. The measured surface concentration is 1×10^{18} atoms/cm^3, and the measured junction depth is 1 μm at a substrate concentration of 1×10^{15}. Calculate the diffusion time and the total dopant in the diffused layer.

°4. To avoid wafer warp due to a sudden reduction in temperature, the temperature in a diffusion furnace is decreased linearly from 1000°C to 500°C in 20 minutes. What is the effective diffusion time at the initial diffusion temperature for a phosphorus diffusion in silicon?

°5. For a low-concentration phosphorus drive-in diffusion in silicon at 1000°C, find the percentage change of surface concentration for 1% variation in diffusion time and temperature.

6. If arsenic is diffused into a thick slice of silicon doped with 10^{15} boron atoms/cm^3 at a temperature of 1100°C for 3 hours, what is the final distribution of arsenic if the surface concentration is held fixed at 4×10^{18} atoms/cm^3? What are the diffusion length and junction depth?

SECTION 6.2: EXTRINSIC DIFFUSION

7. If arsenic is diffused into a thick slice of silicon doped with 10^{15} boron atoms/cm^3 at a temperature of 900°C for 3 hours, what is the final distribution of arsenic if the surface concentration is held fixed at 4×10^{18} atoms/cm^3? What is the junction depth? Assume the following:

$$D_0 = 45.8 \text{ cm}^2/\text{s} \quad E_a = 4.05 \text{ eV} \quad x_j = 1.6\sqrt{Dt}$$

8. Explain the meaning of *intrinsic diffusion* and *extrinsic diffusion*.

SECTION 6.4: DIFFUSION SIMULATION

9. Use SUPREM to perform a drive-in step for 6 hours at 1175°C following the predeposition described in Example 3. Plot the boron profile and give the new junction depth.

°10. After the boron drive-in step in Problem 9, suppose phosphorus is subsequently predeposited and driven in. The phosphorus predeposition occurs at 850°C for 30 minutes, and the drive-in occurs at 1000°C for 30 minutes. Use SUPREM to plot the phosphorus and boron impurity profiles, and determine the junction depth(s).

CHAPTER 7

Ion Implantation

As discussed in Chapter 6, diffusion and ion implantation are the two key methods of impurity doping. Since the early 1970s, many doping operations have been performed by ion implantation, as shown in Figure 7.1. In this process the dopant ions are implanted into the semiconductor by means of an ion beam. The doping concentration has a peak distribution inside the semiconductor, and the profile of the dopant distribution is determined mainly by the ion mass and the implanted-ion energy. This chapter discusses the following topics:

- The process and advantages of ion implantation
- Ion distributions in the crystal lattice and how to remove lattice damage caused by ion implantation
- Implantation-related processes, such as masking, high-energy implantation, and high-current implantation
- The simulation of ion implantation using SUPREM

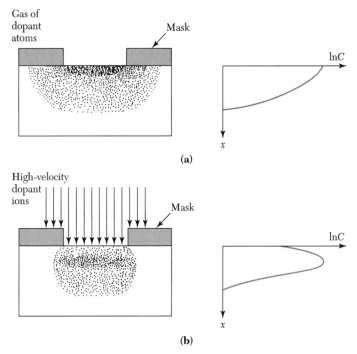

Figure 7.1 Comparison of (*a*) diffusion and (*b*) ion-implantation techniques for the selective introduction of dopants into the semiconductor substrate.

► 7.1 RANGE OF IMPLANTED IONS

Ion implantation is the introduction of energetic, charged particles into a substrate such as silicon. Implantation energies are between 1 keV and 1 MeV, resulting in ion distributions with average depths ranging from 10 nm to 10 μm. Ion doses vary from 10^{12} ions/cm² for threshold voltage adjustment to 10^{18} ions/cm² for the formation of buried insulating layers. Note that the dose is expressed as the number of ions implanted into 1 cm² of the semiconductor surface area. The main advantages of ion implantation are its more precise control and reproducibility of impurity dopings and its lower processing temperature compared with those of the diffusion process.

Figure 7.2 schematically shows a medium-energy ion implantor.[1] The ion source has a heated filament to break up source gases such as BF_3 or AsH_3 into charged ions (B^+ or As^+). An extraction voltage, around 40 kV, causes the charged ions to move out of the ion-source chamber into a mass analyzer. The magnetic field of the analyzer is chosen such that only ions with the desired mass-to-charge ratio can travel through it without being filtered. The selected ions then enter the acceleration tube, where they are accelerated to the implantation energy as they move from high voltage to ground. Apertures ensure that the ion beam is well collimated. The pressure in the implantor is kept below 10^{-4} Pa to minimize ion scattering by gas molecules. The ion beam is then scanned over the wafer surface using electrostatic deflection plates and is implanted into the semiconductor substrate.

The energetic ions lose energy through collisions with electrons and nuclei in the substrate and finally come to rest at some depth within the lattice. The average depth can be controlled by adjusting the acceleration energy. The dopant dose can be controlled by monitoring the ion current during implantation. The principle side effect is disruption or damage of the semiconductor lattice due to ion collisions. Therefore, a subsequent annealing treatment is needed to remove this damage.

7.1.1 Ion Distribution

The total distance that an ion travels in coming to rest is called its *range* (R) and is illustrated in Figure 7.3a.[2] The projection of this distance along the axis of incidence is called the *projected range* (R_p). Because the number of collisions per unit distance and the energy lost per collision are random variables, there will be a spatial distribution of ions having the same mass and the same initial energy. The statistical fluctuations in the projected range are called the *projected straggle* (σ_p). There is also a statistical fluctuation along an axis perpendicular to the axis of incidence, which is called the *lateral straggle* ($\sigma_\perp$).

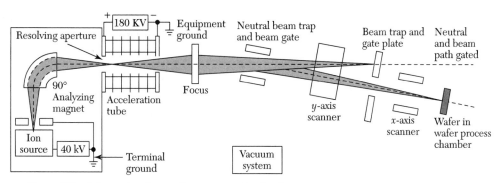

Figure 7.2 Schematic of a medium-current ion implantor.

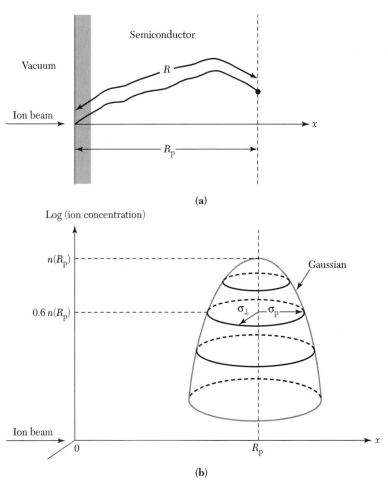

Figure 7.3 (a) Schematic of the ion range R and projected range R_p. (b) Two-dimensional distribution of the implanted ions.[2]

Figure 7.3b shows the ion distribution. Along the axis of incidence, the implanted impurity profile can be approximated by a Gaussian distribution function:

$$n(x) = \frac{S}{\sqrt{2\pi}\sigma_p} \exp\left[-\frac{(x - R_p)^2}{2\sigma_p^2}\right] \tag{1}$$

where S is the ion dose per unit area. This equation is similar to Eq. 14 in Chapter 6 for constant total dopant diffusion, except that the quantity $4Dt$ is replaced by $2\sigma_p^2$ and the distribution is shifted along the x-axis by R_p. Thus, for diffusion, the maximum concentration is at $x = 0$, whereas for ion implantation the maximum concentration is at the projected range. The ion concentration is reduced by 40% from its peak value at $(x - R_p) = \pm\sigma_p$, by one decade at $\pm2\sigma_p$, by two decades at $\pm3\sigma_p$, and by five decades at $\pm4.8\sigma_p$.

Along the axis perpendicular to the axis of incidence, the distribution is also a Gaussian function of the form $\exp(-y^2/2\sigma_\perp^2)$. Because of this distribution, there will be some lateral implantation.[3] However, the lateral penetration from the mask edge (on the order of $\sigma_\perp$) is considerably smaller than that from the thermal diffusion process discussed in Section 6.3.

7.1.2 Ion Stopping

There are two stopping mechanisms by which an energetic ion, on entering a semiconductor substrate (also called the *target*), can be brought to rest. The first is by transferring its energy to the target nuclei. This causes deflection of the incident ion and also dislodges many target nuclei from their original lattice sites. If E is the energy of the ion at any point x along its path, we can define a nuclear stopping power, $S_n(E) \equiv (dE/dx)_n$, to characterize this process. The second stopping mechanism is by the interaction of the incident ion with the cloud of electrons surrounding the target's atoms. The ion loses energy in collisions with electrons through Coulombic interaction. The electrons can be excited to higher energy levels (excitation), or they can be ejected from the atom (ionization). We can define an electronic stopping power, $S_e(E) \equiv (dE/dx)_e$, to characterize this process.

The average rate of energy loss with distance is given by a superposition of the two stopping mechanisms:

$$\frac{dE}{dx} = S_n(E) + S_e(E) \tag{2}$$

If the total distance traveled by the ion before coming to rest is R, then

$$R = \int_0^R dx = \int_0^{E_0} \frac{dE}{S_n(E) + S_e(E)} \tag{3}$$

where E_0 is the initial ion energy. The quantity R has been defined previously as the range.

We can visualize the nuclear stopping process by considering the elastic collision between an incoming hard sphere (energy E_0 and mass M_1) and a target hard sphere (initial energy zero and mass M_2), as illustrated in Figure 7.4. When the spheres collide, momentum is transferred along the centers of the spheres. The deflection angle (θ) and the velocities, v_1 and v_2, can be obtained from the requirements for conservation of momentum and energy. The maximum energy loss is in a head-on collision. For this case, the energy loss by the incident particle M_1, or the energy transferred to M_2, is

$$\frac{1}{2}M_2 v_2^2 = \left[\frac{4M_1 M_2}{(M_1 + M_2)^2}\right] E_0 \tag{4}$$

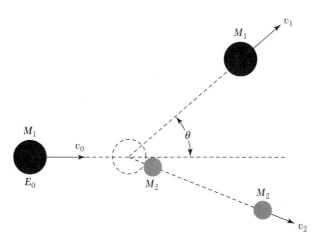

Figure 7.4 Collision of hard spheres.

Since M_2 is usually of the same order of magnitude as M_1, a large amount of energy can be transferred in a nuclear stopping process.

Detailed calculations show that the nuclear stopping power increases linearly with energy at low energies (similar to Eq. 4), and $S_n(E)$ reaches a maximum at some intermediate energy. At high energies, $S_n(E)$ becomes smaller because fast particles may not have sufficient interaction time with the target atoms to achieve effective energy transfer. The calculated values of $S_n(E)$ for arsenic, phosphorus, and boron in silicon at various energies are shown in Figure 7.5 (solid lines, where the superscript indicates the atomic weight).[4] Note that heavier atoms, such as arsenic, have a larger nuclear stopping power, that is, larger energy loss per unit distance.

The electronic stopping power is found to be proportional to the velocity of the incident ion, or

$$S_e(E) = k_e \sqrt{E} \tag{5}$$

where the coefficient k_e is a relatively weak function of atomic mass and atomic number. The value of k_e is approximately 10^7 $(\text{eV})^{1/2}/\text{cm}$ for silicon and 3×10^7 $(\text{eV})^{1/2}/\text{cm}$ for gallium arsenide. The electronic stopping power in silicon is plotted in Figure 7.5 (dotted line). Also shown in the figure are the crossover energies, at which $S_e(E) = S_n(E)$. For boron, which has a relatively low ion mass compared with the target silicon atom, the crossover energy is only 10 keV. This means that over most of the implantation energy range of 1 keV to 1 MeV, the main energy loss mechanism is due to electronic stopping. On the other hand, for arsenic, with its relatively high ion mass, the crossover energy is 700 keV. Thus, nuclear stopping dominates over most of the energy range. For phosphorus,

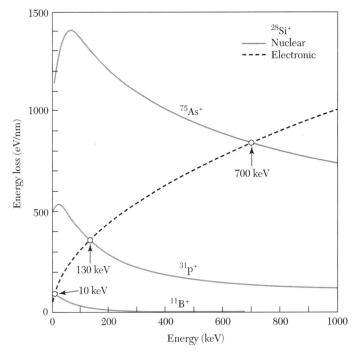

Figure 7.5 Nuclear stopping power, $S_n(E)$, and electronic stopping power, $S_e(E)$, for As, P, and B in Si. The points of intersection of the curves correspond to the energy at which nuclear and electronic stopping are equal.[4]

the crossover energy is 130 keV. For an E_0 less than 130 keV, nuclear stopping will dominate; for higher energies, electronic stopping will take over.

Once $S_n(E)$ and $S_e(E)$ are known, we can calculate the range from Eq. 3. This in turn can give us the projected range and projected straggle with the help of the following approximate equations:[1]

$$R_p \cong \frac{R}{1 + \left(M_2 / 3M_1\right)} \tag{6}$$

$$\sigma_p \cong \frac{2}{3}\left[\frac{\sqrt{M_1 M_2}}{M_1 + M_2}\right] R_p \tag{7}$$

Figure 7.6a shows the projected range (R_p), the projected straggle (σ_p), and the lateral straggle $(\sigma_\perp)$ for arsenic, boron, and phosphorus in silicon.[5] As expected, the larger the energy loss, the smaller the range. Also, the projected range and straggles increase with ion energy. For a given element at a specific incident energy, σ_p and $\sigma_\perp$ are comparable and usually within ±20%. Figure 7.6b shows the corresponding values for hydrogen, zinc, and tellurium in gallium arsenide.[3] If we compare Figures 7.6a and b, we see that most of the popular dopants (except hydrogen) have larger projected ranges in silicon than they have in gallium arsenide.

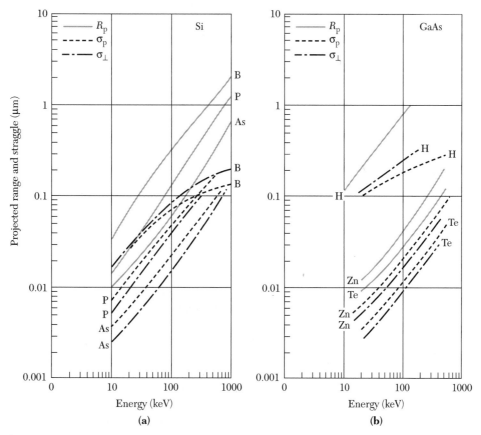

Figure 7.6 Projected range, projected straggle, and lateral straggle for (a) B, P, and As in silicon,[5] and (b) H, Zn, and Te in gallium arsenide.[3]

EXAMPLE 1

Assume 100-keV boron implants on a 200-mm silicon wafer at a dose of 5×10^{14} ions/cm². Calculate the peak concentration and the required ion beam current for 1 minute of implantation.

SOLUTION   From Figure 7.6*a*, we obtain 0.31 and 0.07 μm for the projected range and projected straggle, respectively.

From Eq. 1,

$$n(x) = \frac{S}{\sqrt{2\pi}\sigma_p} \exp\left[-\frac{(x - R_p)^2}{2\sigma_p^2}\right]$$

$$\frac{dn}{dx} = -\frac{S}{\sqrt{2\pi}\sigma_p} \frac{2(x - R_p)}{2\sigma_p^2} \exp\left[\frac{-(x - R_p)^2}{2\sigma_p^2}\right] = 0$$

The peak concentration is at $x = R_p$, and $n(x) = 2.85 \times 10^{19}$ ions/cm³.

The total number of implanted ions, Q, is given by

$$Q = 5 \times 10^{14} \times \pi \times \left(\frac{20}{2}\right)^2 = 1.57 \times 10^{17} \text{ ions / cm}^2$$

The required ion current, I, is given by

$$I = \frac{qQ}{t} = \frac{1.6 \times 10^{-19} \times 1.57 \times 10^{17}}{60} = 4.19 \times 10^{-4} \text{ A} = 0.42 \text{mA}$$ ◀

7.1.3   Ion Channeling

The projected range and straggle of the Gaussian distribution discussed previously give a good description of the implanted ions in amorphous or fine-grain polycrystalline substrates. Both silicon and gallium arsenide behave as if they were amorphous semiconductors, provided the ion beam is misoriented from the low-index crystallographic direction (e.g., <111>). In this situation, the doping profile described by Eq. 1 is followed closely near the peak and extended to one or two decades below the peak value. This is illustrated in Figure 7.7.[2] However, even for a misorientation of 7° from the <111> axis, there still is a tail that varies exponentially with distance as $\exp(-x/\lambda)$, where λ is typically on the order of 0.1 μm.

The exponential tail is related to the *ion channeling* effect. Channeling occurs when incident ions align with a major crystallographic direction and are guided between rows of atoms in a crystal. Figure 7.8 illustrates a diamond lattice viewed along a <110> direction.[6] Ions implanted in the <110> direction will follow trajectories that will not bring them close enough to a target atom to lose significant amounts of energy in nuclear collisions. Thus, for channeled ions, the only energy loss mechanism is electronic stopping, and the range of channeled ions can be significantly larger than it would be in an amorphous target. Ion channeling is particularly critical for low-energy implants and heavy ions.

Channeling can be minimized by several techniques: a blocking amorphous surface layer, misorientation of the wafer, and creating a damaged layer in the wafer surface. The usual blocking amorphous layer is simply a thin layer of grown silicon dioxide (Fig. 7.9*a*). The layer randomizes the direction of the ion beam so that the ions enter the wafer at

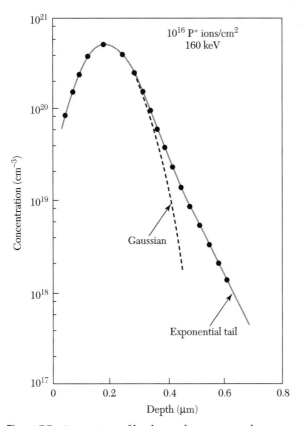

Figure 7.7 Impurity profile obtained in a purposely misoriented target. Ion beam is incident 7° from the <111> axis.[2]

different angles and not directly down the crystal channels. Misorientation of the wafers 5° to 10° off the major plane also has the effect of preventing the ions from entering the channels (Fig. 7.9b). With this method, most implantation machines tilt the wafer by 7° and then apply a 22° twist from the flat to prevent channeling. Pre-damaging the wafer surface with a heavy silicon or germanium implant creates a randomizing layer in the wafer surface (Fig. 7.9c). This method, however, increases the use of the expensive ion implantor.

▶ 7.2 IMPLANT DAMAGE AND ANNEALING

7.2.1 Implant Damage

When energetic ions enter a semiconductor substrate, they lose their energy in a series of nuclear and electronic collisions and finally come to rest. The electronic energy loss can be accounted for in terms of electronic excitations to higher energy levels or in terms of the generation of electron–hole pairs. However, electronic collisions do not displace semiconductor atoms from their lattice positions. Only nuclear collisions can transfer sufficient energy to the lattice so that host atoms are displaced, resulting in implant damage (also called *lattice disorder*).[7] These displaced atoms may possess large fractions of the incident energy, and they in turn cause cascades of secondary displacements of nearby atoms to form a tree of disorder along the ion path. When the displaced atoms per unit volume approach the atomic density of the semiconductor, the material becomes amorphous.

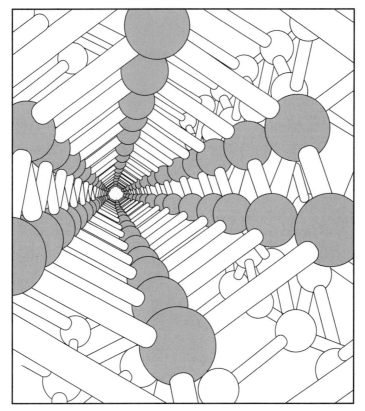

Figure 7.8 Model for a diamond structure, viewed along a <110> axis.[6]

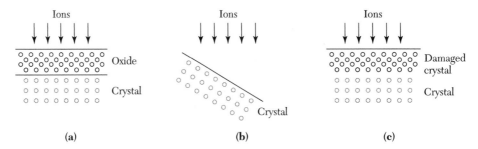

Figure 7.9 Minimizing channeling. (*a*) Implantation through an amorphous oxide layer. (*b*) Misorientation of the beam direction to all crystal axes. (*c*) Pre-damage on the crystal surface.

The tree of disorder for light ions is quite different from that for heavy ions. Much of the energy loss for light ions (e.g., $^{11}B^+$ in silicon) is due to electronic collisions (see Fig. 7.5), which do not cause lattice damage. The ions lose their energies as they penetrate deeper into the substrate. Eventually, the ion energy is reduced below the crossover energy (10 keV for boron) where nuclear stopping becomes dominant. Therefore, most of the lattice disorder occurs near the final ion position. This is illustrated in Figure 7.10*a*.

We can estimate the damage by considering a 100-keV boron ion. Its projected range is 0.31 µm (Fig. 6*a*), and its initial nuclear energy loss is only 3 eV/Å (Fig. 7.5). Since the spacing between lattice planes in silicon is about 2.5 Å, this means that the boron ion will lose 7.5 eV at each lattice plane because of nuclear stopping. The energy required

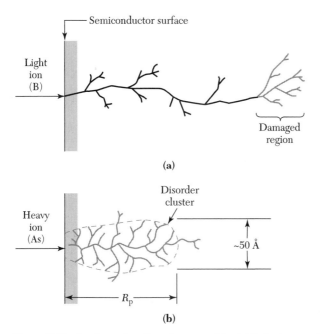

Figure 7.10 Implantation disorder caused by (*a*) light ions and (*b*) heavy ions.

to displace a silicon atom from its lattice position is about 15 eV. Therefore, the incident boron ion does not release enough energy from nuclear stopping to displace a silicon atom when it first enters the silicon substrate. When the ion energy is reduced to about 50 keV (at a depth of 1500 Å), the energy loss due to nuclear stopping increases to 15 eV for each lattice plane (i.e., 6 eV/Å), sufficient to create a lattice disorder. Assuming that 1 atom is displaced per lattice plane for the remaining ion range, we have 600 lattice atoms displaced (i.e., 1500 Å/2.5 Å). If each displaced atom moves roughly 25 Å from its original position, the damage volume is given by $V_D \cong \pi(25\text{Å})^2(1500\text{Å}) = 3 \times 10^{-18}\text{cm}^3$. The damage density is $600/V_D \cong 2 \times 10^{20}$ cm^{-3}, which is only 0.4% of the atoms. Thus, very high doses of light ions are needed to create an amorphous layer.

For heavy ions, the energy loss is primarily due to nuclear collisions; therefore, we expect substantial damage. Consider a 100-keV arsenic ion with a projected range of 0.06 μm, or 60 nm. The average nuclear energy loss over the entire energy range is about 1320 eV/nm (Fig. 7.5). This means that the arsenic ion loses about 330 eV for each lattice plane on the average. Most of the energy is given to one primary silicon atom. Each primary atom will subsequently cause 22 displaced target atoms (i.e., 330 eV/15 eV). The total number of displaced atoms is 5280. Assuming a range of 2.5 nm for the displaced atoms, the damage volume is $V_D \cong \pi(2.5 \text{ nm})^2 (60 \text{ nm}) = 10^{-18}$ cm^3. The damage density is then $5280/V_D \cong 5 \times 10^{21}$ cm^{-3}, or about 10% of the total number of atoms in V_D. As a result of the heavy-ion implantation, the material becomes essentially amorphous. Figure 7.10*b* illustrates the situation in which the damage forms a disordered cluster over the entire projected range.

To estimate the dose required to convert a crystalline material to an amorphous form, we can use the criterion that the energy density is of the same order of magnitude as that needed for melting the material (i.e., 10^{21} keV/cm^3). For 100-keV arsenic ions, the dose required to make amorphous silicon is then

$$S = \frac{\left(10^{21} \text{ keV}/\text{cm}^3\right)R_p}{E_0} = 6 \times 10^{13} \text{ ions}/\text{cm}^2 \tag{8}$$

For 100-keV boron ions, the dose required is 3×10^{14} ions/cm^2 because R_p for boron is five times larger than for arsenic. However, in practice, higher doses (>10^{16} ions/cm^2) are required for boron implantation into a target at room temperature because of the nonuniform distribution of the damage along the ion path.

7.2.2 Annealing

Because of the damaged region and the disorder cluster that result from ion implantation, semiconductor parameters such as mobility and lifetime are severely degraded. In addition, most of the ions as implanted are not located in substitutional sites. To activate the implanted ions and to restore mobility and other material parameters, we must anneal the semiconductor at an appropriate combination of time and temperature.

Conventional annealing uses an open-tube, batch furnace system similar to that used for thermal oxidation. This process requires a long time and high temperature to remove the implant damage. However, conventional annealing may cause substantial dopant diffusion and cannot meet requirements for shallow junctions and narrow doping profiles. *Rapid thermal annealing* (RTA) is an annealing process that employs a variety of energy sources with a wide range of times, from 100 seconds down to nanoseconds—all short compared with conventional annealing. RTA can activate dopants fully with minimal redistribution.

Conventional Annealing of Boron and Phosphorus
Annealing characteristics depend on the dopant type and the dose involved. Figure 7.11 shows the annealing behaviors of boron and phosphorus implantation into silicon substrates.[5] The substrate is held at room temperature (T_s) during implantation. At a given ion dose, the annealing temperature is defined as the temperature at which 90% of the

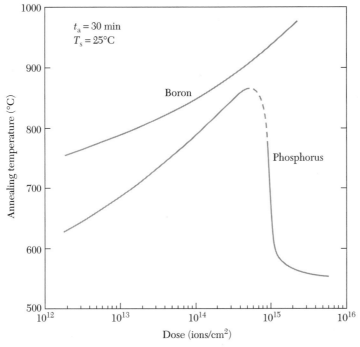

Figure 7.11 Annealing temperature versus dose for 90% activation of boron and phosphorus.[5]

implanted ions are activated by a 30-minute annealing in a conventional annealing furnace. For boron implantation, higher annealing temperatures are needed for higher doses. For phosphorus at lower doses, the annealing behavior is similar to that for boron. However, when the dose is greater than 10^{15} cm^{-2}, the annealing temperature drops to about 600°C. This phenomenon is related to the solid-phase epitaxy process (see Chapter 8). At phosphorus doses greater than 6×10^{14} cm^{-2}, the silicon surface layer becomes amorphous. The single-crystal semiconductor underneath the amorphous layer serves as a seeding area for recrystallization of the amorphous layer. The epitaxial growth rate along the <100> direction is 10 nm/min at 550°C and 50 nm/min at 600°C, with an activation energy at 2.4 eV. Therefore, a 100- to 500-nm amorphous layer can be recrystallized in a few minutes. During the solid-phase epitaxial process, the impurity dopant atoms are incorporated into the lattice sites along with the host atoms. Thus, full activation can be obtained at relatively low temperatures.

Rapid Thermal Annealing

The machine for RTA with transient lamp heating is shown in Figure 7.12. The temperature measured from the heated wafer is usually from 600°C to 1100°C.[8] A wafer is heated quickly under atmospheric conditions or at low pressure under isothermal conditions. Typical lamps in an RTA system are tungsten filaments or arc lamps. The processing chamber is made of either quartz, silicon carbide, stainless steel, or aluminum and has quartz windows through which the optical radiation passes to illuminate the wafer. The wafer holder is often made of quartz and contacts the wafer in a minimum number of places. A measurement system is placed in a control loop to set wafer temperature. The RTA system interfaces with a gas-handling system and a computer that controls system operation. Typically, wafer temperature in an RTA system is measured with a noncontact optical pyrometer that determines temperature from radiated infrared energy.

Table 7.1 compares conventional furnace and RTA technology. To achieve short processing times using RTA, trade-offs must be made in temperature and process uniformity, temperature measurement and control, and wafer stress and throughput. In addition, there are concerns about the introduction of electrically active wafer defects during the

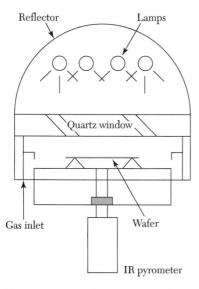

Figure 7.12 Rapid thermal annealing system that is optically heated.

TABLE 7.1 Technology Comparison

Determinant	Conventional Furnace	Rapid Thermal Annealing
Process	Batch	Single-wafer
Furnace	Hot-wall	Cold-wall
Heating rate	Low	High
Cycle time	High	Low
Temperature monitor	Furnace	Wafer
Thermal budget	High	Low
Particle problem	Yes	Minimal
Uniformity and repeatability	High	Low
Throughput	High	Low

very fast (100°–300 °C/s) thermal transients. Rapid heating with temperature gradients in the wafers can cause wafer damage in the form of slip dislocations induced by thermal stress. On the other hand, conventional furnace processing causes significant problems, such as particle generation from the hot walls, limited ambient control in an open system, and a large thermal mass that restricts controlled heating times to tens of minutes. In fact, requirements on contamination, process control, and cost of manufacturing floor space have resulted in a paradigm shift to the RTA process.

▶ 7.3 IMPLANTATION-RELATED PROCESSES

This section considers a few implantation-related processes, such as multiple implantation, masking, tilt-angle implantation, high-energy implantation, and high-current implantation.

7.3.1 Multiple Implantation and Masking

In many applications, doping profiles other than the simple Gaussian distribution are required. One such case is the preimplantation of silicon with an inert ion to make the silicon surface region amorphous. This technique allows close control of the doping profile and permits nearly 100% dopant activation at low temperatures, as discussed previously. In such a case, a deep amorphous region may be required. To obtain this type of region, we must make a series of implants at varying ion energies and doses.

Multiple implantation can also be used to form a flat doping profile, as shown in Figure 7.13. Here, four boron implants into silicon are used to provide a composite doping profile.[9] The measured carrier concentration and that predicted using range theory are shown in the figure. Other doping profiles, unavailable from diffusion techniques, can be obtained by using various combinations of impurity dose and implantation energy. Multiple implants have been used to preserve stoichiometry during the implantation and annealing of GaAs. This approach, whereby equal amounts of gallium and an n-type dopant (or arsenic and a p-type dopant) are implanted prior to annealing, has resulted in higher carrier activation.

To form p–n junctions in selected areas of the semiconductor substrate, an appropriate mask should be used for the implantation. Because implantation is a low-temperature process, a large variety of the masking materials can be used. The minimum thickness of the masking material required to stop a given percentage of incident ions can be estimated from the range parameters for ions. The inset of Figure 7.14 shows a profile of an implant in a masking material. The dose implanted in the region beyond a depth d (shown shaded) is given by integration of Eq. 1 as

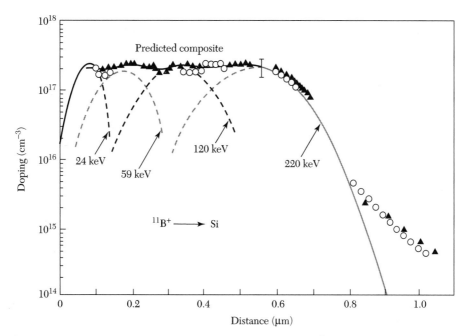

Figure 7.13 Composite doping profile using multiple implants.[9]

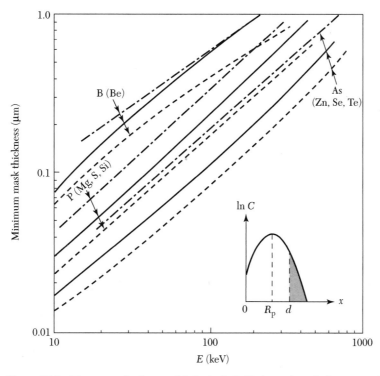

Figure 7.14 Minimum thickness of SiO_2 (—), Si_3N_4 (- - - -), and photoresist (—·—·—) to produce a masking effectiveness of 99.99%.[5,10]

$$S_d = \frac{S}{\sqrt{2\pi}\sigma_p} \int_d^\infty \exp\left[-\left(\frac{x-R_p}{\sqrt{2}\sigma_p}\right)^2\right] dx \tag{9}$$

From Table 6.1, we can derive the expression

$$\int_x^\infty e^{-y^2}\, dy = \frac{\sqrt{\pi}}{2}\,\mathrm{erfc}(x) \tag{10}$$

Therefore, the fraction of the dose that has "transmitted" beyond a depth d is given by the transmission coefficient (T):

$$T \equiv \frac{S_d}{S} = \frac{1}{2}\,\mathrm{erfc}\left(\frac{d-R_p}{\sqrt{2}\sigma_p}\right) \tag{11}$$

Once T is given, we can obtain the mask thickness d from Eq. 11 for any given R_p and σ_p.

The values of d to stop 99.99% of the incident ions ($T = 10^{-4}$) are shown in Figure 7.14 for SiO_2, Si_3N_4, and photoresist as masking materials.[5,10] Mask thicknesses given in this figure are for boron, phosphorus, and arsenic implanted into silicon. These mask thicknesses also can be used as guidelines for impurity masking in gallium arsenide. The dopants are shown in the parentheses. Since both R_p and σ_p vary approximately linearly with energy, the minimum thickness of the masking material also increases linearly with energy. In certain applications, instead of totally stopping the beam, masks can be used as attenuators that can provide an amorphous surface layer to the incident ion beam to minimize the channeling effect.

EXAMPLE 2

When boron ions are implanted at 200 keV, what thickness of SiO_2 will be required to mask 99.996% of the implanted ions ($R_p = 0.53$ μm, $\sigma_p = 0.093$ μm)?

SOLUTION The complementary error function in Eq. 11 can be approximated if the argument is large (see Table 6.1):

$$T \cong \frac{1}{2\sqrt{\pi}}\frac{e^{-u^2}}{u}$$

where the parameter u is given by $(d - R_p)/\sqrt{2}\sigma$. For $T = 10^{-4}$, we solve the above equation to obtain $u = 2.8$. Thus,

$$d = R_p + 3.96\sigma_p = 0.53 + 3.96 \times 0.093 = 0.898 \ \mu m \qquad \blacktriangleleft$$

7.3.2 Tilt-Angle Ion Implantation

In scaling devices to submicron dimensions, it is important also to scale dopant profiles vertically. We need to produce junction depths less than 100 nm, including diffusion during dopant activation and subsequent processing steps. Modern device structures, such as the lightly doped drain (LDD) MOSFET, require precise control of dopant distributions vertically and laterally.

It is the ion velocity perpendicular to the surface that determines the projected range of an implanted ion distribution. If the wafer is tilted at a large angle to the ion beam, then the effective ion energy is greatly reduced. Figure 7.15 illustrates this for 60-keV arsenic ions as a function of the tilt angle, showing that it is possible to achieve extremely shallow distributions using a high tilt angle (86°). In tilt-angle ion implantation, we should

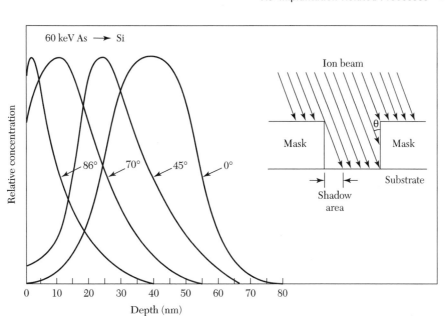

Figure 7.15 60-keV arsenic implanted into silicon, as a function of beam tilt angle. Inset shows the shadow area for tilt-angle ion implantation.

consider the shadow effect (inset in Fig. 7.15) for the patterned wafer. A lower tilt angle leads to a small shadow area. For example, if the height of the patterned mask is 0.5 μm, with vertical sidewall, a 7° incident ion beam will induce a 61-nm shadow. This shadow effect may introduce an unexpected series resistance in the device.

7.3.3 High-Energy and High-Current Implantation

High-energy implantors, capable of energies as high as 1.5 to 5 MeV, are available and have been used for a number of novel applications. The majority of these depend on the ability to dope the semiconductor to many micrometers in depth, without the need for long diffusion times at high temperatures. High-energy implantors can also be used to produce low-resistivity buried layers. For example, a buried layer 1.5 to 3 μm below the surface for a CMOS device can be achieved by high-energy implantation.

High-current implantors (10–20 mA), operating in the 25- to 30-keV range, are routinely used for the predeposition step in diffusion technology because the amount of total dopant can be controlled precisely. After predeposition, the dopant impurities can be driven in by a high-temperature diffusion step at the same time that implant damage at the surface region is annealed out. Another application is the threshold voltage adjustment in MOS devices. A precisely controlled amount of dopant (e.g., boron) is implanted through the gate oxide to the channel region[11] (Fig. 7.16a). Because the projected ranges of boron in silicon and silicon oxide are comparable, if we choose a suitable incident energy, the ions will penetrate just the thin gate oxide, not the thicker field oxide. The threshold voltage will vary approximately linearly with the implanted dose. After boron implantation, polysilicon can be deposited and patterned to form the gate electrode of the MOSFET. The thin oxide surrounding the gate electrode is removed, and the source and drain regions are formed, as shown in Figure 7.16b, by another high-dose arsenic implantation.

High-current implantors with energies in the 150- to 200-keV range are now available. A major use for these machines is to form high-quality silicon films, which are

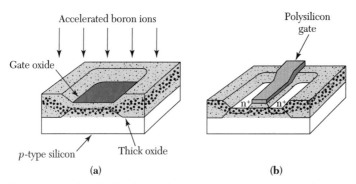

Figure 7.16 Threshold voltage adjustment using boron ion implantation.[11]

insulated from the substrate by implanting oxygen, creating an intervening layer of silicon dioxide. This *separation by implantation of oxygen* (SIMOX) is a key silicon-on-insulator (SOI) technology.

The SIMOX process uses a high-energy O^+ beam, typically in the 150- to 200-keV range, so that the oxygen ions have projected ranges of 100 to 200 nm. Additionally, a heavy dose, 1 to 2×10^{18} ions/cm^2, is used to produce an insulating layer of SiO_2 that is 100 to 500 nm thick. The use of SIMOX material leads to a significant reduction of source/drain capacitances in MOS devices. Moreover, it reduces coupling between devices and thus allows tighter packing without the problem of latchup. As a result, it is widely proposed as the material of choice for advanced, high-speed CMOS circuits.

▶ 7.4 ION IMPLANTATION SIMULATION

SUPREM may be used to simulate ion implantation profiles. Simulated profiles can be implanted and activated using the IMPLANT command, and subsequently driven in using the DIFFUSION command. SUPREM contains the implant parameters for most common dopants, but also allows the user to input range and straggle data for unusual implanted materials. SUPREM can also simulate implants through multiple layers.

EXAMPLE 3

Suppose we want to simulate the implantation of a 2×10^{13} cm^{-2} dose of boron at 30 keV into an *n*-type <100> silicon wafer. The implant is then followed by a drive-in at 950°C for 60 minutes. If the silicon substrate is doped with phosphorus at a level of 10^{15} cm^{-3}, use SUPREM to determine the boron doping profile and the junction depth.

SOLUTION   The SUPREM input listing is as follows:

```
TITLE        Implantation Example
COMMENT      Initialize silicon substrate
INITIALIZE   <100>Silicon Phosphor Concentration=1e15
COMMENT      Implant boron
IMPLANT      Boron Energy=30 Dose=2e13
COMMENT      Diffuse boron
DIFFUSION    Time=60 Temperature=950
PRINT        Layers Chemical Concentration Phosphorus Boron Net
PLOT         Active Net Cmin=1e14
STOP         End implantation example
```

After the simulation is complete, we print and plot the boron concentration as a function of depth into the silicon substrate. The results are shown in Figure 7.17, which indicates a junction depth of 0.4454 μm. ◄

► 7.5 SUMMARY

Ion implantation is a key method for impurity doping. The key parameters for ion implantation are the projected range (R_p) and its standard deviation (σ_p), also called the projected straggle. The implantation profile can be approximated by a Gaussian distribution with its peak located at R_p from the surface of the semiconductor substrate. The advantages of the ion implantation process are more precise control of the amount of dopant, a more reproducible doping profile, and lower processing temperature compared with the diffusion process.

The chapter considered R_p and σ_p for various elements in silicon and gallium arsenide and discussed the channeling effect and ways to minimize this effect. However, implantation may cause severe damage to the crystal lattice. To remove the implant damage and to restore mobility and other device parameters, the semiconductor must be annealed at an appropriate combination of time and temperature. Currently, rapid thermal annealing (RTA) is preferred to conventional furnace annealing because RTA can remove implant damage without thermal broadening of the doping profile.

Ion implantation has wide applications for advanced semiconductor devices. These include (a) multiple implantation to form novel distributions, (b) selection of masking

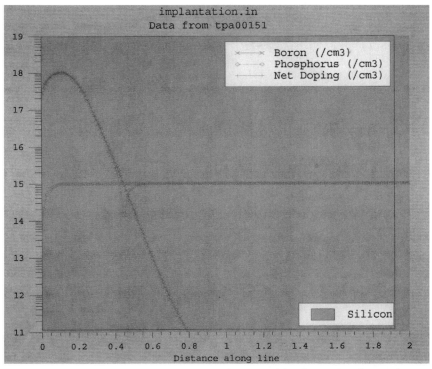

Figure 7.17 Plot of boron concentration as a function of depth into the silicon substrate, using SUPREM.

materials and thickness to stop a given percentage of incident ions from reaching the substrate, (c) tilt-angle implantation to form ultrashallow junctions, (d) high-energy implantation to form buried layers, and (e) high-current implantation for predeposition and threshold voltage adjustment and to form an insulating layer for SOI applications.

▶ REFERENCES

1. I. Brodie and J. J. Murray, *The Physics of Microfabrication*, Plenum, New York, 1982.

2. J. F. Gibbons, "Ion Implantation," in S. P. Keller, Ed., *Handbook on Semiconductors*, Vol. 3, North-Holland, Amsterdam, 1980.

3. S. Furukawa, H. Matsumura, and H. Ishiwara, "Theoretical Consideration on Lateral Spread of Implanted Ions," *Jpn. J. Appl. Phys.*, **11**, 134 (1972).

4. B. Smith, *Ion Implantation Range Data for Silicon and Germanium Device Technologies*, Research Studies, Forest Grove, OR, 1977.

5. K. A. Pickar, "Ion Implantation in Silicon," in R. Wolfe, Ed., *Applied Solid State Science*, Vol. 5, Academic Press, New York, 1975.

6. L. Pauling and R. Hayward, *The Architecture of Molecules*, Freeman, San Francisco, 1964.

7. D. K. Brice, "Recoil Contribution to Ion Implantation Energy Deposition Distribution," *J. Appl. Phys.*, **46**, 3385 (1975).

8. C. Y. Chang and S. M. Sze, Eds., *ULSI Technology*, McGraw-Hill, New York, 1996, Ch. 4.

9. D. H. Lee and J. W. Mayer, "Ion-Implanted Semiconductor Devices," *Proc. IEEE*, **62**, 1241 (1974).

10. G. Deamaley, et al., *Ion Implantation*, North-Holland, Amsterdam, 1973.

11. W. G. Oldham, "The Fabrication of Microelectronic Circuit," in *Microelectronics*, Freeman, San Francisco, 1977.

▶ PROBLEMS

Asterisks denote difficult problems.

SECTION 7.1: RANGE OF IMPLANTED IONS

1. Assume that a 100-mm diameter GaAs wafer is uniformly implanted with 100-keV zinc ions for 5 minutes with a constant ion beam current of 10 μA. What are the ion dose per unit area and the peak ion concentration?

2. A silicon p–n junction is formed by implanting boron ions at 80 keV through a window in an oxide. If the boron dose is 2×10^{15} cm^{-2} and the n-type substrate concentration is 10^{15} cm^{-3}, find the location of the metallurgical junction.

3. A threshold voltage adjustment implantation is made through a 25-nm gate oxide. The substrate is a <100>-oriented p-type silicon with a resistivity of 10 Ω-cm. If the incremental threshold voltage due to a 40-keV boron implantation is 1 V, what is the total implanted dose per unit area? Estimate the location of the peak boron concentration.

°4. For the substrate in Problem 3, what percentage of the total dose is in the silicon?

SECTION 7.2: IMPLANT DAMAGE AND ANNEALING

5. If a 50-keV boron ion is implanted into the silicon substrate, calculate the damage density. Assume silicon atom density is 5.02×10^{22} atoms/cm^3, the silicon displacement energy is 15 eV, the range is 2.5 nm, and the spacing between silicon lattice planes is 0.25 nm.

6. Explain why high-temperature RTA is preferable to low-temperature RTA for defect-free shallow-junction formation.

7. Estimate the implant dose required to reduce a p-channel threshold voltage by 1 V if the gate oxide is 4 nm thick. Assume that the implant voltage is adjusted so that the peak of the distribution occurs at the oxide–silicon interface. Thus, half of the implant goes into the silicon. Further assume that 90% of the implanted ions in the silicon are electrically activated by the annealing process. These assumptions allow 45% of the implanted ions to be used for threshold adjusting. Also assume that all of the charge in the silicon is effectively at the silicon–oxide interface.

SECTION 7.3: IMPLANTATION-RELATED PROCESSES

8. We would like to form 0.1-μm deep, heavily doped junctions for the source and drain regions of a submicron MOSFET. Compare the options that are available to introduce and activate dopant for this application. Which option would you recommend and why?

9. When an arsenic implant at 100 keV is used and the photoresist thickness is 400 nm, find the effectiveness of the resist mask in preventing the transmission of ions ($R_p = 0.6$ μm, $\sigma_p = 0.2$ μm). If the resist thickness is changed to 1 μm, calculate the masking efficiency.

10. With reference to Example 2, what thickness of SiO_2 is required to mask 99.999% of the implanted ions?

SECTION 7.4: ION IMPLANTATION SIMULATION

11. A <100> phosphorus-doped silicon substrate with a doping concentration of 10^{14} cm^{-3} is implanted with boron. The implant energy is 30 keV with a dose of 10^{13} cm^{-2}. Use SUPREM to plot the boron profile. What are (a) the depth of the peak of the implanted profile, (b) the boron concentration at the peak depth, and (c) the junction depth?

*12. Use SUPREM to design an implantation step that gives the same doping profile as the diffusion Example 3 in Chapter 6.

8

Film Deposition

To fabricate discrete devices and integrated circuits, we use many different kinds of thin films. We can classify thin films into five groups: thermal oxides, dielectric layers, epitaxial layers, polycrystalline silicon, and metal films. The growth of thermal oxides was discussed in Chapter 3. This chapter deals with various other techniques for depositing thin films.

Epitaxial growth is closely related to the crystal growth concepts discussed in Chapter 2. It involves the growth of single-crystal semiconductor layers on a single-crystal semiconductor substrate. The word *epitaxy* is derived from the Greek words *epi* (meaning "on") and *taxis* (meaning "arrangement"). The epitaxial layer and the substrate materials may be the same, giving rise to *homoepitaxy*. For example, an n-type silicon can be grown epitaxially on an n^+- silicon substrate. On the other hand, if the epitaxial layer and the substrate are chemically and often crystallographically different, we have *heteroepitaxy*, such as the epitaxial growth of $Al_xGa_{1-x}As$ on GaAs.

Dielectric layers such as silicon dioxide and silicon nitride are used for insulation between conducting layers, for diffusion and ion implantation masks, for capping doped films to prevent the loss of dopants, and for passivation. Polycrystalline silicon, usually referred to as *polysilicon*, is used as a gate electrode material in MOS devices, a conductive material for multilevel metallization, and a contact material for devices with shallow junctions. Metal films such as aluminum and silicides are used to form low-resistance interconnections, ohmic contacts, and rectifying metal-semiconductor barriers.

This chapter covers the following topics:

- Basic techniques of epitaxy, that is, growing a single-crystal layer on a single-crystal substrate
- Structures and defects of lattice-matched and strained-layer epitaxial growth
- Deposition techniques to form low-dielectric-constant and high-dielectric-constant films, as well as polysilicon films
- Deposition techniques to form aluminum and copper interconnections, as well as the related global planarization process
- Characteristics of these thin films and their compatibility with integrated circuit processing

▶ 8.1 EPITAXIAL GROWTH TECHNIQUES

In an epitaxial process, the substrate wafer acts as the seed crystal. Epitaxial processes are differentiated from the melt-growth processes described in Chapter 2 in that the epitaxial layer can be grown at a temperature substantially below the melting point, typically

30% to 50% lower. The common techniques for epitaxial growth are *chemical vapor deposition* (CVD) and *molecular beam epitaxy* (MBE).

8.1.1 Chemical Vapor Deposition

CVD is also known as *vapor-phase epitaxy* (VPE). CVD is a process whereby an epitaxial layer is formed by a chemical reaction between gaseous compounds. CVD can be performed at atmospheric pressure (APCVD) or at low pressure (LPCVD).

Figure 8.1 shows three common susceptors for epitaxial growth. Note that the geometric shape of the susceptor provides the name for the reactor: horizontal, pancake, and barrel. All are made from graphite blocks. Susceptors in epitaxial reactors are analogous to crucibles in the crystal growing furnaces. Not only do they mechanically support the wafer, but in induction-heated reactors, they also serve as the source of thermal energy for the reaction. The mechanism of CVD involves a number of steps: (a) the reactants (gases and dopants) are transported to the substrate region; (b) they are transferred to the substrate surface, where they are adsorbed; (c) a chemical reaction occurs, catalyzed at the surface, followed by growth of the epitaxial layer; (d) the gaseous products are desorbed into the main gas stream; and (e) the reaction products are transported out of the reaction chamber.

CVD for Silicon

Four silicon sources have been used for VPE growth. They are silicon tetrachloride ($SiCl_4$), dichlorosilane (SiH_2Cl_2), trichlorosilane ($SiHCl_3$), and silane (SiH_4). Silicon tetrachloride has been studied the most and has the widest industrial use. The typical reaction temperature is 1200°C. Other silicon sources are used because of lower reaction temperatures. The

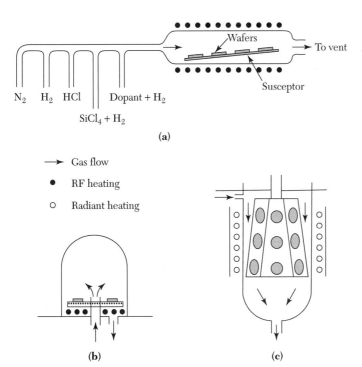

Figure 8.1   Three common susceptors for chemical vapor deposition: (*a*) horizontal, (*b*) pancake, and (*c*) barrel susceptors.

substitution of a hydrogen atom for each chlorine atom from silicon tetrachloride permits about a 50°C reduction in the reaction temperature. The overall reaction of silicon tetrachloride that results in the growth of silicon layers is

$$SiCl_4 \text{ (gas)} + 2H_2 \text{ (gas)} \leftrightarrow Si \text{ (solid)} + 4HCl \text{ (gas)} \tag{1}$$

An additional competing reaction takes place along with that given in Eq. 1:

$$SiCl_4 \text{ (gas)} + Si \text{ (solid)} \leftrightarrow 2SiCl_2 \text{ (gas)} \tag{2}$$

As a result, if the silicon tetrachloride concentration is too high, etching, rather than growth of silicon, will take place. Figure 8.2 shows the effect of the concentration of silicon tetrachloride in the gas on the reaction, where the mole fraction is defined as the ratio of the number of molecules of a given species to the total number of molecules.[1] Note that initially the growth rate increases linearly with an increasing concentration of silicon tetrachloride. As the concentration of silicon tetrachloride is increased, a maximum growth rate is reached. Beyond that, the growth rate starts to decrease, and eventually etching of the silicon will occur. Silicon is usually grown in the low-concentration region, as indicated in Figure 8.2.

The reaction of Eq. 1 is reversible; that is, it can take place in either direction. If the carrier gas entering the reactor contains hydrochloric acid, removal or etching will take place. Actually, this etching operation is used for in situ cleaning of the silicon wafer prior to epitaxial growth.

The dopant is introduced at the same time as the silicon tetrachloride during epitaxial growth (Fig. 8.1a). Gaseous diborane (B_2H_6) is used as the p-type dopant, whereas phosphine (PH_3) and arsine (AsH_3) are used as n-type dopants. Gas mixtures are ordinarily used with hydrogen as the diluent to allow reasonable control of flow rates for the desired doping concentration. The dopant chemistry for arsine is illustrated in Figure 8.3, which shows arsine being adsorbed on the surface, decomposing, and being incorporated

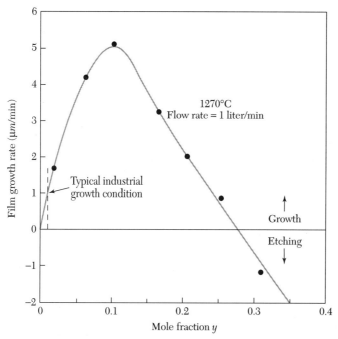

Figure 8.2 Effect of SiCl$_4$ concentration on silicon epitaxial growth.[1]

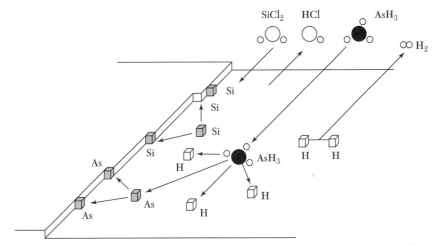

Figure 8.3 Schematic representation of arsenic doping and the growing processes.[2]

into the growing layer. Figure 8.3 also shows the growth mechanisms at the surface, which are based on the surface adsorption of host atoms (silicon) as well as the dopant atoms (e.g., arsenic) and the movement of these atoms toward the ledge sites.[2] To give these adsorbed atoms sufficient mobility for finding their proper positions within the crystal lattice, epitaxial growth requires relatively high temperatures.

CVD for Gallium Arsenide

For gallium arsenide, the basic setup is similar to that shown in Figure 8.1a. Because gallium arsenide decomposes into gallium and arsenic upon evaporation, its direct transport in the vapor phase is not possible. One approach is the use of As_4 for the arsenic component and gallium chloride ($GaCl_3$) for the gallium component. The overall reaction leading to epitaxial growth of gallium arsenide is

$$As_4 + 4GaCl_3 + 6H_2 \rightarrow 4GaAs + 12HCl \tag{3}$$

The As_4 is generated by thermal decomposition of arsine (AsH_3):

$$4AsH_3 \rightarrow As_4 + 6H_2 \tag{3a}$$

and the gallium chloride is generated by the reaction

$$6HCl + 2Ga \rightarrow 2GaCl_3 + 3H_2 \tag{3b}$$

The reactants are introduced into a reactor with a carrier gas (e.g., H_2). The gallium arsenide wafers are typically held within the 650°C to 850°C temperature range. There must be sufficient arsenic overpressure to prevent thermal decomposition of the substrate and the growing layer.

Metalorganic CVD

Metalorganic CVD (MOCVD) is also a VPE process based on pyrolytic reactions. Unlike conventional CVD, MOCVD is distinguished by the chemical nature of the precursor. It is important for those elements that do not form stable hydrides or halides, but that do form stable metalorganic compounds with reasonable vapor pressure. MOCVD has been extensively applied in the heteroepitaxial growth of III-V and II-VI compounds.

To grow GaAs, we can use metalorganic compounds such as trimethylgallium [$Ga(CH_3)_3$] for the gallium component and arsine (AsH_3) for the arsenic component. Both chemicals can be transported in vapor form into the reactor. The overall reaction is

$$AsH_3 + Ga(CH_3)_3 \rightarrow GaAs + 3CH_4 \qquad (4)$$

For Al-containing compounds, such as AlAs, trimethylaluminum [$Al(CH_3)_3$] can be used. During epitaxy, the GaAs is doped by introducing dopants in vapor form. Diethylzinc [$Zn(C_2H_5)_2$] and diethylcadmium [$Cd(C_2H_5)_2$] are typical p-type dopants, and silane (SiH_4) is an n-type dopant for III–V compounds. The hydrides of sulfur and selenium or tetramethyltin are also used for n-type dopants, and chromyl chloride is used to dope chromium into GaAs to form semiinsulating layers. Since these compounds are highly poisonous and often spontaneously inflammable in air, rigorous safety precautions are necessary in the MOCVD process.

Figure 8.4 shows a schematic of an MOCVD reactor.[3] Typically, the metalorganic compound is transported to the quartz reaction vessel by hydrogen carrier gas, where it is mixed with AsH_3 in the case of GaAs growth. The chemical reaction is induced by heating the gases to 600°C to 800°C above a substrate placed on a graphite susceptor using radio frequency heating. A pyrolytic reaction forms the GaAs layer. The advantages of using metalorganics are that they are volatile at moderately low temperatures and that there are no troublesome liquid Ga or In sources in the reactor.

8.1.2 Molecular Beam Epitaxy

MBE is an epitaxial process involving the reaction of one or more thermal beams of atoms or molecules with a crystalline surface under ultrahigh-vacuum conditions (~10^{-8} Pa).[4]

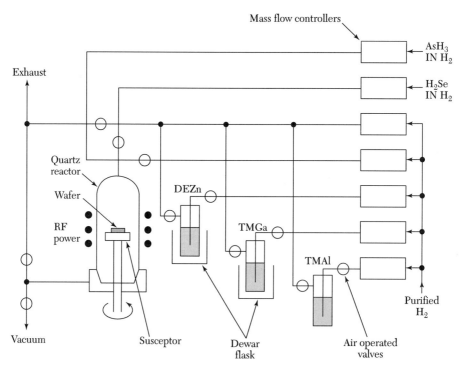

Figure 8.4 Schematic diagram of a vertical atmospheric-pressure MOCVD reactor.[3] DEZn, diethylzinc [$Zn(C_2H_5)_2$]; TMGa, trimethylgallium [$Ga(CH_3)_3$]; TMAl, trimethylaluminum [$Al(CH_3)_3$].

MBE can achieve precise control of both chemical compositions and doping profiles. Single-crystal, multilayer structures with dimensions on the order of atomic layers can be grown using MBE. Thus, the MBE method enables the precise fabrication of semiconductor heterostructures having thin layers from a fraction of a micron down to a monolayer. In general, MBE growth rates are quite low. For GaAs, for example, a value of 1 μm/hr is typical.

Figure 8.5 shows a schematic of an MBE system or gallium arsenide and related III-V compounds, such as $Al_xGa_{1-x}As$. The system represents the ultimate in film deposition control, cleanliness, and in situ chemical characterization capability. Separate effusion ovens made of pyrolytic boron nitride are used for Ga, As, and the dopants. All the effusion ovens are housed in an ultrahigh-vacuum chamber (~10^{-8} Pa). The temperature of each oven is adjusted to give the desired evaporation rate. The substrate holder rotates continuously to achieve uniform epitaxial layers (e.g., ±1% in doping variations and ±0.5% in thickness variations).

To grow GaAs, an overpressure of As is maintained, since the sticking coefficient of Ga to GaAs is unity, whereas that for As is zero, unless there is a previously deposited Ga layer. For a silicon MBE system, an electron gun is used to evaporate silicon. One or more effusion ovens are used for the dopants. Effusion ovens behave like small-area sources and exhibit a cosθ emission, where θ is the angle between the direction of the source and the normal to the substrate surface.

MBE uses an evaporation method in a vacuum system. An important parameter for vacuum technology is the molecular impingement rate; that is, how many molecules impinge on a unit area of the substrate per unit time. The impingement rate (ϕ) is a function of the molecular weight, temperature, and pressure. The rate is derived in Appendix H and can be expressed as[5]

$$\phi = P(2\pi mkT)^{-1/2} \tag{5}$$

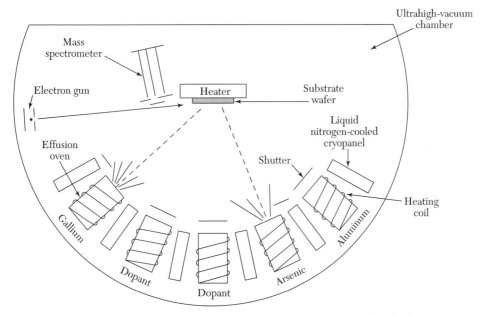

Figure 8.5 Arrangement of the sources and substrate in a conventional molecular beam epitaxy (MBE) system. (Courtesy of M. B. Panish, Bell Laboratories, Lucent Technologies.)

or

$$\phi = 2.64 \times 10^{20} \left(\frac{P}{\sqrt{MT}} \right) \text{ molecules / cm}^2\text{-s} \tag{5a}$$

where P is the pressure in Pa, m is the mass of a molecule in kg, k is Boltzmann's constant in J/K, T is the temperature in degrees Kelvin, and M is the molecular weight. Therefore, at 300 K and 10^{-4} Pa pressure, the impingement rate is 2.7×10^{14} molecules/cm^2-s for oxygen ($M = 32$).

EXAMPLE 1

At 300 K, the molecular diameter of oxygen is 3.64 Å, and the number of molecules per unit area N_s is 7.54×10^{14} cm^{-2}. Find the time required to form a monolayer of oxygen at pressures of 1, 10^{-4}, and 10^{-8} Pa.

SOLUTION The time required to form a monolayer (assuming 100% sticking) is obtained from the impingement rate:

$$t = \frac{N_s}{\phi} \frac{N_s \sqrt{MT}}{2.64 \times 10^{20} P}$$

Therefore,

$$t = 2.8 \times 10^{-4} \approx 0.28 \text{ ms} \quad \text{at 1 Pa}$$
$$= 2.8 \text{ s} \quad \text{at } 10^{-4} \text{ Pa}$$
$$= 7.7 \text{ hr} \quad \text{at } 10^{-8} \text{ Pa}$$

To avoid contamination of the epitaxial layer, it is of paramount importance to maintain ultrahigh-vacuum conditions (~10^{-8} Pa) for the MBE process. ◀

During molecular motion, molecules will collide with one another. The average distance traversed by all the molecules between successive collisions with each other is defined as the *mean free path*. It can be derived from a simple collision theory. A molecule having a diameter d and a velocity v will move a distance $v\delta t$ in the time δt. The molecule suffers a collision with another molecule if its center is anywhere within the distance d of the center of another molecule. Therefore, it sweeps out (without collision) a cylinder of diameter $2d$. The volume of the cylinder is

$$\delta V = \frac{\pi}{4}(2d)^2 v\delta t = \pi d^2 v\delta t \tag{6}$$

Since there are n molecules/cm^3, the volume associated with one molecule is on the average $1/n$ cm^3. When the volume δV is equal to $1/n$, it must contain on the average one other molecule. Thus, a collision would have occurred. Setting $\tau = \delta t$ as the average time between collisions,

$$\frac{1}{n} = \pi d^2 v\tau \tag{7}$$

The mean free path (λ) is then

$$\lambda = v\tau = \frac{1}{\pi n d^2} = \frac{kT}{\pi P d^2} \tag{8}$$

A more rigorous derivation gives

$$\lambda = \frac{kT}{\sqrt{2}\pi P d^2} \tag{9}$$

and

$$\lambda = \frac{0.66}{P(\text{in Pa})}\, \text{cm} \tag{10}$$

for air molecules (equivalent molecular diameter of 3.7 Å) at room temperature. Therefore, at a system pressure of 10^{-8} Pa, λ would be 660 km.

EXAMPLE 2

Assume an effusion oven geometry of area $A = 5$ cm^2 and a distance L between the top of the oven and the gallium arsenide substrate of 10 cm. Calculate the MBE growth rate for the effusion oven filled with gallium arsenide at 900°C. The surface density of gallium atoms is 6×10^{14} cm^{-2}, and the average thickness of a monolayer is 2.8 Å.

SOLUTION   On heating gallium arsenide, the volatile arsenic vaporizes first, leaving a gallium-rich solution. Therefore, only the pressures marked "Ga-rich" in Figure 2.11 are of interest. The pressure at 900°C is 5.5×10^{-2} Pa for gallium and 1.1 Pa for arsenic (As$_2$). The arrival rate can be obtained from the impingement rate (Eq. 5a) by multiplying it by $A/\pi L^2$:

$$\text{Arrival rate} = 2.64 \times 10^{20} \left(\frac{P}{\sqrt{MT}}\right)\left(\frac{A}{\pi L^2}\right) \text{molecules}/\text{cm}^2\text{-s}$$

The molecular weight (M) is 69.72 for Ga and 74.92 × 2 for As$_2$. Substituting values of P, M, and T (1173 K) into the above equation gives

$$\text{Arrival rate} = 8.2 \times 10^{14} / \text{cm}^2\text{-s for Ga}$$
$$= 1.1 \times 10^{16} / \text{cm}^2\text{-s for As}^2$$

The growth rate of gallium arsenide is governed by the arrival rate of gallium. The growth rate is

$$\frac{8.2 \times 10^{14} \times 2.8}{6 \times 10^{14}} \approx 3.8 \text{A/s} = 23 \text{ nm/min}$$

Note that the growth rate is relatively low compared with that of VPE.   ◀

There are two ways to clean a surface in situ for MBE. High-temperature baking can decompose the native oxide and remove other adsorbed species by evaporation or diffusion into the wafer. Another approach is to use a low-energy ion beam of an inert gas to sputter-clean the surface, followed by a low-temperature annealing to reorder the surface lattice structure.

MBE can use a wide variety of dopants (compared with CVD and MOCVD), and the doping profile can be exactly controlled. However, the doping process is similar to the vapor-phase growth process: A flux of evaporated dopant atoms arrives at a favorable lattice site and is incorporated along the growing interface. Fine control of the doping profile is achieved by adjusting the dopant flux relative to the flux of silicon atoms (for silicon epitaxial films) or gallium atoms (for gallium arsenide epitaxial films). It is also possible to dope the epitaxial film using a low-current, low-energy ion beam to implant the dopant (see Chapter 7).

The substrate temperatures for MBE range from 400°C to 900°C, and the growth rates range from 0.001 to 0.3 μm/min. Because of low-temperature processing and the low growth rate, many unique doping profiles and alloy compositions not obtainable from conventional CVD can be produced in MBE. Many novel structures have been made using MBE. These include the *superlattice*, which is a periodic structure consisting of alternating ultrathin layers with a period less than the electron mean free path (e.g., GaAs/Al$_x$Ga$_{1-x}$As, with each layer 10 nm or less in thickness), and heterojunction field-effect transistors.

A further development in MBE has replaced the group III elemental sources by metalorganic compounds such as trimethylgallium (TMG) or triethylgallium (TEG). This approach is called *metalorganic molecular beam epitaxy* (MOMBE) and is also referred to as *chemical beam epitaxy* (CBE). Although closely related to MOCVD, it is considered a special form of MBE. The metalorganics are sufficiently volatile that they can be admitted directly into the MBE growth chamber as a beam and are not decomposed before forming the beam. The dopants are generally elemental sources, typically Be for *p*-type and Si or Sn for *n*-type GaAs epitaxial layers.

▶ 8.2 STRUCTURES AND DEFECTS IN EPITAXIAL LAYERS

8.2.1 Lattice-Matched and Strained-Layer Epitaxy

For conventional homoepitaxial growth, a single-crystal semiconductor layer is grown on a single-crystal semiconductor substrate. The semiconductor layer and the substrate are the same material having the same lattice constant. Therefore, homoepitaxy is, by definition, a lattice-matched epitaxial process. The homoepitaxial process offers one important means of controlling the doping profiles so that device and circuit performance can be optimized. For example, an *n*-type silicon layer with a relatively low doping concentration can be grown epitaxially on an *n*$^+$ silicon substrate. This structure substantically reduces the series resistance associated with the substrate.

For heteroepitaxy, the epitaxial layer and the substrate are two different semiconductors, and the epitaxial layer must be grown in such a way that an idealized interfacial structure is maintained. This implies that atomic bonding across the interface must be continuous. Therefore, the two semiconductors must either have the same lattice spacing or be able to deform to adopt a common spacing. These two cases are referred to as *lattice-matched epitaxy* and *strained-layer epitaxy*, respectively.

Figure 8.6*a* shows a lattice-matched epitaxy where the substrate and the film have the same lattice constant. An important example is the epitaxial growth of Al$_x$Ga$_{1-x}$As on a GaAs substrate where for any *x* between 0 and 1, the lattice constant of Al$_x$Ga$_{1-x}$As differs from that of GaAs by less than 0.13%.

For the lattice-mismatched case, if the epitaxial layer has a larger lattice constant and is flexible, it will be compressed in the plane of growth to conform to the substrate spacing. Elastic forces then compel it to dilate in a direction perpendicular to the interface. This type of structure is called strained-layer epitaxy and is illustrated in Figure 8.6*b*.[6] On the other hand, if the epitaxial layer has a smaller lattice constant, it will be dilated in the plane of growth and compressed in a direction perpendicular to the interface. In the above strained-layer epitaxy, as the strained-layer thickness increases, the total number of atoms under strain of the distorted atomic bonds grows, and at some point, misfit dislocations are nucleated to relieve the homogeneous strain energy. This thickness is referred to as the *critical layer thickness* for the system. Figure 8.6*c* shows the case in which there are edge dislocations at the interface.

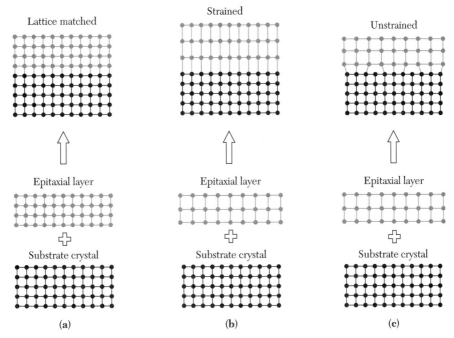

Figure 8.6 Schematic illustration of (*a*) lattice-matched, (*b*) strained,[6] and (*c*) relaxed heteroepitaxial structures. Homoepitaxy is structurally identical to the lattice-matched heteroepitaxy.

The critical layer thicknesses for two material systems are shown in Figure 8.7.[7] The upper curve is for the strained-layer epitaxy of a Ge_xSi_{1-x} layer on a silicon substrate, and the lower curve is for a $Ga_{1-x}In_xAs$ layer on a GaAs substrate. For example, for $Ge_{0.3}Si_{0.7}$ on silicon, the maximum epitaxial thickness is about 70 nm. For thicker films, edge dislocations will occur.

A related heteroepitaxial structure is the *strained-layer superlattice* (SLS). A superlattice is an artificial one-dimensional periodic structure constituted by different materials with a period of about 10 nm. Figure 8.8 shows[4] an SLS having two semiconductors with different equilibrium lattice constants $a_1 > a_2$, grown in a structure with a common inplane lattice constant b, where $a_1 > b > a_2$. For sufficiently thin layers, the lattice mismatch is accommodated by uniform strains in the layers. Under these conditions, no misfit dislocations are generated at the interfaces, so high-quality crystalline materials can be obtained. These artificially structured materials can be grown by MBE. These materials provide a new area in semiconductor research and permit new solid-state devices, especially for high-speed and photonic applications.

8.2.2 Defects in Epitaxial Layers

Defects in epitaxial layers will degrade device properties. For example, defects can result in reduced mobility or increased leakage current. The defects in epitaxial layers can be categorized into five groups:

1. *Defects from the substrates.* These defects may propagate from the substrate into the epitaxial layer. To avoid these defects, dislocation-free semiconductor substrates are required.

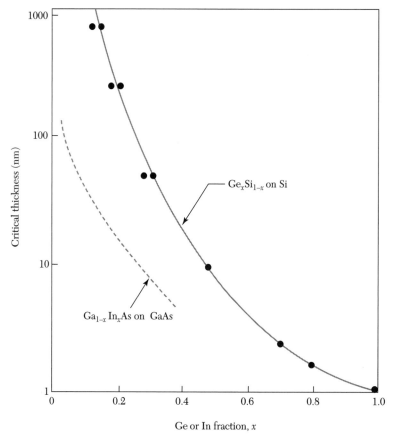

Figure 8.7   Experimentally determined critical layer thickness for defect-free, strained-layer epitaxy of Ge$_x$Si$_{1-x}$ on Si, and Ga$_{1-x}$In$_x$As on GaAs.[7]

2. *Defects from the interface.* Oxide precipitates or any contamination at the interface of the epitaxial layer and substrate may cause the formation of misoriented clusters or nuclei containing stacking faults. These clusters and stacking faults may coalesce with normal nuclei and grow into the film in the shape of an inverted pyramid. To avoid these defects, the surface of the substrate must be thoroughly cleaned. In addition, an in situ etchback may be used, such as the reversable reaction of Eq. 1.

3. *Precipitates or dislocation loops.* Their formation is due to supersaturation of impurities or dopants. Epitaxial layers containing very high intentional or unintentional dopants or impurity concentrations are susceptible to such defects.

4. *Low-angle grain boundaries and twins.* Any misoriented areas of an epitaxial film during growth may meet and coalesce to form these defects.

5. *Edge dislocations.* These are formed in the heteroepitaxy of two lattice-mismatched semiconductors. If both lattices are rigid, they will retain their fundamental lattice spacings, and the interface will contain rows of misbonded atoms described as misfit or edge dislocations. The edge dislocations can also form in a strained layer when the layer thickness becomes larger than the critical layer thickness.

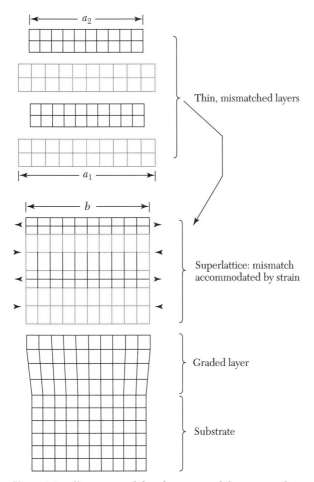

Figure 8.8 Illustration of the elements and formation of a strained-layer superlattice.[4] Arrows show the direction of the strain.

▶ 8.3 DIELECTRIC DEPOSITION

Deposited dielectric films are used mainly for insulation and passivation of discrete devices and integrated circuits. There are three commonly used deposition methods: atmospheric-pressure CVD, low-pressure CVD (LPCVD), and plasma-enhanced chemical vapor deposition (PECVD, or plasma deposition). PECVD is an energy-enhanced CVD method in which plasma energy is added to the thermal energy of a conventional CVD system. Considerations in selecting a deposition process are the substrate temperature, the deposition rate and film uniformity, the morphology, the electrical and mechanical properties, and the chemical composition of the dielectric films.

The reactor for atmospheric-pressure CVD is similar to the one shown in Figure 3.2, except that different gases are used at the gas inlet. In a hot-wall, reduced-pressure reactor like the one shown in Figure 8.9a, the quartz-tube is heated by a three-zone furnace, and gas is introduced at one end and pumped out at the opposite end. The semiconductor wafers are held vertically in a slotted quartz boat.[8] The quartz-tube wall is hot because it is adjacent to the furnace, in contrast to a cold-wall reactor such as the horizontal epitaxial reactor that uses radio frequency (rf) heating.

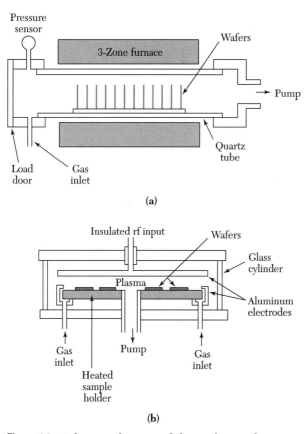

Figure 8.9   Schematic diagrams of chemical vapor deposition reactors. (*a*) Hot-wall, reduced-pressure reactor. (*b*) Parallel-plate plasma deposition reactor. rf, radio frequency.

The parallel-plate, radial-flow PECVD reactor shown in Figure 8.9*b* consists of a cylindrical glass or aluminum chamber sealed with aluminum endplates. Inside are two parallel aluminum electrodes. An rf voltage is applied to the upper electrode, whereas the lower electrode is grounded. The rf voltage causes a plasma discharge between the electrodes. Wafers are placed on the lower electrode, which is heated between 100°C and 400°C by resistance heaters. The reaction gases flow through the discharge from outlets located along the circumference of the lower electrode. The main advantage of this reactor is its low deposition temperature. However, its capacity is limited, especially for large-diameter wafers, and the wafers may become contaminated if loosely adhering deposits fall on them.

8.3.1   Silicon Dioxide

CVD silicon dioxide cannot replace thermally grown oxides, because the best electrical properties are obtained with thermally grown films. CVD oxides are used instead to complement thermal oxides. A layer of undoped silicon dioxide is used to insulate multilevel metallization, to mask ion implantation and diffusion, and to increase the thickness of thermally grown field oxides. Phosphorus-doped silicon dioxide is used both as an insulator between metal layers and as a final passivation layer over devices. Oxides doped with phosphorus, arsenic, or boron are used occasionally as diffusion sources.

Deposition Methods

Silicon dioxide films can be deposited by several methods. For low-temperature deposition (300–500°C), the films are formed by reacting silane, dopant, and oxygen. The chemical reactions for phosphorus-doped oxides are

$$SiH_4 + O_2 \xrightarrow{450°C} SiO_2 + 2H_2 \tag{11}$$

$$4PH_3 + 5O_2 \xrightarrow{450°C} 3P_2O_5 + 6H_2 \tag{12}$$

The deposition process can be performed either at atmospheric pressure in a CVD reactor or at reduced pressure in an LPCVD reactor (Fig. 8.9a). The low deposition temperature of the silane–oxygen reaction makes it a suitable process when films must be deposited over a layer of aluminum.

For intermediate-temperature deposition (500–800°C), silicon dioxide can be formed by decomposing tetraethylorthosilicate, $Si(OC_2H_5)_4$, in an LPCVD reactor. The compound, abbreviated TEOS, is vaporized from a liquid source. The TEOS compound decomposes as follows:

$$Si(OC_2H_5)_4 \xrightarrow{700°C} SiO_2 + by\text{-}products \tag{13}$$

forming both SiO_2 and a mixture of organic and organosilicon by-products. Although the higher temperature required for the reaction prevents its use over aluminum, it is suitable for polysilicon gates requiring a uniform insulating layer with good step coverage. The good step coverage is a result of enhanced surface mobility at higher temperatures. The oxides can be doped by adding small amounts of the dopant hydrides (phosphines, arsine, or diborane), similar to the process in epitaxial growth.

The deposition rate as a function of temperature varies as $\exp(-E_a/kT)$, where E_a is the activation energy. The E_a of the silane–oxygen reaction is quite low: about 0.6 eV for undoped oxides and almost zero for phosphorus-doped oxide. In contrast, E_a for the TEOS reaction is much higher: about 1.9 eV for undoped oxide and 1.4 eV when phosphorus doping compounds are present. The dependence of the deposition rate on TEOS partial pressure is proportional to $(1-e^{-P/P_0})$, where P is the TEOS partial pressure and P_0 is about 30 Pa. At low TEOS partial pressures, the deposition rate is determined by the rate of the surface reaction. At high partial pressures, the surface becomes nearly saturated with adsorbed TEOS, and the deposition rate becomes essentially independent of TEOS pressure.[8]

Recently, atmospheric-pressure and low-temperature CVD processes using TEOS and ozone (O_3) have been proposed,[9] as shown in Figure 8.10. This CVD technology produces oxide films with high conformality and low viscosity under low deposition temperature. The shrinkage of oxide film during annealing is also a function of ozone concentration, as shown in Figure 8.11. Because of their porosity, O_3–TEOS CVD oxides are often accompanied by plasma-assisted oxides to permit planarization in ULSI processing.

For high-temperature deposition (900°C), silicon dioxide is formed by reacting dichlorosilane, $SiCl_2H_2$, with nitrous oxide at reduced pressure:

$$SiCl_2H_2 + 2N_2O \xrightarrow{900°C} SiO_2 + 2N_2 + 2HCl \tag{14}$$

This deposition gives excellent film uniformity and is sometimes used to deposit insulating layers over polysilicon.

Properties of Silicon Dioxide

Table 8.1 lists deposition methods and properties of silicon dioxide films.[8] In general, there is a direct correlation between deposition temperature and film quality. At higher

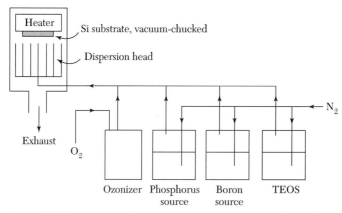

Figure 8.10   Experimental apparatus for the O_3–TEOS chemical vapor deposition system.

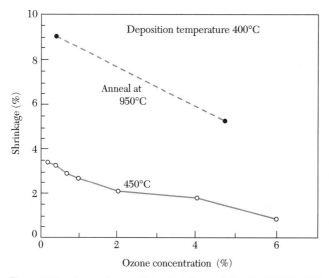

Figure 8.11   Dependence of the shrinkage of the O_3–TEOS CVD film on ozone concentration using annealing. (Courtesy of SAMCO Company, Japan.)

TABLE 8.1   Properties of Silicon Dioxide Films

Property	Thermally Grown at 1000°C	$SiH_4 + O_2$ at 450°C	TEOS at 700°C	$SiCl_2H_2 + N_2O$ at 900°C
Composition	SiO_2	SiO_2 (H)	SiO_2	SiO_2(Cl)
Density (g/cm³)	2.2	2.1	2.2	2.2
Refractive index	1.46	1.44	1.46	1.46
Dielectric strength (10^6 V/cm)	>10	8	10	10
Etch rate (Å/min) (100:1 H_2O:HF)	30	60	30	30
Etch rate (Å/min) (buffered HF)	440	1200	450	450
Step coverage	—	Nonconformal	Conformal	Conformal

temperatures, deposited oxide films are structurally similar to silicon dioxide that has been thermally grown.

The lower densities occur in films deposited below 500°C. Heating deposited silicon dioxide at temperatures between 600°C and 1000°C causes densification, during which the oxide thickness decreases, whereas the density increases to 2.2 g/cm³. The refractive index of silicon dioxide is 1.46 at a wavelength of 0.6328 µm. Oxides with lower indices are porous, such as the oxide resulting from silane–oxygen deposition, which has a refractive index of 1.44. The porous nature of the oxide also is responsible for the lower dielectric strength, which is the applied electric field that will cause a high current to flow in the oxide film. The etch rates of oxides in a hydrofluoric acid solution depend on deposition temperature, annealing history, and dopant concentration. Usually, higher-quality oxides are etched at lower rates.

Step Coverage

Step coverage relates the surface topography of a deposited film to the various steps on the semiconductor substrate. In the illustration of ideal, or *conformal*, step coverage shown in Figure 8.12a, film thickness is uniform along all surfaces of the step. The uniformity of the film thickness, regardless of topography, is due to the rapid migration of reactants after adsorption on the step surfaces.[10]

Figure 8.12b shows an example of nonconformal step coverage, which results when the reactants adsorb and react without significant surface migration. In this instance, the

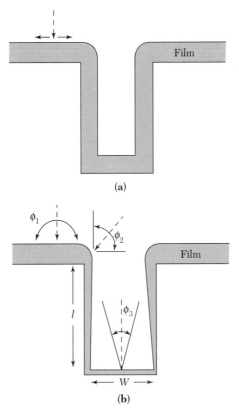

Figure 8.12 Step coverage of deposited films.[10] (*a*) Conformal step coverage. (*b*) Nonconformal step coverage.

deposition rate is proportional to the arrival angle of the gas molecules. Reactants arriving along the top horizontal surface come from many different angles, and ϕ_1, the arrival angle, varies in two dimensions, from 0° to 180°, whereas reactants arriving at the top of a vertical wall have an arrival angle (ϕ_2) that varies from 0° to 90°. Thus, the film thickness on the top surface is double that of a wall surface. Further down the wall, ϕ_3 is related to the width of the opening, and the film thickness is proportional to

$$\phi_3 \cong \arctan\frac{W}{l} \tag{15}$$

where l is the distance from the top surface and W is the width of the opening. This type of step coverage is thin along the vertical walls, with a possible crack at the bottom of the step caused by self-shadowing.

Silicon dioxide formed by TEOS decomposition at reduced pressure gives a nearly conformal coverage due to rapid surface migration. Similarly, the high-temperature dichlorosilane–nitrous oxide reaction also results in conformal coverage. However, during silane–oxygen deposition, no surface migration takes place, and the step coverage is determined by the arrival angle. Most evaporated or sputtered materials have a step coverage similar to that in Figure 8.12b.

P-Glass Flow

A smooth topography is usually required for the deposited silicon dioxide used as an insulator between metal layers. If the oxide used to cover the lower metal layer is concave, circuit failure may result from an opening that may occur in the upper metal layer during deposition. Because phosphorus-doped silicon dioxide (P-glass) deposited at low temperatures becomes soft and flows upon heating, it provides a smooth surface and is often used to insulate adjacent metal layers. This process is called *P-glass flow*.

Figure 8.13 shows four cross sections of scanning electron microscope photographs of P-glass covering a polysilicon step.[10] All samples are heated in steam at 1100°C for 20 minutes. Figure 8.13a shows a sample of glass that contains a negligibly small amount of phosphorus and does not flow. Note the concavity of the film and that the corresponding angle (θ) is about 120°. Figures 8.13b, c, and d show samples of P-glass with progressively higher phosphorus contents, up to 7.2 wt% (weight percent). In these samples, the decreasing step angles of the P-glass layer indicate how flow increases with phosphorus concentration. P-glass flow depends on annealing time, temperature, phosphorus concentration, and the annealing ambient.[10]

The angle θ as a function of weight percent of phosphorus as shown in Figure 8.13 can be approximated by

$$\theta \cong 120°\left(\frac{10 - wt\%}{10}\right) \tag{16}$$

If we want an angle smaller than 45°, we require a phosphorus concentration larger than 6 wt%. However, at concentrations above 8 wt%, the metal film (e.g., aluminum) may be corroded by the acid products formed during the reaction between the phosphorus in the oxide and atmospheric moisture. Therefore, the P-glass flow process uses phosphorus concentrations of 6 to 8 wt%.

8.3.2 Silicon Nitride

It is difficult to grow silicon nitride by thermal nitridation (e.g., with ammonia, NH_3) because of its slow growth rate and high growth temperature. However, silicon nitride

 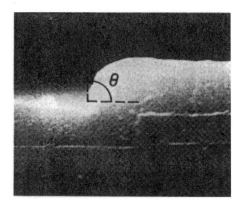

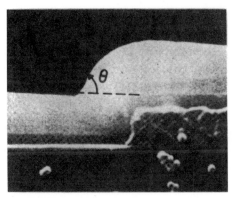 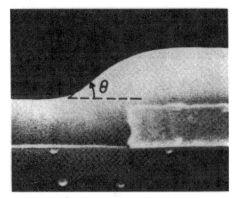

Figure 8.13 Scanning electron micrographs (10,000×) of samples annealed in steam at 1100°C for 20 minutes for the following weight percent of phosphorus:[10] (*a*) 0 wt%, (*b*) 2.2 wt%, (*c*) 4.6 wt%, and (*d*) 7.2 wt%.

films can be deposited by an intermediate-temperature (750°C) LPCVD process or a low-temperature (300°C) plasma-assisted CVD process.[11,12] The LPCVD films are of stoichiometric composition (Si_3N_4) with high density (2.9–3.1 g/cm^3). These films can be used to passivate devices because they serve as good barriers to the diffusion of water and sodium. The films also can be used as masks for the selective oxidation of silicon because silicon nitride oxidizes very slowly and prevents the underlying silicon from oxidizing. The films deposited by plasma-assisted CVD are not stoichiometric and have a lower density (2.4–2.8 g/cm^3). Because of the low deposition temperature, silicon nitride films can be deposited over fabricated devices and serve as their final passivation. Plasma-deposited nitride provides excellent scratch protection, serves as a moisture barrier, and prevents sodium diffusion.

In the LPCVD process, dichlorosilane and ammonia react at reduced pressure to deposit silicon nitride at temperatures between 700°C and 800°C. The reaction is

$$3SiCl_2H_2 + 4NH_2 \xrightarrow{\sim 750°C} Si_3N_4 + 6HCl + 6H_2 \qquad (17)$$

Good film uniformity and high wafer throughput (i.e., wafers processed per hour) are advantages of the reduced-pressure process. As in the case of oxide deposition, silicon nitride deposition is controlled by temperature, pressure, and reactant concentration. The activation energy for deposition is about 1.8 eV. The deposition rate increases with increasing total pressure or dichlorosilane partial pressure and decreases with an increasing ammonia-to-dichlorosilane ratio.

Silicon nitride deposited by LPCVD is an amorphous dielectric containing up to 8 atomic percent (at%) hydrogen. The etch rate in buffered HF is less than 1 nm/min. The film has a very high tensile stress of approximately 10^{10} dynes/cm^2, which is nearly 10 times that of TEOS-deposited SiO$_2$. Films thicker than 200 nm may crack because of the very high stress. The resistivity of silicon nitride at room temperature is about 10^{16} Ω-cm. Its dielectric constant is 7, and its dielectric strength is 10^7 V/cm.

In the plasma-assisted CVD process, silicon nitride is formed either by reacting silane and ammonia in an argon plasma or by reacting silane in a nitrogen discharge. The reactions are as follows:

$$\text{SiH}_4 + \text{NH}_3 \xrightarrow{\;300^\circ C\;} \text{SiNH} + 3\text{H}_2 \tag{18a}$$

$$2\text{SiH}_4 + \text{N}_2 \xrightarrow{\;300^\circ C\;} 2\text{SiNH} + 3\text{H}_2 \tag{18b}$$

The products depend strongly on deposition conditions. The radial-flow, parallel-plate reactor (Fig. 8.9b) is used to deposit the films. The deposition rate generally increases with increasing temperature, power input, and reactant gas pressure.

Large concentrations of hydrogen are contained in plasma-deposited films. The plasma nitride (also referred to as SiN) used in semiconductor processing generally contains 20 to 25 at% hydrogen. Films with low tensile stress (~2×10^9 dynes/cm^2) can be prepared by plasma deposition. Film resistivities range from 10^5 to 10^{21} Ω-cm, depending on silicon-to-nitrogen ratio, whereas dielectric strengths are between 1×10^6 and 6×10^6 V/cm.

8.3.3 Low-Dielectric-Constant Materials

As device sizes continue to shrink down to the deep submicron region, they require a multilevel interconnection architecture to minimize the time delay due to parasitic resistance (R) and capacitance (C). The gain in device speed at the gate level is offset by the propagation delay at the metal interconnects because of the increased RC time constant, as shown in Figure 8.14. For example, in devices with gate length of 250 nm or less, up to 50% of the time delay is due to the RC delay of long interconnections.[13] Therefore, the device interconnection network becomes a limiting factor in determining chip performance metrics such as device speed, cross talk, and power consumption of ULSI circuits.

Reducing the RC time constant of ULSI circuits requires interconnection materials with low resistivity and interlayer films with low capacitance. Note that $C = \varepsilon_i A/d$, where ε_i is the dielectric permittivity, A is the area, and d is the thickness of the dielectric film. Regarding the low-capacitance issue, it is not easy to lower the parasitic capacitance by increasing thickness of the interlayer dielectric (which makes gap filling more difficult) or decreasing wiring height and area (which results in the increase of interconnect resistance). Therefore, materials with low dielectric constant (low k) are required. The dielectric permittivity is equal to the product of k and ε_0, where k and ε_0 are the dielectric constant and permittivity of free space, respectively.

The properties of the interlayer dielectric film and how it is formed have to meet the following requirements: low dielectric constant, low residual stress, high planarization

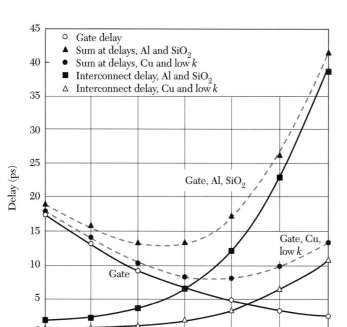

Figure 8.14 Calculated gate and interconnect delay versus technology generation. The dielectric constant for the low-k material is 2.0. Both Al and Cu interconnects are 0.8 μm thick and 43 μm long.

capability, high capability for gap filling, low deposition temperature, simplicity of process, and ease of integration. A substantial number of low-k materials have been synthesized for the intermetal dielectric in ULSI circuits. Some of the promising low-k materials are shown in Table 8.2. These materials can be either inorganic or organic and can be deposited by either CVD or spin-on techniques.[13]

TABLE 8.2   Low-k Materials

Determinant	Materials	Dielectric Constant
Vapor-phase deposition polymers	Fluorosilicate glass (FSG)	3.5–4.0
	Parylene N	2.6
	Parylene F	2.4–2.5
	Black diamond (C-doped oxide)	2.7–3.0
	Fluorinated hydrocarbon	2.0–2.4
	Teflon-AF	1.93
Spin-on polymers	HSQ/MSQ	2.8–3.0
	Polyimide	2.7–2.9
	SiLK (aromatic hydrocarbon polymer)	2.7
	PAE [poly(arylene ethers)]	2.6
	Fluorinated amorphous carbon	2.1
	Xerogels (porous silica)	1.1–2.0

EXAMPLE 3

Estimate the intrinsic RC value of two parallel Al wires 0.5 μm × 0.5 μm in cross section, 1 mm in length, and separated by a polyimide ($k \sim 2.7$) dielectric layer that is 0.5 μm thick. The resistivity of Al is 2.7 μΩ-cm.

SOLUTION Let t_m be the cross-sectional dimension of the wires. Resistance is equal to ρl divided by the cross-sectional area of the wire, so

$$RC = \left(\rho \frac{l}{t_m^2} \right) \times \left(\varepsilon_i \frac{A}{\text{Spacing width}} \right)$$

$$= \left(2.7 \times 10^{-6} \times \frac{10^{-1}}{0.25 \times 10^{-8}} \right) \times \left(8.85 \times 10^{-14} \times 2.7 \times \frac{0.5 \times 10^{-4} \times 10^{-1}}{0.5 \times 10^{-4}} \right) = 2.57 \text{ ps} \quad ◀$$

8.3.4 High-Dielectric-Constant Materials

High-k materials are also required for ULSI circuits, especially for dynamic random access memory (DRAM) circuits. The storage capacitor in a DRAM has to maintain a certain value of capacitance for proper operation (e.g., 40 fF). For a given capacitance, a minimum d is usually selected to meet the conditions of the maximum allowed leakage current and the minimum required breakdown voltage. The area of the capacitor can be increased by using stacked or trench structures. These structures are considered in Chapter 9. However, for a planar structure, area is reduced with increasing DRAM density. Therefore, the dielectric constant of the film must be increased.

Several high-k materials have been proposed, such as barium strontium titanate (BST) and lead zirconium titanate (PZT). These materials are shown in Table 8.3. In addition, there are titanates doped with one or more acceptors, such as alkaline earth metals, or doped with one or more donors, such as rare earth elements. Tantalum oxide (Ta_2O_5) has a dielectric constant in the range of 20 to 30. As a reference, the dielectric constant of Si_3N_4 is in the range of 6 to 7, and that for SiO_2 is 3.9. A Ta_2O_5 film can be deposited by a CVD process using gaseous $TaCl_5$ and O_5 as the starting materials.

TABLE 8.3   High-k Materials

	Materials	Dielectric Constant
Binary	Ta_2O_5	25
	TiO_2	40
	Y_2O_3	17
	Si_3N_4	7
Paraelectric perovskite	$SrTiO_3$ (STO)	140
	$(Ba_{1-x}Sr_x) TiO_3$(BST)	300–500
	$Ba(Ti_{1-x}Zr_x)O_3$(BZT)	300
	$(Pb_{1-x}La_x)(Zr_{1-y}Ti_y)O_3$(PLZT)	800–1000
	$Pb(Mg_{1/3}Nb_{2/3})O_3$(PMN)	1000–2000
Ferroelectric perovskite	$Pb(Zr_{0.47}Ti_{0.53})O_3$(PZT)	>1000

EXAMPLE 4

A DRAM capacitor has the following parameters: C = 40 fF, cell size (A) = 1.28 μm^2, and k = 3.9 for silicon dioxide. If we replace SiO_2 with Ta_2O_5 (k = 25) without changing thickness, what is the equivalent cell area of the capacitor?

SOLUTION

$$C = \frac{\varepsilon_i A}{d}$$

so

$$\frac{3.9 \times 1.28}{d} = \frac{25 \times A}{d}$$

Therefore, the equivalent cell size is

$$A = \frac{3.9}{25} \times 1.28 = 0.2 \ \mu m^2$$

◀

▶ 8.4 POLYSILICON DEPOSITION

Using polysilicon as the gate electrode in MOS devices is a significant development in MOS technology. One important reason is that polysilicon surpasses aluminum for electrode reliability. Figure 8.15 shows the maximum time to breakdown for capacitors with both polysilicon and aluminum electrodes.[14] Polysilicon is clearly superior, especially for

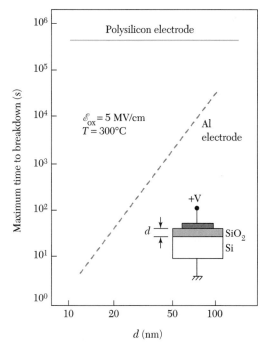

Figure 8.15 Maximum time to breakdown versus oxide thickness for a polysilicon electrode and an aluminum electrode.[14]

thinner gate oxides. The inferior time to breakdown of aluminum electrodes is due to the migration of aluminum atoms into the thin oxide under an electrical field. Polysilicon is also used as a diffusion source to create shallow junctions and to ensure ohmic contact to crystalline silicon. Additional uses include the manufacture of conductors and high-value resistors.

A low-pressure reactor (Fig. 8.9a) operated between 600°C and 650°C is used to deposit polysilicon by pyrolyzing silane according to the reaction

$$SiH_4 \xrightarrow{600°C} Si + 2H_2 \qquad (19)$$

Of the two most common low-pressure processes, one operates at a pressure of 25 to 130 Pa using 100% silane, whereas the other involves a diluted mixture of 20% to 30% silane in nitrogen at the same total pressure. Both processes can deposit polysilicon on hundreds of wafers per run with good uniformity (i.e., thicknesses within ± 5%).

Figure 8.16 shows the deposition rate at four deposition temperatures. At low silane partial pressure, the deposition rate is proportional to the silane pressure.[8] At higher silane concentrations, saturation of the deposition rate occurs. Deposition at reduced pressure is generally limited to temperatures between 600°C and 650°C. In this temperature range, the deposition rate varies as $exp(-E_a/kT)$, where E_a is 1.7 eV, which is essentially independent of the total pressure in the reactor. At higher temperatures, gas-phase reactions that result in a rough, loosely adhering deposit become significant, and silane depletion occurs, causing poor uniformity. At temperatures much lower than 600°C, the deposition rate is too slow to be practical.

Process parameters that affect the polysilicon structure are deposition temperature, dopants, and the heat cycle applied following the deposition step. A columnar structure results when polysilicon is deposited at a temperature of 600°C to 650°C. This structure consists of polycrystalline grains ranging in size from 0.03 to 0.3 μm at a preferred orientation of (110). When phosphorus is diffused at 950°C, the structure changes to crystallite, and grain size increases to a size between 0.5 and 1.0 μm. When temperature is increased to 1050°C during oxidation, the grains reach a final size of 1 to 3 μm. Although the initially deposited film appears amorphous when deposition occurs below 600°C, growth characteristics similar to the polycrystalline-grain columnar structure are observed after doping and heating.

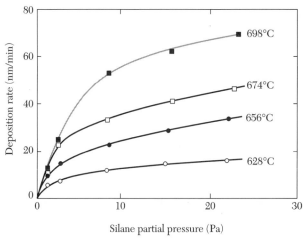

Figure 8.16 Effect of silane concentration on the polysilicon deposition rate.[8]

Polysilicon can be doped by diffusion, ion implantation, or the addition of dopant gases during deposition, referred to as in situ *doping*. The implantation method is most commonly used because of its lower processing temperatures. Figure 8.17 shows the sheet resistance of single-crystal silicon and of 500-nm polysilicon doped with phosphorus and antimony using ion implantation.[15] The ion implantation process was considered in Chapter 7. Implant dose, annealing temperature, and annealing time all influence the sheet resistance of implanted polysilicon. Carrier traps at the grain boundaries cause a very high resistance in the lightly implanted polysilicon. As Figure 8.17 illustrates, resistance drops rapidly, approaching that of implanted single-crystal silicon, as the carrier traps become saturated with dopants.

▶ 8.5 METALLIZATION

8.5.1 Physical Vapor Deposition

The most common methods of physical vapor deposition (PVD) of metals are evaporation, e-beam evaporation, plasma spray deposition, and sputtering. Metals and metal compounds such as Ti, Al, Cu, TiN, and TaN can be deposited by PVD. Evaporation occurs when a source material is heated above its melting point in an evacuated chamber. The evaporated atoms then travel at high velocity in straight-line trajectories. The source can be melted by resistance heating, by rf heating, or with a focused electron beam. Evaporation and e-beam evaporation were used extensively in earlier generations of integrated circuits, but they have been replaced by sputtering for ULSI circuits.

In ion beam sputtering, a source of ions is accelerated toward the target and impinges on its surface. Figure 8.18*a* shows a standard sputtering system. The sputtered material deposits on a wafer that is placed facing the target. The ion current and energy can be independently adjusted. Since the target and wafer are placed in a chamber that has lower pressure, more target material and less contamination are transferred to the wafer.

One method to increase the ion density and, hence, the sputter-deposition rate is to use a third electrode that provides more electrons for ionization. Another method is to use

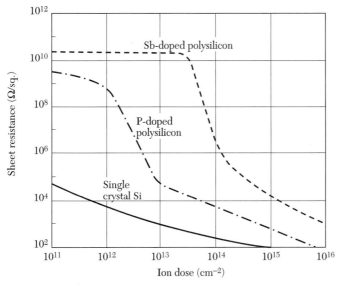

Figure 8.17 Sheet resistance versus ion dose into 500-nm polysilicon at 30 keV.[15]

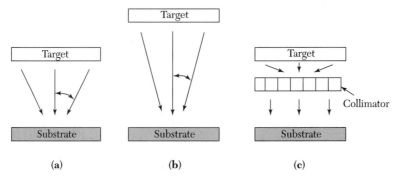

Figure 8.18 (*a*) Standard sputtering. (*b*) Long-throw sputtering. (*c*) Sputtering with a collimator.

a magnetic field, such as *electron cyclotron resonance* (ECR), to capture and spiral electrons, increasing their ionizing efficiency in the vicinity of the sputtering target. This technique, referred to as *magnetron sputtering*, has found widespread applications for the deposition of aluminum and its alloys at a rate that can approach 1 μm/min.

Long-throw sputtering is another technique used to control the angular distribution. Figure 8.18*b* shows a long-throw sputtering system. In standard sputtering configurations, there are two primary reasons for a wide angular distribution of incident flux at the surface: the use of a small target-to-substrate separation, d_{ts}; and scattering of the flux by the working gas as the flux travels from the target to the substrate. These two factors are linked because a small d_{ts} is needed to achieve good throughput, uniformity, and film properties when there is substantial gas scattering. A solution to this problem is to sputter at very low pressures, a capability that has been developed using a variety of systems that can sustain the magnetron plasma under more rarefied conditions. These systems allow for sputtering at working pressures of less than 0.1 Pa. At these pressures, gas scattering is less important, and the target-to-substrate distance can be greatly increased. From a simple geometrical argument, this allows the angular distribution to be greatly narrowed, which permits more deposition at the bottom of high-aspect features such as contact holes.

Contact holes with large aspect ratio are difficult to fill with material, mainly because scattering events cause the top opening of the hole to seal before appreciable material has deposited on its floor. This problem can be overcome by collimating the sputtered atoms by placing an array of collimating tubes just above the wafer to restrict the depositing flux to normal ±5°. Sputtering with a collimator is shown in Figure 8.18*c*. Atoms whose trajectory is more than 5° from normal are deposited on the inner surface of the collimators.

8.5.2 Chemical Vapor Deposition

CVD is attractive for metallization because it offers coatings that are conformal, has good step coverage, and can coat a large number of wafers at a time. The basic CVD setup is the same as that used for deposition of dielectrics and polysilicon (see Fig. 8.9*a*). Low-pressure CVD is capable of producing conformal step coverage over a wide range of topographical profiles, often with lower electrical resistivity than that from PVD.

One of the major new applications of CVD metal deposition for integrated circuit production is in the area of refractory metal deposition. For example, tungsten's low electrical resistivity (5.3 μΩ-cm) and refractory nature make it a desirable metal for use in integrated circuit fabrication.

CVD Tungsten

Tungsten is used both as a contact plug and as a first-level metal. Tungsten can be deposited by using WF_6 as the source gas, since it is a liquid that boils at room temperature. WF_6 can be reduced by silicon, hydrogen, or silane. The basic chemistry for CVD W is as follows:

$$WF_6 + 3H_2 \rightarrow W + 6HF \text{ (hydrogen reduction)} \tag{20}$$

$$2WF_6 + 3Si \rightarrow 2W + 3SiF_4 \text{ (silicon reduction)} \tag{21}$$

$$2WF_6 + 3SiH_4 \rightarrow 2W + 3SiF_4 + 6H_2 \text{ (silane reduction)} \tag{22}$$

On a Si contact, the selective process starts from a silicon reduction process. This process provides a nucleation layer of W grown on Si but not on SiO_2. The hydrogen reduction process can deposit W rapidly on the nucleation layer, forming the plug. The hydrogen reduction process provides excellent conformal coverage of the topography. This process, however, does not have perfect selectivity, and the HF gas by-product of the reaction is responsible for the encroachment of the oxide, as well as for the rough surface of deposited W films.

The silane reduction process gives a high deposition rate and much smaller W grain size than that obtained with the hydrogen reduction process. In addition, the problems of encroachment and a rough W surface are eliminated because there is no HF by-product generation. Usually, a silane reduction process is used as the first step in blanket W deposition to serve as a nucleation layer and to reduce junction damage. After the silane reduction, hydrogen reduction is used to grow the blanket W layer.

CVD TiN

TiN is widely used as a diffusion barrier metal layer in metallization and can be deposited by sputtering from a compound target or by CVD. CVD TiN can provide better step coverage than PVD methods in deep submicron technology. CVD TiN can be deposited[16-18] using $TiCl_4$ with NH_3, H_2/N_2, or NH_3/H_2:

$$6TiCl_4 + 8NH_3 \rightarrow 6 \text{ TiN} + 24HCl + N_2 \tag{23}$$

$$2TiCl_4 + N_2 + 4H_2 \rightarrow 2TiN + 8HCl \tag{24}$$

$$2TiCl_4 + 2NH_3 + H_2 \rightarrow 2 \text{ TiN} + 8HCl \tag{25}$$

The deposition temperature is about 400°C to 700°C for NH_3 reduction and is higher than 700°C for the N_2/H_2 reaction. The higher the deposition temperature, the better the TiN film, and the less Cl incorporated in TiN (~5%).

8.5.3 Aluminum Metallization

Aluminum and its alloys are used extensively for metallization in integrated circuits. The Al film can be deposited by PVD or CVD. Since aluminum and its alloys have low resistivities (2.7 $\mu\Omega$-cm for Al and up to 3.5 $\mu\Omega$-cm for its alloys), these metals satisfy the low-resistance requirements. Aluminum also adheres well to silicon dioxide. However, the use of aluminum in integrated circuits with shallow junctions often creates problems, such as spiking and electromigration. This section considers the problems of aluminum metallization and their solutions.

Junction Spiking

Figure 8.19 shows the phase diagram of the Al–Si system at 1 atm.[19] The phase diagram relates these two components as a function of temperature. The Al–Si system exhibits

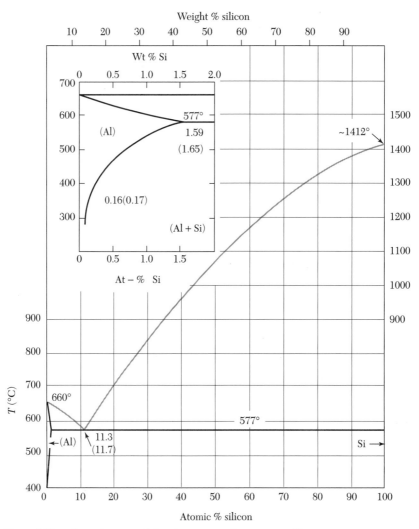

Figure 8.19 Phase diagram of the aluminum–silicon system.[19]

eutectic characteristics; that is, the addition of either component lowers the system's melting point below that of either metal. Here, the minimum melting temperature, called *eutectic temperature*, is 577°C, corresponding to a 11.3% Si and 88.7% Al composition. The melting points of pure aluminum and pure silicon are 660°C and 1412°C, respectively. Because of the eutectic characteristics, during aluminum deposition the temperature on the silicon substrate must be limited to less than 577°C.

The inset of Figure 8.19 shows the solid solubility of silicon in aluminum. For example, the solubility of silicon in aluminum is 0.25 wt% at 400°C, 0.5 wt% at 450°C, and 0.8 wt% at 500°C. Therefore, wherever aluminum contacts silicon, the silicon will dissolve into the aluminum during annealing. The amount of silicon dissolved will depend not only on the solubility at the annealing temperature, but also on the volume of aluminum to be saturated with silicon. Consider a long aluminum metal line in contact with an area ZL of silicon, as shown in Figure 8.20. After an annealing time t, the silicon will diffuse a distance of approximately $\sqrt{Dt}$ along the aluminum line from the edge of the contact, where D is the diffusion coefficient given by $4 \times 10^{-2} \exp(-0.92/kT)$ for silicon

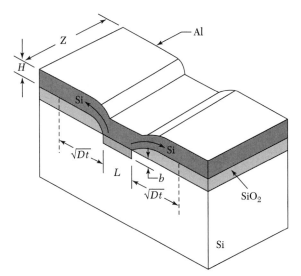

Figure 8.20   Diffusion of silicon in aluminum metallization.[20]

diffusion in deposited aluminum films. Assuming that this length of aluminum is completely saturated with silicon, the volume of silicon consumed is then

$$\text{Vol} \cong 2\sqrt{Dt}\,(HZ)S\left(\frac{\rho_{Al}}{\rho_{Si}}\right) \tag{26}$$

where ρ_{Al} and ρ_{Si} are the densities of aluminum and silicon, respectively, and S is the solubility of silicon in aluminum at the annealing temperature.[20] If the consumption takes place uniformly over the contact area A (where $A = ZL$ for uniform dissolution), the depth to which silicon would be consumed is

$$b \cong 2\sqrt{Dt}\left(\frac{HZ}{A}\right)S\left(\frac{\rho_{Al}}{\rho_{Si}}\right) \tag{27}$$

EXAMPLE 5

Let $T = 500°C$, $t = 30$ min, $ZL = 16$ μm^2, $Z = 5$ μm, and $H = 1$ μm. Find the depth b, assuming uniform dissolution.

SOLUTION   The diffusion coefficient of silicon in aluminum at 500°C is about 2×10^{-8} cm²/s; thus, $\sqrt{Dt}$ is 60 μm. The density ratio is 2.7/2.33 = 1.16. At 500°C, S is 0.8 wt%. From Eq. 27, we have

$$b = 2 \times 60\left(\frac{1 \times 5}{16}\right)0.8\% \times 1.16 = 0.35 \text{ } \mu m$$

Aluminum will fill a depth of $b = 0.35$ μm from which silicon is consumed. If at the contact point there is a shallow junction whose depth is less than b, the diffusion of silicon into aluminum can short-circuit the junction.   ◀

In a practical situation, the dissolution of silicon does not take place uniformly, but rather at only a few points. The effective area in Eq. 27 is less than the actual contact area; hence, b is much larger. Figure 8.21 illustrates the actual situation in the p–n junction area of aluminum penetrating the silicon at only the few points where spikes are formed. One way to minimize aluminum spiking is to add silicon to the aluminum by co-evaporation

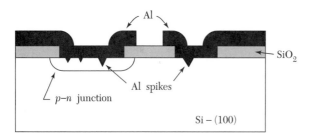

Figure 8.21 Schematic view of aluminum films contacting silicon. Note the aluminum spiking in the silicon.

until the amount of silicon contained by the alloy satisfies the solubility requirement. Another method is to introduce a barrier metal layer between the aluminum and the silicon substrate (Fig. 8.22). This barrier metal layer must meet the following requirements: It must form low contact resistance with silicon, it must not react with aluminum, and its deposition and formation must be compatible with the overall process. Barrier metals such as titanium nitride (TiN) have been evaluated and found to be stable for contact annealing temperatures of up to 550°C for 30 minutes.

Electromigration

As devices become smaller, the corresponding current density becomes larger. High current densities can cause device failure due to electromigration. The term *electromigration* refers to the transport of mass (i.e., atoms) in metals under the influence of current. It occurs by the transfer of momentum from the electrons to the positive metal ions. When a high current passes through thin metal conductors in integrated circuits, metal ions in some regions will pile up, and voids will form in other regions. This pileup can short-circuit adjacent conductors, whereas the voids can result in an open circuit.

The *mean time to failure* (MTF) of a conductor due to electromigration can be related to the current density (J) and the activation energy by

$$\text{MTF} \sim \frac{1}{J^2} \exp\left(\frac{E_a}{kT}\right) \tag{28}$$

Experimentally, a value of $E_a \cong 0.5$ eV is obtained for deposited aluminum. This indicates that low-temperature grain-boundary diffusion is the primary vehicle of material transport, since an $E_a \cong 1.4$ eV would characterize the self-diffusion of single-crystal aluminum. The electromigration resistance of aluminum conductors can be increased by

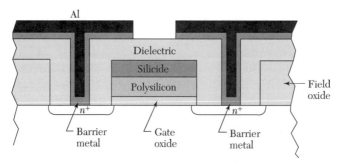

Figure 8.22 Cross-sectional view of a MOSFET with a barrier metal between the aluminum and silicon and a composite gate electrode of silicide and polysilicon.

using several techniques. These techniques include alloying with copper (e.g., Al with 0.5% Cu), encapsulating the conductor in a dielectric, or incorporating oxygen during film deposition.

8.5.4 Copper Metallization

It is well known that both high-conductivity wiring and low-dielectric-constant insulators are required to lower the RC time delay of the interconnect network. Copper is the obvious choice for a new interconnection metallization because it has higher conductivity and higher electromigration resistance than aluminum. Copper can be deposited by PVD, CVD, and electrochemical methods. However, the use of Cu as an alternative material to Al in ULSI circuits has drawbacks, such as its tendency to corrode under standard chip manufacturing conditions, its lack of a feasible dry etching method or a stable self-passivating oxide similar to Al_2O_3 on Al, and its poor adhesion to dielectric materials, such as SiO_2 and low-k polymers. This section discusses copper metallization techniques.

Several different techniques for fabrication of multilevel Cu interconnects have been reported.[21, 22] The first method is a conventional method to pattern the metal lines, followed by dielectric deposition. The second method is to pattern the dielectric layer first and fill copper metal into trenches. This step is followed by *chemical mechanical polishing*, discussed later, to remove the excess metal on the top surface of the dielectric and leave Cu material in the holes and trenches. This method is also known as a *damascene* process.

Damascene Technology

The approach for fabricating a copper/low-k dielectric interconnect structure is by the damascene or dual damascene process. Figure 8.23 shows the dual damascene sequence for an advanced Cu interconnection structure. For a typical damascene structure, trenches for metal lines are defined and etched in the interlayer dielectric (ILD), followed by metal deposition of TaN/Cu. The TaN layer serves as a diffusion barrier layer and prevents copper from penetrating the low-k dielectric. The excess copper metal on the surface is removed to obtain a planar structure with metal inlays in the dielectric.

For the dual damascene process, the vias and trenches in the dielectric are defined using two lithography and reactive ion etching (RIE) steps before depositing the Cu metal (Figs. 8.23a–c). Then a Cu chemical mechanical polishing process is used to remove the metal on the top surface, leaving the planarized wiring and via imbedded in the insulator.[23] One special benefit of the dual damascene technique is that the via plug is now of the same material as the metal line and the risk of via electromigration failure is reduced.

EXAMPLE 6

If we replace Al with Cu wire associated with some low-k dielectric (k = 2.6) instead of a SiO_2 layer, what percentage of reduction in the RC time constant will be achieved? The resistivity of Al is 2.7 $\mu\Omega$-cm, and the resistivity of Cu is 1.7 $\mu\Omega$-cm.

SOLUTION

$$\frac{1.7}{2.7} \times \frac{2.6}{3.9} \times 100\% = 42\%$$ ◀

Chemical Mechanical Polishing

In recent years, the development of chemical mechanical polishing (CMP) has become increasingly important for multilevel interconnection because it is the only technology

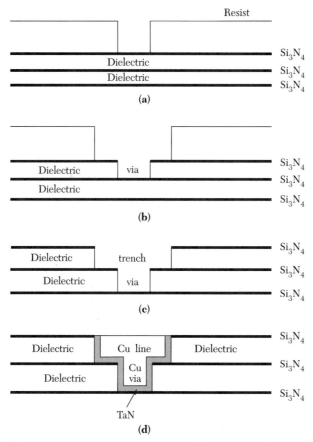

Figure 8.23 Process sequence used to fabricate a Cu line-stud structure using dual damascene. (*a*) Resist stencil applied. (*b*) Reactive ion etching dielectric and resist patterning. (*c*) Trench and via definition. (*d*) Cu depositions followed by chemical mechanical polishing.

that allows global planarization (i.e., makes a flat surface across the whole wafer). It offers many advantages over other types of technologies, including better global planarization over large or small structures, reduced defect density, and the avoidance of plasma damage. Three CMP approaches are summarized in Table 8.4.

The CMP process consists of moving the sample surface against a pad that carries slurry between the sample surface and the pad. Abrasive particles in the slurry cause mechanical damage on the sample surface, loosening the material for enhanced chemical attack or fracturing off the pieces of surface into a slurry where they dissolve or are swept away. The process is tailored to provide an enhanced material removal rate from high points on surfaces, thus affecting the planarization because most chemical actions are isotropic. Mechanical grinding alone may theoretically achieve the desired planarization,

TABLE 8.4   Three Methods of Chemical Mechanical Polishing (CMP)

Method	Wafer Facing	Platen Movement	Slurry Feeding
Rotary CMP	Down	Rotary against rotating wafer carrier	Dripping to pad surface
Orbital CMP	Down	Orbital against rotating wafer carrier	Through the pad surface
Linear CMP	Down	Linear against rotating wafer carrier	Dripping to pad surface

but is not desirable because of extensive associated damage to the material surfaces. There are three main parts of the process: (a) the surface to be polished; (b) the pad, which is the key medium enabling the transfer of mechanical action to the surface being polished; and (c) the slurry, which provides both chemical and mechanical effects. Figure 8.24 shows the CMP setup.[24]

EXAMPLE 7

The oxide removal rate and the removal rate of a layer underneath the oxide (called a stop layer) are $1r$ and $0.1r$, respectively. To remove 1 μm of oxide and a 0.01 μm stop layer, the total removal time is 5.5 minutes. Find the oxide removal rate.

SOLUTION

$$\frac{1}{1r} + \frac{0.01}{0.1r} = 5.5$$
$$r = 0.2 \ \mu m/min$$

◀

8.5.5   Silicide

Silicon forms many stable metallic and semiconducting compounds with metals. Several metal silicides show low resisivity and high thermal stability, making them suitable for ULSI application. Silicides such as $TiSi_2$ and $CoSi_2$ have reasonably low resistivities and are generally compatible with integrated circuit processing. Silicides become important metallization materials as devices become smaller. One important application of silicide is for the MOSFET gate electrode, either alone or with doped polysilicon (polycide) above the gate oxide. Table 8.5 shows a comparison of titanium silicide and cobalt silicide.

Metal silicides have been used to reduce the contact resistance of the source and drain, the gate electrodes, and interconnections. Self-aligned metal silicide technology

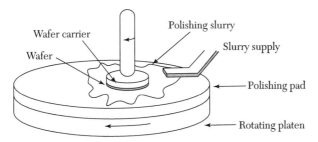

Figure 8.24   Schematic of a CMP polisher.

TABLE 8.5   A Comparison of TiSi$_2$ and CoSi$_2$ Films

Properties	TiSi$_2$	CoSi$_2$
Resistivity	13–16	22–28
Silicide/metal ratio	2.37	3.56
Silicide/Si ratio	1.04	0.97
Reactive to native oxide	Yes	No
Silicidation temperature (°C)	800–850	550–900
Film stress (dyne/cm2)	1.5×1010	1.2×10^{10}

(*salicide*) has been proven to be a highly attractive technique for improving the performance of submicron devices and circuits. The self-aligned process uses the silicide gate electrode as the mask to form the source and drain electrodes of a MOSFET (e.g., by ion implantation, considered in Chapter 7). This process can minimize the overlap of these electrodes and thus reduce parasitic capacitances.

Figure 8.25 shows the polycide and salicide processes. A typical polycide formation sequence is shown in Figure 8.25a. For sputter deposition, a high-temperature, high-purity compound target is used to ensure the quality of the silicide. The most commonly used silicides for the polycide process are WSi_2, $TaSi_2$, and $MoSi_2$. They are all refractory, thermally stable, and resistant to processing chemicals. A self-aligned silicide process is illustrated in Figure 8.25b. In the process, the polysilicon gate is patterned without any silicide, and a sidewall spacer (silicon oxide or silicon nitride) is formed to prevent shorting the gate to the source and drain during the silicidation process. A metal layer, either Ti or Co, is blanket-sputtered on the entire structure, followed by silicide sintering. Silicide is formed, in principle, only where the metal is in contact with Si. A wet chemical wash then rinses off the unreacted metal, leaving only the silicide. This technique eliminates the need to pattern the composite polycide gate structure and adds silicide to the source/drain area to reduce the contact resistance.

Silicides are promising materials for ULSI circuits because of their low resistivity and excellent thermal stability. Cobalt silicide has been widely investigated because of its low resistivity and high-temperature thermal stability. However, cobalt is sensitive to native oxides as well as an oxygen-containing environment, and a large amount of silicon is consumed during silicidation.

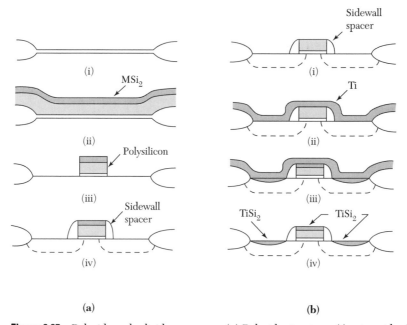

Figure 8.25 Polycide and salicide processes. (*a*) Polycide structure: (i) gate oxide, (ii) polysilicon and silicide deposition, (iii) pattern polycide, and (iv) lightly doped drain (LDD) implant, sidewall formation, and source/drain implant. (*b*) Salicide structure: (i) gate patterning (polysilicon only), LDD, sidewall, and source/drain implant; (ii) metal (Ti, Co) deposition; (iii) anneal to form salicide; and (iv) selective (wet) etch to remove unreacted metal.

EXAMPLE 8

Calculate the thickness of cobalt silicide for a desired sheet resistance of 0.6 $\Omega/\square$. The resistivity is 18 $\mu\Omega$-cm.

SOLUTION The resistivity is equal to the product of the sheet resistance and the film thickness:

$$\rho = R_s \times t$$

Then

$$t = \frac{\rho}{R_s} = \frac{18 \times 10^{-6}}{0.6} = 3 \times 10^{-5} \text{ cm} = 300 \text{ nm}$$ ◀

▶ 8.6 DEPOSITION SIMULATION

SUPREM may be used to simulate the deposition process. Like etch simulation, deposition modeling is very straightforward. Simulation is executed using the **DEPOSITION** command, which deposits a given amount of user-specified material on top of the current structure. The material deposited may be either undoped or uniformly doped. If single-crystal silicon is deposited, then the crystal orientation must also be specified. If polysilicon is deposited, the temperature must be specified for SUPREM to determine the proper polysilicon grain size.

EXAMPLE 9

Suppose we want to simulate the deposition of 800Å CVD silicon nitride on top of a dry oxide layer approximately 400Å thick. If the p-type silicon substrate is doped with boron at a level of 10^{15} cm^{-3}, use SUPREM to determine the final oxide and nitride layer thicknesses, as well as the boron doping profile in the oxide and nitride layers.

SOLUTION The SUPREM input listing is as follows:

```
TITLE           Deposition Example
COMMENT         Initialize silicon substrate
INITIALIZE      <100> Silicon Boron Concentration=1e15
COMMENT         Grow 400A oxide
DIFFUSION       Time=40 Temperature=1000 DryO2
COMMENT         Deposit 800A CVD nitride
DEPOSITION      Nitride Thickness=0.08
PRINT           Layers
PLOT            Chemical Boron Net
STOP            End Deposition Example
```

After simulation is complete, the results are shown in Figure 8.26, which indicates final oxide and nitride layer thicknesses of 379 and 800Å, respectively, and depicts the boron incorporation in the oxide layer. ◀

▶ 8.7 SUMMARY

Modern semiconductor device fabrication requires the use of thin films. In the epitaxial growth process, the substrate wafer is the seed. High-quality, single-crystal films can be grown at a temperature 30% to 50% lower than the melting point. The common techniques for epitaxial growth are chemical vapor deposition, metalorganic CVD, and molecular beam

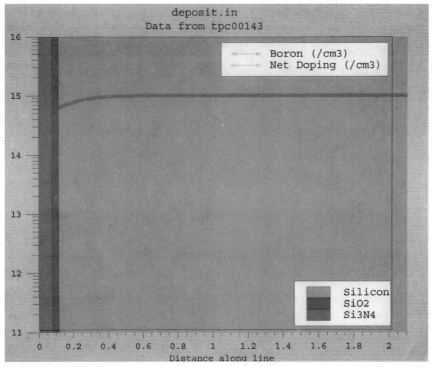

Figure 8.26 Plot of boron incorporation into the oxide layer, using SUPREM.

epitaxy. CVD and MOCVD are chemical deposition processes. Gases and dopants are transported in vapor form to the substrate, where a chemical reaction occurs that results in the deposition of the epitaxial layer. Inorganic compounds are used for CVD, whereas metalorganic compounds are used for MOCVD. MBE, on the other hand, is a physical deposition process. It is done by the evaporation of a species in an ultrahigh-vacuum system. Because it is a low-temperature process that has a low growth rate, MBE can grow single-crystal, multilayer structures with dimensions on the order of atomic layers.

In addition to conventional homoepitaxy, such as n-type silicon on an n^+ silicon substrate, this chapter also considered heteroepitaxy, which includes lattice-matched and strained-layer structures. For strained-layer epitaxy, there is a critical layer thickness above which edge dislocations will nucleate to relieve the strain energy.

Besides the edge dislocations in an epitaxial layer, defects from the substrate, defects from the interface, precipitates, and low-angle grain boundaries and twins exist. These defects degrade device performance. Various means have been presented to minimize, or even eliminate, these defects so that a defect-free semiconductor layer can be grown either homoepitaxially or heteroepitaxially.

Aside from epitaxial layers, there are four other important types of films: thermal oxides, dielectric layers, polycrystalline silicon, and metal films. The major issues related to film formation are low-temperature processing, step coverage, selective deposition, uniformity, film quality, planarization, throughput, and large wafer capacity.

Thermal oxidation offers the best quality for the Si–SiO$_2$ interface and has the lowest interface trap density (see Chapter 3). Therefore, it is used to form gate and field oxides. LPCVD of dielectrics and polysilicon offers conformal step coverage. In contrast, PVD and atmospheric-pressure CVD generally result in noncomformal step coverage. CMP offers global planarization and reduces defect density. Conformal step coverage and planarization are also required for precise pattern transfer at the deep-submicron lithography level.

To minimize the *RC* time delay due to parasitic resistance and capacitance, the silicide process for ohmic contacts, copper metallization for interconnects, and low-dielectric-constant materials for interlayer films are extensively used to meet the requirements of the multilevel interconnect structures of ULSI circuits. In addition, this chapter discussed high-dielectric-constant materials to improve the gate insulator performance and to increase the capacitance per unit area for DRAM.

▶ REFERENCES

1. A. S. Grove, *Physics and Technology of Semiconductor Devices,* Wiley, New York, 1967.

2. R. Reif, T. I. Kamins, and K. C. Saraswat, "A Model for Dopant Incorporation into Growing Silicon Epitaxial Films," *J. Electrochem. Soc.,* **126,** 644, 653 (1979).

3. R. D. Dupuis, "Metalorganic Chemical Vapor Deposition of III-V Semiconductors," *Science,* **226,** 623 (1984).

4. M. A. Herman and H. Sitter, *Molecular Beam Epitaxy,* Springer-Verlag, Berlin, 1996.

5. A. Roth, *Vacuum Technology,* North-Holland, Amsterdam, 1976.

6. M. Ohring, *The Materials Science of Thin Films,* Academic Press, New York, 1992.

7. J. C. Bean, "The Growth of Novel Silicon Materials," *Physics Today,* **39,** 10, 36 (1986).

8. For a discussion on film deposition, see, for example, A. C. Adams, "Dielectric and Polysilicon Film Deposition," in S. M. Sze, Ed., *VLSI Technology,* McGraw-Hill, New York, 1983.

9. K. Eujino, et al., "Doped Silicon Oxide Deposition by Atmospheric Pressure and Low Temperature Chemical Vapor Deposition Using Tetraethoxysilane and Ozone," *J. Electrochem. Soc.,* **138,** 3019 (1991).

10. A. C. Adams and C. D. Capio, "Planarization of Phosphorus-Doped Silicon Dioxide," *J. Electrochem. Soc.,* **127,** 2222 (1980).

11. T. Yamamoto, et al., "An Advanced 2.5nm Oxidized Nitride Gate Dielectric for Highly Reliable 0.25 μm MOSFETs," *Symp. VLSI Technol. Dig. Tech. Pap.,* p. 45 (1997).

12. K. Kumar, et al., "Optimization of Some 3 nm Gate Dielectrics Grown by Rapid Thermal Oxidation in a Nitric Oxide Ambient," *Appl. Phys. Lett.,* **70,** 384 (1997).

13. T. Homma, "Low Dielectric Constant Materials and Methods for Interlayer Dielectric Films in Ultralarge-Scale Integrated Circuit Multilevel Interconnects," *Mater. Sci. Eng.,* **23,** 243 (1998).

14. H. N. Yu, et al., "1 μm MOSFET VLSI Technology. Part I—An Overview," *IEEE Trans. Electron Devices,* ED-26, 318 (1979).

15. J. M. Andrews, "Electrical Conduction in Implanted Polycrystalline Sillicon," *J. Electron. Mater.,* **8,** 3, 227 (1979).

16. M. J. Buiting, A. F. Otterloo, and A. H. Montree, "Kinetical Aspects of the LPCVD of Titanium Nitride from Titanium Tetrachloride and Ammonia," *J. Electrochem. Soc.,* **138,** 500 (1991).

17. R. Tobe, et al., "Plasma-Enhanced CVD of TiN and Ti Using Low-Pressure and High-Density Helicon Plasma," *Thin Solid Film,* **281–282,** 155 (1996).

18. J. Hu, et al., "Electrical Properties of Ti/TiN Films Prepared by Chemical Vapor Deposition and Their Applications in Submicron Structures as Contact and Barrier Materials," *Thin Solid Film,* **308,** 589 (1997).

19. M. Hansen and A. Anderko, *Constitution of Binary Alloys,* McGraw-Hill, New York, 1958.

20. D. Pramanik and A. N. Saxena, "VLSI Metallization Using Aluminum and Its Alloys," *Solid State Tech.,* **26**(1), 127 (1983); **26**(3), 131 (1983).

21. C. L. Hu and J. M. E. Harper, "Copper Interconnections and Reliability," *Mater. Chem. Phys.,* **52,** 5 (1998).

22. P. C. Andricacos, et al., "Damascene Copper Electroplating for Chip Interconnects," *193rd Meet. Electrochem. Soc.,* p. 3 (1998).

23. J. M. Steigerwald, et al., *Chemical Mechanical Planarization of Microelectronic Materials,* Wiley, New York, 1997.

24. L. M. Cook, et al., "Theoretical and Practical Aspects of Dielectric and Metal CMP," *Semicond. Int.,* p. 141 (1995).

▶ PROBLEMS

Asterisks denote difficult problems.

SECTION 8.1: EPITAXIAL GROWTH TECHNIQUES

°**1.** Find the average molecular velocity of air at 300 K (the molecular weight for air is 29).

2. The distance between source and wafer in a deposition chamber is 15 cm. Estimate the pressure at which this distance become 10% of the mean free path of source molecules.

°**3.** Find the number of atoms per unit area, N_s, needed to form a monolayer under close-packing condition (i.e., each atom is in contact with its six neighboring atoms), assuming the diameter d of the atom is 4.68 Å.

°**4.** Assume an effusion oven geometry of $A = 5$ cm^2 and $L = 12$ cm. (a) Calculate the arrival rate of gallium and the MBE growth rate for the effusion oven filled with gallium arsenide at 970°C. (b) For a tin effusion oven operated at 700°C under the same geometry, calculate the doping concentration (assuming tin atoms are fully incorporated in the gallium arsenide grown at the aforementioned rate). The molecular weight for tin is 118.69, and the pressure at 700°C for tin is 2.66×10^{-6} Pa.

SECTION 8.2: STRUCTURES AND DEFECTS IN EPITAXIAL LAYERS

5. Find the maximum percentage of In (i.e., the x value for Ga$_x$In$_{1-x}$ As) film grown on GaAs substrate without formation of misfit dislocation, if the final film thickness is 10 nm.

6. The lattice misfit, f, of a film is defined as $f + [a_0(s) - a_0(f)]/a_0(f) = \Delta a_0/a_0$ where $a_0(s)$ and $a_0(f)$ are the unstrained lattice constants of the substrate and the film, respectively. Find the f values for InAs–GaAs and Ge–Si systems.

SECTION 8.3: DIELECTRIC DEPOSITION

7. (a) In a plasma-deposited silicon nitride that contains 20 at% hydrogen and has a silicon-to-nitrogen ratio (Si/N) of 1.2, find x and y in the empirical formula of SiN$_x$H$_y$. (b) If the variation of film resistivity with Si/N ratio is given by 5×10^{28} exp(-33.3γ) for $2 > \gamma > 0.8$, where γ is the ratio, find the resistivity of the film in (a).

8. The dielectric constants of SiO$_2$, Si$_3$N$_4$, and Ta$_2$O$_5$ are about 3.9, 7.6, and 25, respectively. What is the capacitance ratio for the capacitors with the Ta$_2$O$_5$ and oxide/nitride/oxide dielectrics for the same dielectric thickness, provided the oxide/nitride/oxide has thickness ratio 1:1:1 for the oxide to the nitride?

9. In Problem 8, if BST with a dielectric constant of 500 is chosen to replace Ta$_2$O$_5$, calculate the area reduction ratio to maintain the same capacitance if the two films have the same thickness.

10. In Problem 8, calculate the equivalent thickness of the Ta$_2$O$_5$ in terms of SiO$_2$ thickness if both have the same capacitance. Assume the actual thickness of Ta$_2$O$_5$ is 3μm.

11. In a silane–oxygen reaction to deposit undoped SiO$_2$ film, the deposition rate is 15 nm/min at 425°C. What temperature is required to double the deposition rate?

12. The P-glass flow process requires temperatures above 1000°C. As device dimensions become smaller in ULSI, we must use lower temperatures. Suggest methods to obtain a smooth topography at < 900°C for deposited silicon dioxide that can be used as an insulator between metal layers.

SECTION 8.4: POLYSILICON DEPOSITION

13. Why is silane more often used for polysilicon deposition than silicon chloride?

14. Explain why the deposition temperature for polysilicon films is moderately low, usually between 600°C and 650°C.

SECTION 8.5: METALLIZATION

15. An e-beam evaporation system is used to deposit aluminum to form MOS capacitors. If the flatband voltage of the capacitance is shifted by 0.5 V because of e-beam radiation, find the number of fixed oxide charges (the silicon dioxide thickness is 50 nm). How can these charges be removed?

16. A metal line ($L = 20\ \mu m$, $W = 0.25\ \mu m$) has a sheet resistance of 5 $\Omega/\square$. Calculate the resistance of the metal line.

17. Calculate the thicknesses of the $TiSi_2$ and $CoSi_2$, where the initial Ti and Co film thicknesses are 30 nm.

18. Compare the advantages and disadvantages of $TiSi_2$ and $CoSi_2$ for salicide applications.

19. A dielectric material is placed between the two parallel metal lines. The length $L = 1$ cm, width $W = 0.28\ \mu m$, thickness $T = 0.3\ \mu m$, and spacing $S = 0.36\ \mu m$. (a) Calculate the RC time delay. The metal is Al with a resistivity of 2.67 $\mu\Omega$-cm, and the dielectric is oxide with dielectric constant 3.9. (b) Calculate the RC time delay. The metal is Cu with a resistivity of 1.7 $\mu\Omega$-cm, and the dielectric is organic polymer with dielectric constant 2.8. (c) Compare the results in (a) and (b). How much can we decrease the RC time delay?

20. Repeat Problems 19 (a) and (b) if the fringing factor for the capacitors is 3. The fringing factor is due to the spreading of the electric-field lines beyond the length and the width of the metal lines.

°21. To avoid electromigration problems, the maximum allowed current density in an aluminum runner is about 5×10^5 A/cm^2. If the runner is 2 mm long, 1 μm wide, and nominally 1 μm thick, and if 20% of the runner length passes over steps and is only 0.5 μm thick there, find the total resistance of the runner if the resistivity is 3×10^{-6} Ω-cm. Find the maximum voltage that can be applied across the runner.

°22. To use Cu for wiring, one must overcome several obstacles: the diffusion of Cu through SiO_2, adhesion of Cu to SiO_2, and corrosion of Cu. One way to overcome these obstacles is to use a cladding/adhesion layer (e.g., Ta or TiN) to protect the Cu wires. Consider a cladded Cu wire with a square cross section of 0.5 $\mu m \times$ 0.5 μm and compare it with a layered TiN/Al/TiN wire of the same size, with the top and bottom TiN layers 40 nm and 60 nm thick, respectively. What is the maximum thickness of the cladding layer if the resistances of the cladded Cu wire and the TiN/Al/TiN wire are the same?

CHAPTER 9

Process Integration

Microwave, photonic, and power applications generally employ discrete devices. For example, an IMPATT diode is used as a microwave generator, an injection laser as an optical source, and a thyristor as a high-power switch. However, most electronic systems are built on the *integrated circuit*, which is an ensemble of both active (e.g., transistor) and passive (e.g., resistor, capacitor, and inductor) devices formed on and within a single-crystal semiconductor substrate and interconnected by a metallization pattern.[1] ICs have enormous advantages over discrete devices connected by wire bonding. The advantages include (a) reduction of the interconnection parasitics, because an IC with multilevel metallization can substantially reduce the overall wiring length; (b) full utilization of a semiconductor wafer's area, because devices can be closely packed within an IC chip; and (c) drastic reduction in processing cost, because wire bonding is a time-consuming and error-prone operation.

This chapter discusses combinations of the basic processes described in the previous chapters to fabricate active and passive components in an IC. Because the key element of an IC is the transistor, specific processing sequences are developed to optimize its performance. The chapter considers three major IC technologies associated with the three transistor families: the bipolar transistor, the MOSFET, and the MESFET. In addition, it discusses the fabrication of microelectromechanical systems by micromachining techniques. Specifically, it covers the following topics:

- The design and fabrication of IC resistors, capacitors, and inductors
- The processing sequence for standard bipolar transistor and advanced bipolar devices
- The processing sequence for MOSFETs, with special emphasis on CMOS and memory devices
- The processing sequence for high-performance MESFETs and monolithic microwave ICs
- The major challenges for future microelectronics, including ultrashallow junction, ultrathin oxide, new interconnection materials, low power dissipation, and isolation
- Microelectromechanical systems formed by orientation-dependent etching, sacrificial etching, or LIGA (lithography, electroplating, and molding) processes
- The simulation of IC fabrication processes using SUPREM

Figure 9.1 illustrates the interrelationship between the major process steps used for IC fabrication. Polished wafers with a specific resistivity and orientation are used as the starting material. The film formation steps include thermally grown oxide films (Chapter 3) and deposited polysilicon, dielectric, and metal films (Chapter 8). Film formation is often

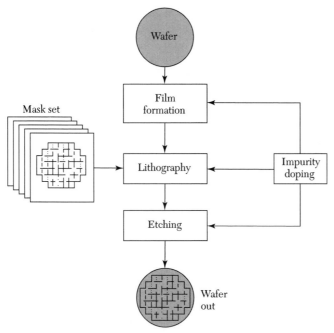

Figure 9.1 Schematic flow diagram of integrated circuit fabrication.

followed by lithography (Chapter 4) or impurity doping (Chapters 6 and 7). Lithography is generally followed by etching (Chapter 5), which in turn is often followed by another impurity doping or film formation. The final IC is made by sequentially transferring the patterns from each mask, level by level, onto the surface of the semiconductor wafer.

After processing, each wafer contains hundreds of identical rectangular chips (or *dice*), typically between 1 and 20 mm on each side, as shown in Figure 9.2*a*. The chips are separated by sawing or laser cutting; Figure 9.2*b* shows a separated chip. Schematic top views of a single MOSFET and a single bipolar transistor are shown in Figure 9.2*c* to give some perspective of the relative size of a component in an IC chip. Prior to chip separation, each chip is electrically tested (see Chapter 10). Defective chips are usually marked with

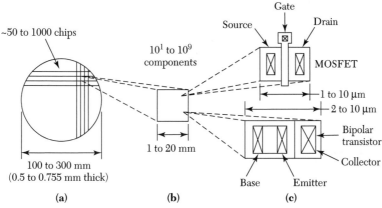

Figure 9.2 Size comparison of a wafer to individual components. (*a*) Semiconductor wafer. (*b*) Chip. (*c*) MOSFET and bipolar transistor.

a dab of black ink. Good chips are selected and packaged to provide an appropriate thermal, electrical, and interconnection environment for electronic applications.[2]

IC chips may contain a few components (transistors, diodes, resistors, capacitors, etc.) or as many as a billion or more. Since the invention of the monolithic IC in 1959, the number of components on a state-of-the-art IC chip has grown exponentially. We usually refer to the complexity of an IC as its level of integration. *Small-scale integration* (SSI) refers to up to 100 components per chip, *medium-scale integration* (MSI) to up to 1000 components per chip, *large-scale integration* (LSI) to up to 100,000 components per chip, *very large-scale integration* (VLSI) to up to 10^7 components per chip, and *ultralarge-scale integration* (ULSI) to larger numbers of components per chip. Section 9.3 describes two ULSI chips: a 32-bit microprocessor chip, which contains over 42 million components, and a 1-Gb dynamic random access memory (DRAM) chip, which contains over 2 billion components.

▶ 9.1 PASSIVE COMPONENTS

9.1.1 The Integrated Circuit Resistor

To form an IC resistor, we can deposit a resistive layer on a silicon substrate, then pattern the layer by photolithography and etching. We can also define a window in a silicon dioxide layer grown thermally on a silicon substrate and then implant (or diffuse) impurities of the opposite conductivity type into the wafer. Figure 9.3 shows the top and cross-sectional views of two resistors formed by the latter approach: One has a meander shape, and the other has a bar shape.

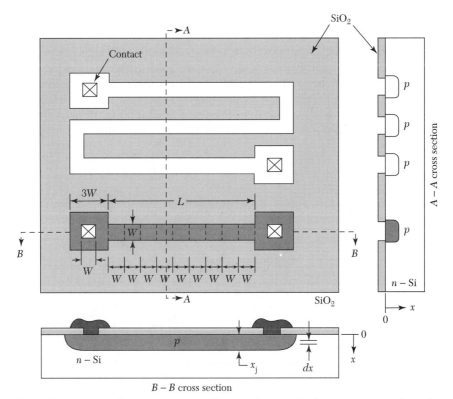

Figure 9.3 Integrated circuit resistors. All narrow lines in the large square area have the same width, W, and all contacts are the same size.

Consider the bar-shaped resistor first. The differential conductance dG of a thin layer of the p-type material that is of thickness dx parallel to the surface and at a depth x (as shown by the B–B cross section) is

$$dG = q\mu_p p(x)\frac{W}{L}dx \tag{1}$$

where W is the width of the bar, L is the length of the bar (we neglect the end contact areas for the time being), μ_p is the mobility of a hole, and $p(x)$ is the doping concentration. The total conductance of the entire implanted region of the bar is given by

$$G = \int_0^{x_j} dG = q\frac{W}{L}\int_0^{x_j} \mu_p p(x)dx \tag{2}$$

where x_j is the junction depth. If the value of μ_p (which is a function of the hole concentration), and the distribution of $p(x)$ are known, the total conductance can be evaluated from Eq. 2. We can write

$$G \equiv g\frac{W}{L} \tag{3}$$

where $g \equiv q\int_0^{x_j} \mu_p p(x)dx$ is the conductance of a square resistor pattern; that is, $G = g$ when $L = W$.

The resistance is therefore given by

$$R \equiv \frac{1}{G} = \frac{L}{W}\left(\frac{1}{g}\right) \tag{4}$$

where $1/g$ usually is defined by the symbol $R_\square$ and is called the sheet resistance. The sheet resistance has units of ohms but is conventionally specified in units of ohms per square ($\Omega/\square$).

Many resistors in an integrated circuit are fabricated simultaneously by defining different geometric patterns in the mask, such as those shown in Figure 9.3. Since the same processing cycle is used for all these resistors, it is convenient to separate the resistance into two parts: the sheet resistance $R_\square$, determined by the implantation (or diffusion) process; and the ratio L/W, determined by the pattern dimensions. Once the value of $R_\square$ is known, the resistance is given by the ratio L/W, or the number of squares (each square has an area of $W \times W$) in the resistor pattern. The end contact areas will introduce additional resistance to the IC resistors. For the type shown in Figure 9.3, each end contact corresponds to approximately 0.65 squares. For the meander-shape resistor, the electric-field lines at the bends are not spaced uniformly across the width of the resistor but are crowded toward the inside corner. A square at the bend does not contribute exactly 1 square, but rather 0.65 squares.

EXAMPLE 1

Find the value of a resistor 90 μm long and 10 μm wide, such as the bar-shaped resistor in Figure 9.3. The sheet resistance is 1 kΩ/$\square$.

SOLUTION　The resistor contains 9 squares. The two end contacts correspond to 1.3 $\square$. The value of the resistor is $(9 + 1.3) \times 1$ kΩ/$\square$ = 10.3 kΩ. ◀

9.1.2　The Integrated Circuit Capacitor

Basically, two types of capacitors are used in integrated circuits: MOS capacitors and p–n junctions. The MOS capacitor can be fabricated by using a heavily doped region (such as an emitter region) as one plate, the top metal electrode as the other plate, and the intervening oxide layer as the dielectric. The top and cross-sectional views of a MOS capacitor are shown in Figure 9.4a. To form a MOS capacitor, a thick oxide layer is thermally grown on a silicon substrate. Next, a window is lithographically defined and etched in the oxide. Diffusion or ion implantation is used to form a p^+ region in the window area, whereas the surrounding thick oxide serves as a mask. A thin oxide layer is then thermally grown in the window area, followed by a metallization step. The capacitance per unit area is given by

$$C = \frac{\varepsilon_{ox}}{d} \tag{5}$$

where ε_{ox} is the dielectric permittivity of silicon dioxide (the dielectric constant $\varepsilon_{ox}/\varepsilon_0$ is 3.9) and d is the oxide thickness. To increase the capacitance further, insulators with higher dielectric constants are being studied, such as Si_3N_4 and Ta_2O_5, which have dielectric constants of 7 and 25, respectively. The MOS capacitance is essentially independent of the applied voltage, because the lower plate of the capacitor is made of heavily doped material. This also reduces the asociated series resistance.

A p–n junction is sometimes used as a capacitor in an integrated circuit. The top and cross-sectional views of an n^+–p junction capacitor are shown in Figure 9.4b. The detailed fabrication process is considered in Section 9.2, because this structure forms part of a bipolar transistor. As a capacitor, the device is usually reverse biased; that is, the p region is reverse biased with respect to the n^+ region. The capacitance is not a constant, but varies as $(V_R + V_{bi})^{-1/2}$, where V_R is the applied reverse voltage and V_{bi} is the built-in potential of the junction. The series resistance is considerably higher than that of a MOS capacitor because the p region has higher resistivity than does the p^+ region.

EXAMPLE 2

What is the stored charge and the number of electrons on a MOS capacitor with an area of 4 µm^2 for a dielectric of (a) 10-nm thick SiO_2 and (b) 5-nm thick Ta_2O_5? The applied voltage is 5 V for both cases.

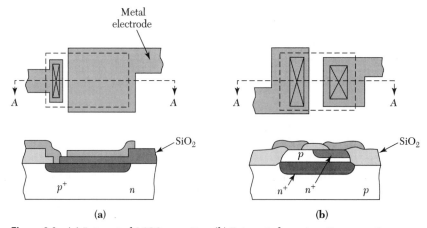

(a) (b)

Figure 9.4 (a) Integrated MOS capacitor. (b) Integrated p–n junction capacitor.

SOLUTION

(a)
$$Q = \varepsilon_{ox} \times A \times \frac{V_s}{d}$$

$$= 3.9 \times 8.85 \times 10^{-14} \, \text{F/cm} \times 4 \times 10^{-8} \, \text{cm}^2 \times \frac{5V}{10^{-6} \, \text{cm}}$$

$$= 6.9 \times 10^{-14} \, \text{C}$$

$$\text{Number of electrons} = 6.9 \times 10^{-14} \, \text{C} / q = 4.3 \times 10^5$$

(b) Changing the dielectric constant from 3.9 to 25 and the thickness from 10 nm to 5 nm, we obtain $Q_s = 8.85 \times 10^{-13}$ C, and number of electrons = 8.85×10^{-13} C$/q = 5.53 \times 10^6$. ◀

9.1.3 The Integrated Circuit Inductor

IC inductors have been widely used in III-V–based monolithic microwave integrated circuits (MMICs).[3] With the increased speed of silicon devices and advancement in multilevel interconnection technology, IC inductors have started to receive more and more attention in silicon-based radio frequency and high-frequency applications. Many kinds of inductors can be fabricated using IC processes. The most popular method is the thin-film spiral inductor. Figures 9.5a and b show the top view and the cross section of a silicon-based, two-level-metal spiral inductor. To form a spiral inductor, a thick oxide is thermally grown or deposited on a silicon substrate. The first metal is then deposited and defined as one end of the inductor. Next, another dielectric is deposited onto metal 1. A via hole is defined lithographically and etched in the oxide. Metal 2 is deposited and the via hole is filled. The spiral pattern can be defined and etched on metal 2 as the second end of the inductor.

To evaluate the inductor, an important figure of merit is the quality factor, Q, which is defined as $L\omega/R$, where L, R, and ω are the inductance, resistance, and frequency, respectively. The higher the Q value, the lower the loss from resistance, and hence the better the performance. Figure 9.5c shows the equivalent circuit model for an IC inductor. R_1 is the inherent resistivity of the metal, C_{p1} and C_{p2} are the coupling capacitances between the metal lines and the substrate, and R_{sub1} and R_{sub2} are the resistances of the silicon substrate associated with the metal lines, respectively. The Q value increases linearly with frequency initially and then drops at higher frequencies because of parasitic resistances and capacitances.

Some approaches exist for improving the Q value. The first is to use low-dielectric-constant materials (< 3.9) to reduce C_p. The other is to use a thick-film metal or low-resistivity metals (e.g., Cu, Au to replace Al) to reduce R_1. The third approach uses an insulating substrate (e.g., silicon on sapphire, silicon on glass, or quartz) to reduce R_{sub}.

To obtain the exact value of a thin-film inductor, complicated simulation tools must be employed for both circuit simulation and inductor optimization. The model for thin-film inductors must take into account the resistance of the metal, the capacitance of the oxide, line-to-line capacitance, the resistance of the substrate, the capacitance to the substrate, and the inductance and mutual inductance of the metal lines. Hence, it is more difficult to calculate the integrated inductance than to calculate integrated capacitance or resistance. However, a simple equation to estimate the square planar spiral inductor is given as[3]

$$L \approx \mu_0 n^2 r \approx 1.2 \times 10^{-6} n^2 r \tag{6}$$

where μ_0 is the permeability in vacuum ($4\pi \times 10^{-7}$ H/m), L is in henries, n is the number of turns, and r is the radius of the spiral in meters.

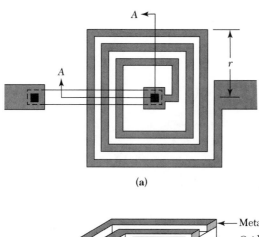

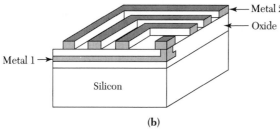

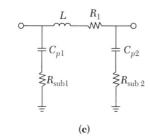

Figure 9.5 (a) Schematic view of a spiral inductor on a silicon substrate. (b) Perspective view along A–A'. (c) An equivalent circuit model for an integrated inductor.

EXAMPLE 3

For an integrated inductor with an inductance of 10 nH, what is the required radius if the number of turns is 20?

SOLUTION According to Eq. 6,

$$r = \frac{10 \times 10^{-9}}{1.2 \times 10^{-6} \times 20^2} = 2.08 \times 10^{-5} \text{ m} = 20.8 \text{ μm}$$ ◀

▶ 9.2 BIPOLAR TECHNOLOGY

For IC applications, especially for VLSI and ULSI, the size of bipolar transistors must be reduced to meet high density requirements. Figure 9.6 illustrates the reduction in the size of the bipolar transistor in recent years.[4] The main differences in a bipolar transistor in an IC, compared with a discrete transistor, are that all electrode contacts are

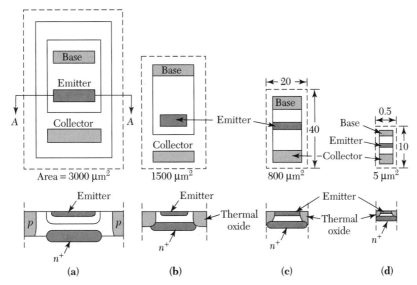

Figure 9.6 Reduction of the horizontal and vertical dimensions of a bipolar transistor. (*a*) Junction isolation. (*b*) Oxide isolation. (*c* and *d*) Scaled oxide isolation.[4]

located on the top surface of the IC wafer, and each transistor must be electrically isolated to prevent interactions between devices. Prior to 1970, both the lateral and vertical isolations were provided by *p–n* junctions (Fig. 9.6*a*), and the lateral *p* isolation region was always reverse biased with respect to the *n*-type collector. In 1971, thermal oxide was used for lateral isolation, resulting in a substantial reduction in device size (Fig. 9.6*b*) because the base and collector contacts abut the isolation region. In the mid-1970s, the emitter extended to the walls of the oxide, resulting in an additional reduction in area (Fig. 9.6*c*). At the present time, all lateral and vertical dimensions have been scaled down, and emitter stripe widths have dimensions in the submicron region (Fig. 9.6*d*).

9.2.1 The Basic Fabrication Process

The majority of bipolar transistors used in ICs are of the *n–p–n* type because the higher mobility of minority carriers (electrons) in the base region results in higher-speed performance than can be obtained with *p–n–p* types. Figure 9.7 shows a perspective view of an *n-p-n* bipolar transistor, in which lateral isolation is provided by oxide walls and vertical isolation is provided by the n^+–*p* junction. The lateral oxide isolation approach reduces not only the device size, but also the parasitic capacitance because of the smaller dielectric constant of silicon dioxide (3.9, compared with 11.9 for silicon). This section now considers the major process steps that are used to fabricate the device shown in Figure 9.7.

For an *n–p–n* bipolar transistor, the starting material is a *p*-type, lightly doped ($\sim 10^{15}$ cm^{-3}), <111>- or <100>-oriented, polished silicon wafer. Because the junctions are formed inside the semiconductor, the choice of crystal orientation is not as critical as for MOS devices (see Section 9.3). The first step is to form a buried layer. The main purpose of this layer is to minimize the series resistance of the collector. A thick oxide (0.5–1 μm) is thermally grown on the wafer, and a window is then opened in the oxide. A precisely controlled amount of low-energy arsenic ions (~ 30 keV, $\sim 10^{15}$ cm^{-2}) is implanted into the window region to serve as a predeposit (Fig. 9.8*a*). Next, a high-temperature (~ 1100°C) drive-in step forms the n^+ buried layer, which has a typical sheet resistance of approximately 20 Ω/□.

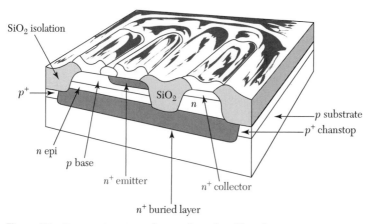

SiO₂ isolation

p^+

SiO₂

n

p substrate

p^+ chanstop

n epi

p base

n^+ emitter

n^+ collector

n^+ buried layer

Figure 9.7 Perspective view of an oxide-isolated bipolar transistor.

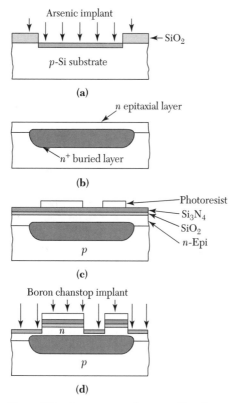

Arsenic implant

SiO₂

p-Si substrate

(a)

n epitaxial layer

n^+ buried layer

(b)

Photoresist

Si₃N₄

SiO₂

n-Epi

p

(c)

Boron chanstop implant

n

p

(d)

Figure 9.8 Cross-sectional views of bipolar transistor fabrication. (a) Buried-layer implantation. (b) Epitaxial layer. (c) Photoresist mask. (d) Chanstop implant.

The second step is to deposit an n-type epitaxial layer. The oxide is removed and the wafer is placed in an epitaxial reactor for epitaxial growth. The thickness and the doping concentration of the epitaxial layer are determined by the ultimate use of the device. Analog circuits (with their higher voltages for amplification) require thicker layers ($\sim 10\ \mu$m) and lower dopings ($\sim 5 \times 10^{15}\ \text{cm}^{-3}$), whereas digital circuits (with their lower voltages for switching) require thinner layers ($\sim 3\ \mu$m) and higher dopings ($\sim 2 \times 10^{16}\ \text{cm}^{-3}$). Figure 9.8$b$ shows a cross-sectional view of the device after the epitaxial process. Note that there is some outdiffusion from the buried layer into the epitaxial layer. To minimize outdiffusion, a low-temperature epitaxial process should be employed, and low-diffusivity impurities (e.g., As) should be used in the buried layer.

The third step is to form the lateral oxide isolation region. A thin oxide pad (~50 nm) is thermally grown on the epitaxial layer, followed by a silicon–nitride deposition (~100 nm). If nitride is deposited directly onto the silicon without the thin oxide pad, the nitride may cause damage to the silicon surface during subsequent high-temperature steps. Next, the nitride–oxide layers and about half of the epitaxial layer are etched using a photoresist as mask (Figs. 9.8c and d). Boron ions are then implanted into the exposed silicon areas (Fig. 9.8d).

The photoresist is removed, and the wafer is placed in an oxidation furnace. Since the nitride layer has a very low oxidation rate, thick oxides will be grown only in the areas not protected by the nitride layer. The isolation oxide is usually grown to a thickness such that the top of the oxide becomes coplanar with the original silicon surface to minimize the surface topography. This oxide isolation process is called *local oxidation of silicon* (LOCOS). Figure 9.9a shows the cross section of the isolation oxide after the removal of the nitride layer. Because of segregation effects, most of the implanted boron ions are pushed underneath the isolation oxide to form a p^+ layer. This is called the p^+ *channel stop* (or *chanstop*) because the high concentration of p-type semiconductor will prevent

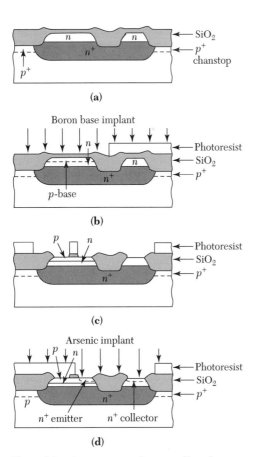

Figure 9.9 Cross-sectional views of bipolar transistor fabrication. (*a*) Oxide isolation. (*b*) Base implant. (*c*) Removal of thin oxide. (*d*) Emitter and collector implant.

surface inversion and eliminate possible high-conductivity paths (or channels) among neighboring buried layers.

The fourth step is to form the base region. A photoresist is used as a mask to protect the right half of the device. Then, boron ions ($\sim 10^{12}$ cm^{-2}) are implanted to form the base regions, as shown in Figure 9.9b. Another lithographic process removes all the thin pad oxide except a small area near the center of the base region (Fig. 9.9c).

The fifth step is to form the emitter region. As shown in Figure 9.9d, the base contact area is protected by a photoresist mask. Then, a low-energy, high-arsenic-dose ($\sim 10^{16}$ cm^{-2}) implantation forms the n^+ emitter and the n^+ collector contact regions. The photoresist is removed, and a final metallization step forms the contacts to the base, emitter, and collector, as shown in Figure 9.7.

In this basic bipolar process, there are six film formation operations, six lithographic operations, four ion implantations, and four etching operations. Each operation must be precisely controlled and monitored. Failure of any one of the operations generally will render the wafer useless.

The doping profiles of the completed transistor along a coordinate perpendicular to the surface and passing through the emitter, base, and collector are shown in Figure 9.10. The emitter profile is abrupt because of the concentration-dependent diffusivity of arsenic. The base doping profile beneath the emitter can be approximated by a Gaussian distribution for limited source diffusion. The collector doping is given by the epitaxial doping level ($\sim 2 \times 10^{16}$ cm^{-3}) for a representative switching transistor. However, at larger depths, the collector doping concentration increases because of outdiffusion from the buried layer.

9.2.2 Dielectric Isolation

In the isolation scheme described previously for the bipolar transistor, the device is isolated from other devices by the oxide layer around its periphery and is isolated from its common substrate by a $n^+ p$ junction (buried layer). In high-voltage applications, a different

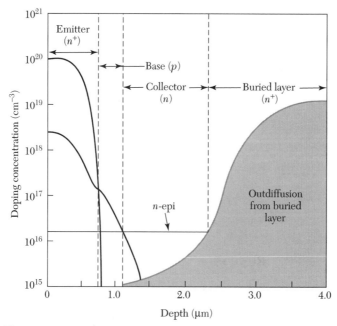

Figure 9.10 n–p–n transistor doping profiles.

approach, called *dielectric isolation*, is used to form insulating tubs to isolate a number of pockets of single-crystal semiconductors. In this approach, the device is isolated from both its common substrate and its surrounding neighbors by a dielectric layer.

A process sequence for dielectric isolation is shown in Figure 9.11. An oxide layer is formed inside a <100>-oriented n-type silicon substrate using high-energy oxygen ion implantation (Fig. 9.11a). Next, the wafer undergoes a high-temperature annealing process so that the implanted oxygen will react with silicon to form the oxide layer. The damage resulting from implantation is also annealed out in this process (Fig. 9.11b). After this, we obtain an n silicon layer that is fully isolated on an oxide (namely, a silicon-on-insulator, or SOI, layer). This process is called SIMOX (separation by implanted oxygen). Since the top silicon is so thin, the isolation region is easily formed by the LOCOS process illustrated in Figure 9.8c or by etching a trench (Fig. 9.11c) and refilling it with oxide (Fig. 9.11d). The other processes are almost the same as those in Figures 9.8c through 9.9 to form the p-type base, n^+ emitter, and n collector.

The main advantage of this technique is its high breakdown voltage between the emitter and the collector, which can be in excess of several hundred volts. This technique is also compatible with modern CMOS integration (Section 9.3.3). This CMOS-compatible process is very useful for mixed high-voltage and high-density ICs.

9.2.3 Self-Aligned Double-Polysilicon Bipolar Structures

The process shown in Figure 9.9c needs another lithographic step to define an oxide region to separate the base and emitter contact regions. This gives rise to a large inactive device area within the isolated boundary, which increases not only the parasitic capacitances, but also the resistance that degrades transistor performance. The most effective way to reduce these effects is by using a *self-aligned* structure.

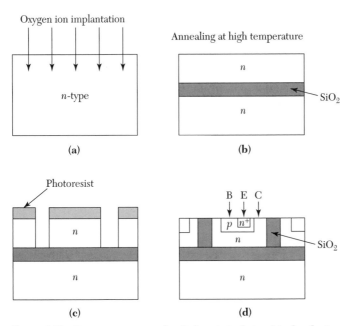

Figure 9.11 Process sequence for dielectric isolation bipolar device using silicon-on-insulator for high-voltage application. (a) Oxygen ion implantation. (b) Annealing at high temperature to form the isolation dielectric. (c) Trench isolation formed by a dry etching process. (d) Base, emitter, and collector formation.

The most widely used self-aligned structure is the double-polysilicon structure with advanced isolation provided by a trench refilled with polysilicon,[5] shown in Figure 9.12. Figure 9.13 shows the detailed sequence of steps for the self-aligned double-polysilicon (n–p–n) bipolar structure.[6] The transistor is built on an n-type epitaxial layer. A trench approximately 5.0 μm in depth is etched by reactive ion etching through the n^+ subcollector region into the p^- substrate region. A thin layer of thermal oxide is then grown and serves as the screen oxide for the channel stop implant of boron at the bottom of the trench. The trench is then filled with undoped polysilicon and capped by a thick planar field oxide.

The first polysilicon layer is deposited and heavily doped with boron. The p^+ polysilicon (called poly 1) will be used as a solid-phase diffusion source to form the extrinsic base region and the base electrode. This layer is covered with a chemical vapor deposition oxide and nitride (Fig. 9.13a). The emitter mask is used to pattern the emitter regions, and a dry etch process is used to produce an opening in the CVD oxide and poly 1 (Fig. 9.13b). A thermal oxide is then grown over the etched structure, and a relatively thick oxide (approximately 0.1–0.4 μm) is grown on the vertical sidewalls of the heavily doped poly. The thickness of this oxide determines the spacing between the edges of the base and emitter contacts. The extrinsic p^+ base regions are also formed during the thermal oxide growth step as a result of the outdiffusion of boron from the poly 1 into the substrate (Fig. 9.13c). Because boron diffuses laterally as well as vertically, the extrinsic base region will be able to make contact with the intrinsic base region that is formed next, under the emitter contact.

Following the oxide growth step, the intrinsic base region is formed using ion implantation of boron (Fig. 9.13d). This serves to self-align the intrinsic and extrinsic base regions. After the contact is cleaned to remove any oxide layer, the second polysilicon layer is deposited and implanted with As or P. The n^+ polysilicon (called poly 2) is used as a solid-phase diffusion source to form the emitter region and the emitter electrode. A shallow emitter region is then formed through dopant outdiffusion from poly 2. A rapid thermal anneal for the base and emitter outdiffusion steps facilitates the formation of shallow emitter–base and collector–base junctions. Finally, Pt film is deposited and sintered to form PtSi over the n^+ polysilicon emitter and the p^+ polysilicon base contact (Fig. 9.13e).

This self-aligned structure allows the fabrication of emitter regions smaller than the minimum lithographic dimension. When the sidewall-spacer oxide is grown, it fills the contact hole to some degree because the thermal oxide occupies a larger volume than the original volume of polysilicon. Thus, an opening 0.8 μm wide will shrink to about 0.4 μm if sidewall oxide 0.2 μm thick is grown on each side.

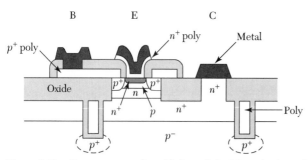

Figure 9.12 Cross section of a self-aligned double-polysilicon bipolar transistor with advanced trench isolation.[5]

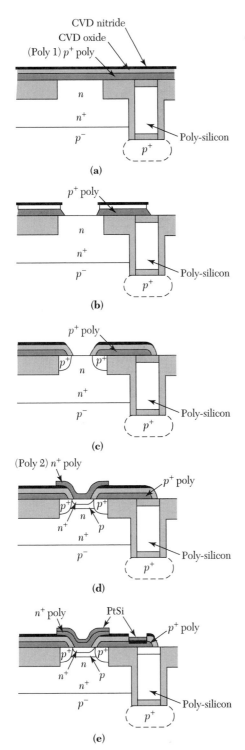

Figure 9.13 Process sequence for fabricating double-polysilicon self-aligned *n–p–n* transistors.[6]

▶ 9.3 MOSFET TECHNOLOGY

At present, the MOSFET is the dominant device used in ULSI circuits because it can be scaled to smaller dimensions than other types of devices. The dominant technology for MOSFET is *CMOS* (complementary MOSFET) technology, in which both *n*-channel and *p*-channel devices (called NMOS and PMOS, respectively) are provided on the same chip. CMOS technology is particular attractive for ULSI circuits because it has the lowest power consumption of all IC technology.

Figure 9.14 shows the reduction in the size of the MOSFET in recent years. In the early 1970s, the gate length was 7.5 µm and the corresponding device area was about 6000 µm². As the device is scaled down, there is a drastic reduction in the device area. For a MOSFET with a gate length of 0.5 µm, the device area shrinks to less than 1% of the early MOSFET. Device miniaturization is expected to continue. The gate length will probably decrease to less than 0.10 µm in the early twenty-first century. The future trends of MOSFET devices are considered briefly in Section 9.7.

9.3.1 The Basic Fabrication Process

Figure 9.15 shows a perspective view of an *n*-channel MOSFET prior to final metallization.[7] The top layer is a phosphorus-doped silicon dioxide (P-glass) that is used as an insulator between the polysilicon gate and the gate metallization and also as a gettering layer for mobile ions. Compare Figure 9.15 with Figure 9.7 for the bipolar transistor and note that a MOSFET is considerably simpler in its basic structure. Although both devices use

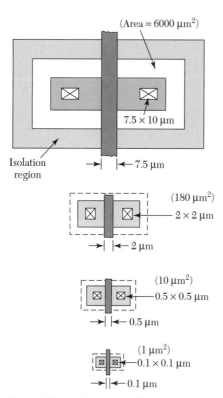

Figure 9.14 Reduction in the area of the MOSFET as the gate length (minimum feature length) is reduced.

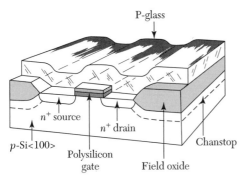

Figure 9.15 Perspective view of an *n*-channel MOSFET.[7]

lateral oxide isolation, there is no need for vertical isolation in the MOSFET, whereas a buried-layer n^+–p junction is required in the bipolar transistor. The doping profile in a MOSFET is not as complicated as that in a bipolar transistor, and control of the dopant distribution is also less critical. This section now considers the major process steps that are used to fabricate the device in Figure 9.15.

To process an *n*-channel MOSFET (NMOS), the starting material is a *p*-type, lightly doped ($\sim 10^{15}$ cm^{-3}), <100>-oriented, polished silicon wafer. The <100> orientation is preferred over <111> because it has an interface trap density that is about one-tenth that of <111>. The first step is to form the oxide isolation region using LOCOS technology. The process sequence for this step is similar to that for the bipolar transistor. A thin pad oxide ($\sim$35 nm) is thermally grown, followed by a silicon nitride ($\sim$150 nm) deposition (Fig. 9.16*a*).[7] The active device area is defined by a photoresist mask and a boron chanstop layer and is then implanted through the composite nitride–oxide layer (Fig. 9.16*b*). The nitride layer not covered by the photoresist mask is subsequently removed by etching. After stripping the photoresist, the wafer is placed in an oxidation furnace to grow an oxide (called the *field oxide*), where the nitride layer is removed, and to drive in the boron implant. The thickness of the field oxide is typically 0.5 to 1 μm.

The second step is to grow the gate oxide and to adjust the threshold voltage. The composite nitride–oxide layer over the active device area is removed, and a thin gate oxide layer (less than 10 nm) is grown. For an enhancement-mode *n*-channel device, boron ions are implanted in the channel region, as shown in Figure 9.16*c*, to increase the threshold voltage to a predetermined value (e.g., +0.5 V). For a depletion-mode *n*-channel device, arsenic ions are implanted in the channel region to decrease the threshold voltage (e.g., –0.5 V).

The third step is to form the gate. A polysilicon is deposited and is heavily doped by diffusion or implantation of phosphorus to a typical sheet resistance of 20 to 30 Ω/□. This resistance is adequate for MOSFETs with gate lengths larger than 3 μm. For smaller devices, polycide (a composite layer of metal silicide and polysilicon, such as W-polycide) can be used as the gate material to reduce the sheet resistance to about 1 Ω/□.

The fourth step is to form the source and drain. After the gate is patterned (Fig. 9.16*d*), it serves as a mask for the arsenic implantation ($\sim$30 keV, $\sim 5 \times 10^{15}$ cm^{-2}) to form the source and drain (Fig. 9.17*a*), which are self-aligned with respect to the gate.[7] At this stage, the only overlapping of the gate is due to lateral straggling of the implanted ions (for 30-keV As, $\sigma_\perp$ is only 5 nm). If low-temperature processes are used for subsequent steps to minimize lateral diffusion, the parasitic gate-drain and gate-source coupling capacitances can be much smaller than the gate-channel capacitance.

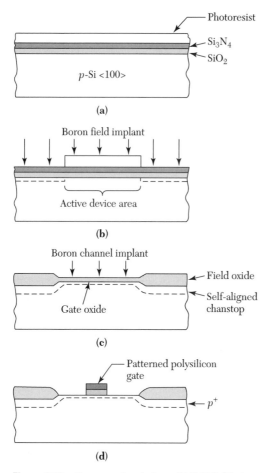

(a)

(b)

(c)

(d)

Figure 9.16 Cross-sectional view of NMOS fabrication sequence.[7] (a) Formation of SiO_2, Si_3N_4, and photoresist layers. (b) Boron implant. (c) Field oxide. (d) Gate.

The last step is metallization. A phosphorus-doped oxide (P-glass) is deposited over the entire wafer and is flowed by heating the wafer to give a smooth surface topography (Fig. 9.17b). Contact windows are defined and etched in the P-glass. A metal layer, such as aluminum, is then deposited and patterned. A cross-sectional view of the completed MOSFET is shown in Figure 9.17c, and the corresponding top view is shown in Figure 9.17d. The gate contact is usually made outside the active device area to avoid possible damage to the thin gate oxide.

EXAMPLE 4

What is the maximum gate-to-source voltage that a MOSFET with a 5-nm gate oxide can withstand? Assume that the oxide breaks down at 8 MV/cm and that the substrate voltage is zero.

SOLUTION

$$V = \xi \times d = 8 \times 10^6 \times 5 \times 10^{-7} = 4\,\text{V}$$ ◀

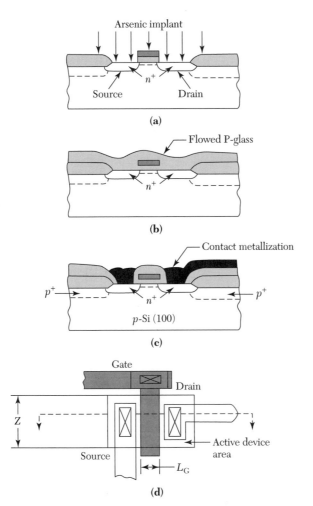

Figure 9.17 NMOS fabrication sequence.[7] (*a*) Source and drain. (*b*) P-glass deposition. (*c*) Cross section of the MOSFET. (*d*) Top view of the MOSFET.

9.3.2 Memory Devices

Memories are devices that can store digital information (or data) in terms of *bits* (binary digits). Various memory chips have been designed and fabricated using NMOS technology. For most large memories, the random access memory (RAM) organization is preferred. In a RAM, memory cells are organized in a matrix structure, and data can be accessed (i.e., stored, retrieved, or erased) in random order, independent of its physical location. A static random access memory (SRAM) can retain stored data indefinitely as long as the power supply is on. The SRAM is basically a flip-flop circuit that can store one bit of information. An SRAM cell has four enhancement-mode MOSFETs and two depletion-mode MOSFETs. The depletion-mode MOSFETs can be replaced by resistors formed in undoped polysilicon to minimize power consumption.[8]

To reduce the cell area and power consumption, the dynamic random access memory (DRAM) has been developed. Figure 9.18*a* shows the circuit diagram of the one-transistor DRAM cell, in which the transistor serves as a switch and one bit of information

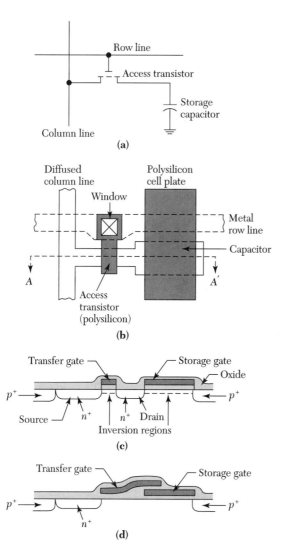

Figure 9.18 Single-transistor DRAM cell with a storage capacitor.[8] (a) Circuit diagram. (b) Cell layout. (c) Cross section through A–A'. (d) Double-level polysilicon.

can be stored in the storage capacitor. The voltage level on the capacitor determines the state of the cell. For example, +1.5 V may be defined as logic 1, and 0 V defined as logic 0. The stored charge will be removed typically in a few milliseconds, mainly because of the leakage current of the capacitors. Thus, dynamic memories require periodic "refreshing" of the stored charge.

Figure 9.18b shows the layout of a DRAM cell, and Figure 9.18c shows the corresponding cross section through A–A'. The storage capacitor uses the channel region as one plate, the polysilicon gate as the other plate, and the gate oxide as the dielectric. The row line is a metal track to minimize the delay due to parasitic resistance (R) and parasitic capacitance (C), the product of which is the RC delay. The column line is formed by n^+ diffusion. The internal drain region of the MOSFET serves as a conductive link between the inversion layers under the storage gate and the transfer gate. The drain region can be eliminated using the double-level polysilicon approach shown in Figure 9.18d.

The second polysilicon electrode is separated from the first polysilicon capacitor plate by an oxide layer that is thermally grown on the first-level polysilicon before the second electrode has been defined. The charge from the column line can therefore be transmitted directly to the area under the storage gate by the continuity of inversion layers under the transfer and storage gates.

To meet the requirements of high-density DRAMs, the DRAM structure has been extended to the third dimension with stacked or trench capacitors. Figure 9.19*a* shows a simple trench cell structure.[9] The advantage of the trench type is that the capacitance of the cell can be increased by increasing the depth of the trench without increasing the surface area of silicon occupied by the cell. The main difficulties of making trench-type cells are the etching of the deep trench, which needs a rounded bottom corner, and the growth of a uniform thin dielectric film on trench walls. Figure 9.19*b* shows a stacked cell structure. The storage capacitance increases as a result of stacking the storage capacitor on top of the access transistor. The dielectric is formed using the thermal oxidation or CVD nitride methods between the two polysilicon plates. Hence, the stacked cell process is easier than the trench-type process.

Figure 9.20 shows a 1-Gb DRAM chip. This memory chip uses 0.18-μm design rules. Trench capacitors and its peripheral circuits are in CMOS, which is discussed in Section 9.3.3. The memory chip has an area of 390 mm² (14.3 mm × 27.3 mm) that contains over 2 billion components and operates at 2.5 V. This 1-Gb DRAM is mounted in an 88-pin ceramic package, which can provide adequate heat dissipation.

Both SRAM and DRAM are volatile memories; that is, they lose their stored data when power is switched off. Nonvolatile memories, on the other hand, can retain their data. Figure 9.21*a* shows a floating-gate nonvolatile memory, which is basically a conventional MOSFET that has a modified gate electrode. The composite gate has a regular (control) gate and a floating gate that is surrounded by insulators. When a large positive voltage is applied to the control gate, charge will be injected from the channel region through the gate oxide into the floating gate. When the applied voltage is removed, the injected charge can be stored in the floating gate for a long time. To remove this charge, a large negative voltage must be applied to the control gate so that the charge will be injected back into the channel region.

Another version of the nonvolatile memory is the metal-nitride-oxide-semiconductor (MNOS) type shown in Figure 9.21*b*. When a positive gate voltage is applied, electrons can tunnel through the thin oxide layer (~2 nm), be captured by the traps at the oxide–nitride interface, and thus become stored charges there. The equivalent circuit for both types of nonvolatile memories can be represented by two capacitors in series for the

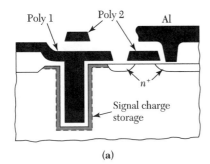

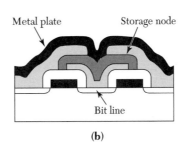

(a)   (b)

Figure 9.19 (*a*) DRAM with a trench cell structure.[9] (*b*) DRAM with a single-layer stacked-capacitor cell.

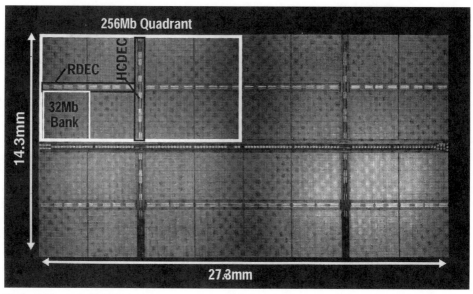

Figure 9.20 A 1-Gb DRAM that contains over 2 billion components. (Photograph courtesy of IBM/Siemens, 1999 IEEE Int. Solid State Circuit Conference.)

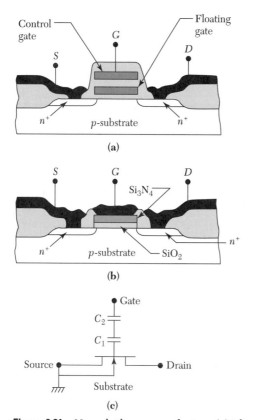

Figure 9.21 Nonvolatile memory devices. (*a*) Floating-gate nonvolatile memory. (*b*) MNOS non-volatile memory. (*c*) Equivalent circuit of either type of nonvolatile memory.

gate structure, as illustrated in Figure 9.21c. The charge stored in the capacitor (C_1) causes a shift in the threshold voltage, and the device remains at the higher threshold voltage state (logic 1). For a well-designed memory device, the charge retention time can be over 100 years. To erase the memory (i.e., the stored charge) and return the device to a lower threshold voltage state (logic 0), a gate voltage or other means (such as ultraviolet light) can be used.

Nonvolatile semiconductor memory (NVSM) has been extensively used in portable electronics systems, such as cellular phones and digital cameras. Another interesting application is the chip card, also called the IC card. The photo in Figure 9.22 is an IC card. The diagram at the bottom of Figure 9.22 illustrates the nonvolatile memory device that stores the data, which can be read and written through the bus to a central processing unit (CPU). In contrast to the limited volume (1 kB) inside a conventional magnetic tape card, the size of the nonvolatile memory can be increased to 16 kB, 64 kB, or even larger depending on the application (e.g., one could store personal photos or fingerprints). Through the IC card read/write machines, the data can be used in numerous applications, including telecommunications (card telephone, mobile radio), payment transactions (electronic purse, credit card), pay television, transport (electronic ticket, public transport), health care (patient-data card), and access control. The IC card will likely play a central role in the global information and service society of the future.[10]

9.3.3 CMOS Technology

Figure 9.23a shows a CMOS inverter. The gate of the upper PMOS device is connected to the gate of the lower NMOS device. Both devices are enhancement-mode MOSFETs

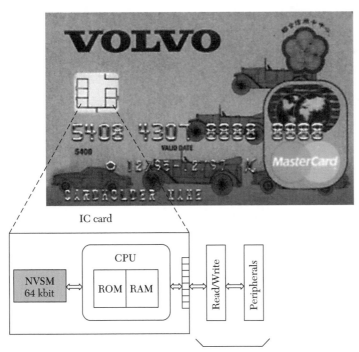

Figure 9.22 An IC card. The data stored in the NVSM can be accessed through the bus of the central processing unit (CPU). There are several metal pads connecting to the read/write machine. (Photograph courtesy of Retone Information System Co., LTD.)

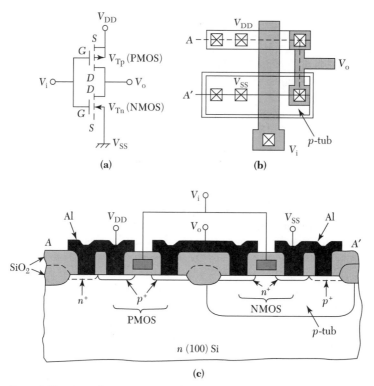

Figure 9.23 Complementary MOS (CMOS) inverter. (*a*) Circuit diagram. (*b*) Circuit layout. (*c*) Cross section along dotted A–A′ line of (*b*).

with the threshold voltage V_{T_p} less than zero for the PMOS device and V_{T_n} greater than zero for the NMOS device (typically the threshold voltage is about 1/4 V_{DD}). When the input voltage (V_i) is at ground or at small positive values, the PMOS device is turned on (the gate-to-ground potential of PMOS is $-V_{DD}$, which is more negative than V_{T_p}), and the NMOS device is off. Hence, the output voltage (V_o) is very close to V_{DD} (logic 1). When the input is at V_{DD}, the PMOS (with $V_{GS} = 0$) is turned off, and the NMOS is turned on ($V_i = V_{DD} > V_{T_n}$). Therefore, V_o equals zero (logic 0).

The CMOS inverter has a unique feature: In either logic state, one device in the series path from V_{DD} to ground is nonconductive. The current that flows in either steady state is a small leakage current, and only when both devices are on during switching does a significant current flow through the CMOS inverter. Thus, the average power dissipation is small, on the order of nanowatts. As the number of components per chip increases, power dissipation becomes a major limiting factor. Low power consumption is the most attractive feature of the CMOS circuit.

Figure 9.23*b* shows a layout of the CMOS inverter, and Figure 9.23*c* shows the device cross section along the A–A′ line. In the processing, a *p* tub (also called a *p well*) is first implanted and subsequently driven into the *n* substrate. The *p*-type dopant concentration must be high enough to overcompensate the background doping of the *n* substrate. The subsequent processes for the *n*-channel MOSFET in the *p* tub are identical to those described previously. For the *p*-channel MOSFET, $^{11}B^+$ or $^{49}(BF_2)^+$ ions are implanted into the *n* substrate to form the source and drain regions. A channel implant of $^{75}As^+$ ions may be used to adjust the threshold voltage, and an n^+ chanstop is formed underneath the field oxide around the *p*-channel device. Because of the *p* tub and the additional steps needed to make the p-channel MOSFET, the number of steps to make a CMOS circuit is essentially double that to make an NMOS circuit. Thus, there is a trade-off between the complexity of processing and the reduction in power consumption.

Instead of the p tub just described, an alternate approach is to use an n tub formed in p-type substrate, as shown in Figure 9.24a. In this case, the n-type dopant concentration must be high enough to overcompensate for the background doping of the p substrate (i.e., $N_D > N_A$). In both the p-tub and the n-tub approach, the channel mobility will be degraded because mobility is determined by the total dopant concentration ($N_A + N_D$). An approach using two separated tubs implanted into a lightly doped substrate is shown in Figure 9.24b. This structure is called a *twin tub*.[1] Because no overcompensation is needed in either of the twin tubs, higher channel mobility can be obtained.

All CMOS circuits have the potential for a troublesome problem called *latchup* that is associated with parasitic bipolar transistors. These parasitic devices consist of the n–p–n transistor formed by the NMOS source/drain regions, p tub, and n-type substrate, as well as the p–n–p transistor formed by the PMOS source/drain regions, n-type substrate, and p tub. Under appropriate conditions, the collector of the p–n–p device supplies base current to the n–p–n, and vice versa in a positive feedback arrangement. This latchup current can have serious negative repercussions in a CMOS circuit.

An effective processing technique to eliminate the latchup problem is to use deep-trench isolation, as shown in Figure 9.24c.[11] In this technique, a trench with a depth deeper than the well is formed in the silicon by anisotropic reactive sputter etching. An oxide layer is thermally grown on the bottom and walls of the trench, which is then refilled by deposited polysilicon or silicon dioxide. This technique can eliminate latchup because the n-channel and p-channel devices are physically isolated by the refilled trench. The detailed steps for trench isolation and some related CMOS processes are discussed next.

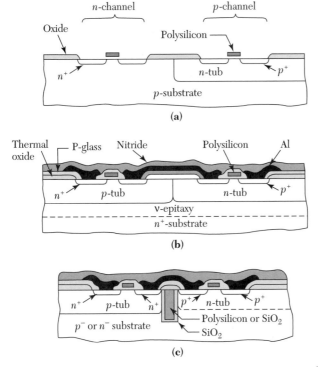

Figure 9.24 Various CMOS structures. (a) n tub. (b) Twin tub.[1] (c) Refilled trench.[11]

Well Formation Technology

The well of a CMOS can be a single well, a twin well, or a retrograde well. The twin-well process exhibits some disadvantages. For example, it needs high-temperature processing (above 1050°C) and a long diffusion time (longer than 8 hours) to achieve the required well depth of 2 to 3 μm. In this process, the doping concentration is highest at the surface and decreases monotonically with depth.

To reduce the process temperature and time, high-energy implantation is used (i.e., implanting the ion to the desired depth instead of diffusion from the surface). Since the depth is determined by the implantation energy, we can design the well depth with different implantation energy. The profile of the well in this case can have a peak at a certain depth in the silicon substrate. This is called a *retrograde well*. Figure 9.25 shows a comparison of the impurity profiles in the retrograde well and the conventional thermally diffused well.[12] The energy for the n- and p-type retrograde wells is around 700 keV and 400 keV, respectively. As mentioned previously, the advantage of high-energy implantation is that it can form the well under low-temperature and short-time conditions. Hence, it can reduce the lateral diffusion and increase the device density. The retrograde well offers some additional advantages over the conventional well: (a) Because of high doping near the bottom, the well resistivity is lower than that of the conventional well, and the latchup problem can be minimized; (b) the chanstop can be formed at the same time as the retrograde well implantation, reducing processing steps and time; and (c) higher well doping in the bottom can reduce the chance of punchthrough from the drain to the source.

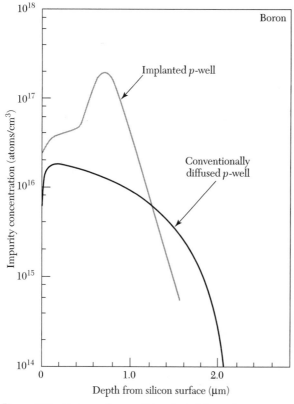

Figure 9.25   Retrograde p-well implanted impurity concentration profile. Also shown is a conventionally diffused well.[12]

Advanced Isolation Technology

The conventional isolation process (Section 9.3.1) has some disadvantages that make it unsuitable for deep-submicron (0.25 μm features and smaller) fabrication. The high-temperature oxidation of silicon and long oxidation time result in the encroachment of the chanstop implantation (usually boron for NMOS) to the active region and cause V_T shift. The area of the active region is reduced because of the lateral oxidation. In addition, the field-oxide thickness in submicron isolation spacings is significantly less than the thickness of field oxide grown in wider spacings. Trench isolation technology can avoid these problems and has become the mainstream technology for isolation.

Figure 9.26 shows the process sequence for forming a deep (larger than 3 μm) but narrow (less than 2 μm) trench isolation structure. There are four steps: patterning the area; trench etching and oxide growth; refilling with dielectric materials, such as oxide or undoped polysilicon; and planarization. This deep-trench isolation can be used in both advanced CMOS and bipolar devices and for the trench-type DRAM. Since the isolation material is deposited by CVD, it does not need a long time or a high-temperature process, and it eliminates lateral oxidation and boron encroachment problems.

Another example is shallow-trench (depth less than 1 μm) isolation for CMOS, shown in Figure 9.27. After patterning (Fig. 9.27a), the trench area is etched (Fig. 9.27b) and then refilled with oxide (Fig. 9.27c). Before refilling, a chanstop implantation can be performed. Since the oxide has overfilled the trench, the oxide on the nitride should be removed. Chemical mechanical polishing, which was discussed in Section 8.5.4, is used to remove the oxide on the nitride and to get a flat surface (Fig. 9.27d). Because of its high resistance to polishing, the nitride acts as a stop layer for the CMP process. After

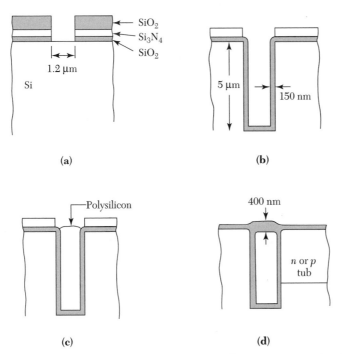

Figure 9.26 Process sequence for forming a deep, narrow-trench isolation structure. (a) Trench mask patterning. (b) Trench etching and oxide growth. (c) Polysilicon deposition to fill the trench. (d) Planarization.

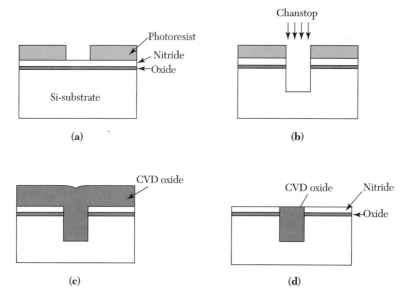

Figure 9.27 Shallow-trench isolation for CMOS. (*a*) Patterning with photoresist on nitride/oxide films. (*b*) Dry etching and chanstop implantation. (*c*) Chemical vapor deposition (CVD) oxide to refill. (*d*) Planar surface after chemical mechanical polishing.

the polishing, the nitride layer and the oxide layer can be removed by H_3PO_4 and HF, respectively. This initial planarization step is helpful for the subsequent polysilicon patterning and planarizations of the multilevel interconnection processes.

Gate Engineering Technology

If we use n^+ polysilicon for both PMOS and NMOS gates, the threshold voltage for PMOS ($V_{T_p} \cong -0.5$ to -1.0 V) has to be adjusted by boron implantation. This makes the channel of the PMOS a buried type, as shown in Figure 9.28*a*. The buried-type PMOS suffers serious short-channel effects as the device size shrinks below 0.25 μm. The most noticeable phenomena for short-channel effects are the V_T roll-off, drain-induced barrier lowering (DIBL), and the large leakage current at the off state, so that even with the gate voltage at zero, leakage current flows through source and drain. To alleviate this problem, one can change n^+ polysilicon to p^+ polysilicon for PMOS. Due to the work function difference (1.0 eV from n^+ to p^+ polysilicon), one can obtain a surface p-type channel device without the boron V_T adjustment implantation. Hence, as the technology shrinks to 0.25 μm and less, dual-gate structures are required: p^+ polysilicon gate for PMOS, and n^+ polysilicon for NMOS (Fig. 9.28*b*). A comparison of V_T for the surface channel and the buried channel is shown in Figure 9.29. Note that the V_T of the surface channel rolls off more slowly in the deep-submicron regime than in the buried-channel device. This makes the surface-channel device with p^+ polysilicon suitable for deep-submicron device operation.

To form the p^+ polysilicon gate, ion implantation of BF_2^+ is commonly used. However, boron penetrates easily from the polysilicon through the oxide into the silicon substrate at high temperatures, resulting in a V_T shift. This penetration is enhanced in the presence of an F atom. There are methods to reduce this effect: use of rapid thermal annealing to reduce the time at high temperatures and, consequently, the diffusion of boron; use of nitrided oxide to suppress the boron penetration, since boron can easily combine with nitrogen and becomes less mobile; and the making of a multilayer of polysilicon to trap the boron atoms at the interface of the two layers.

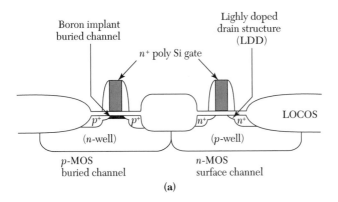

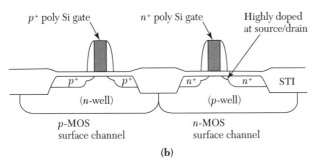

Figure 9.28 (*a*) A conventional long-channel CMOS structure with a single polysilicon gate (n^+). (*b*) Advanced CMOS structures with dual polysilicon gates.

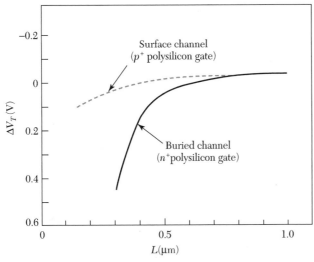

Figure 9.29 The V_T roll-off for a buried-type channel and for a surface-type channel. The V_T drops very quickly as the channel length becomes less than 0.5 μm.

Figure 9.30 shows a microprocessor chip (Pentium 4) that has an area of about 200 mm^2 and contains 42 million components. This ULSI chip is fabricated using 0.18-μm CMOS technology with a six-level aluminum metallization.

9.3.4 BiCMOS Technology

BiCMOS is a technology that combines both CMOS and bipolar device structures in a single IC. The reason to combine these two different technologies is to create an IC chip that has the advantages of both CMOS and bipolar devices. CMOS exhibits advantages in power dissipation, noise margin, and packing density, whereas bipolar technology shows advantages in switching speed, current drive capability, and analog capability. As a result, for a given design rule, BiCMOS can have a higher speed than CMOS, better performance in analog circuits than CMOS, a lower power dissipation than bipolar, and a higher component density than bipolar. Figure 9.31 compares a BiCMOS and a CMOS logic gate. For a CMOS inverter, the current to drive (or to charge) the next load, C_L, is the drain current I_{DS}. For a BiCMOS inverter, the current is $h_{fe}I_{DS}$, where h_{fe} is the current gain of the bipolar transistor and I_{DS} is the base current of the bipolar transistor and is equal to the drain current of M_2 in the CMOS. Since h_{fe} is much larger than 1, the speed can be substantially enhanced.

Figure 9.30   Micrograph of a 32-bit Pentium 4 microprocessor chip. (Photograph courtesy of Intel Corporation.)

BiCMOS has been widely used in many applications. In the early days, it was used in SRAM circuits. At the present time, BiCMOS technology has been successfully developed for transceiver, amplifier, and oscillator applications in wireless communication equipment. Most of the BiCMOS processes are based on the CMOS process, with some modifications, such as adding masks for bipolar transistor fabrication. The following example (Fig. 9.32) is for a high-performance BiCMOS process based on the twin-well CMOS process.[13]

The initial material is a p-type silicon substrate. An n^+ buried layer is then formed to reduce collector resistance. The buried p layer is formed by ion implantation to increase the doping level to prevent punchthrough. A lightly doped n-epi layer is grown on the wafer, and a twin-well process for the CMOS is performed. To achieve high performance for the bipolar transistor, four additional masks are needed. They are the buried n^+ mask, the collector deep n^+ mask, the base p mask, and the poly-emitter mask. In other processing steps, the p^+ region for base contact can be formed with the p^+ implant in the source/drain implantation of the PMOS, and the n^+ emitter can be formed with the source/drain implantation of the NMOS. The additional masks and longer processing time compared with a standard CMOS are the main drawbacks of BiCMOS. The additional cost should be justified by the enhanced performance.

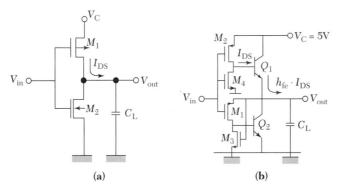

Figure 9.31 (a) CMOS logic gate. (b) Bipolar CMOS (BiCMOS) logic gate.

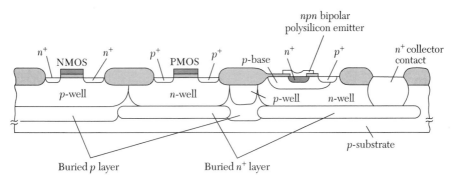

Figure 9.32 Optimized BiCMOS device structure. Key features include self-aligned p and n^+ buried layers for improved packing density, separately optimized n and p wells (twin-well CMOS) formed in an epitaxial layer with intrinsic background doping, and a polysilicon emitter for improved bipolar performance.[13]

▶ 9.4 MESFET TECHNOLOGY

Advances in gallium arsenide processing techniques, in conjunction with new fabrication and circuit approaches, have made possible the development of "silicon-like" gallium arsenide IC technology. Gallium arsenide has three inherent advantages compared with silicon: higher electron mobility, which results in lower series resistance for a given device geometry; higher drift velocity at a given electric field, which improves device speed; and the ability to be made semiinsulating, which can provide a lattice-matched dielectric-insulated substrate. However, gallium arsenide also has three disadvantages: a very short minority-carrier lifetime; lack of a stable, passivating native oxide; and crystal defects that are many orders of magnitude higher than in silicon. The short minority-carrier lifetime and the lack of high-quality insulating films have prevented the development of bipolar devices and delayed MOS technology using gallium arsenide. Thus, the emphasis of gallium arsenide IC technology is in the MESFET area, in which our main concerns are the majority carriers' transport and the metal-semiconductor contact.

A typical fabrication sequence[14] for a high-performance MESFET is shown in Figure 9.33. A layer of GaAs is epitaxially grown on a semiinsulating GaAs substrate, followed by an n^+ contact layer (Fig. 9.33a). A mesa etch step is performed for isolation (Fig. 9.33b), and a metal layer is evaporated for the source and drain ohmic contacts (Fig. 9.33c). A channel recess etch is followed by a gate recess etch and gate evaporation (Figs. 9.33d and e). After a liftoff process that removes the photoresist, shown in Figure 9.33e, the MESFET is completed (Fig. 9.33f).

The n^+ contact layer reduces the source and drain ohmic contact resistances. Note that the gate is offset toward the source to minimize the source resistance. The epitaxial layer is thick enough to minimize the effect of surface depletion on the source and drain resistance. The gate electrode has maximal cross-sectional area with a minimal footprint, which provides low gate resistance and minimal gate length. In addition, the length (L_{GD}) is designed to be greater than the depletion width at gate-drain breakdown.

A representative fabrication sequence for a MESFET integrated circuit is shown in Figure 9.34.[15] In this process, n^+ source and drain regions are self-aligned to the gate of each MESFET. A relatively light channel implant is used for the enhancement-mode switching device, and a heavier implant is used for the depletion-mode load device. A gate recess is usually not used for such digital IC fabrication because the uniformity of each depth is difficult to control, leading to unacceptable variation in the threshold voltage. This process sequence can also be used for a *monolithic microwave integrated circuit* (MMIC). Note that gallium arsenide MESFET processing technology is similar to the silicon-based MOSFET processing technology.

Gallium arsenide ICs with complexities up to the large-scale integration level (~10,000 components per chip) have been fabricated. Because of the higher drift velocity (~20% higher than silicon), gallium arsenide ICs have a 20% higher speed than silicon ICs that use the same design rules. However, substantial improvements in crystal quality and processing technology are needed before gallium arsenide can seriously challenge the preeminent position of silicon in ULSI applications.

▶ 9.5 MEMS TECHNOLOGY

There has been rapidly growing interest in *microelectromechanical systems* (MEMS) since the late 1980s, when a spinning micromotor made from polysilicon was fabricated on a silicon chip.[16,17] The manufacture of silicon MEMS adopted many of the highly developed technologies of silicon integrated circuits. This approach has enabled MEMS products to

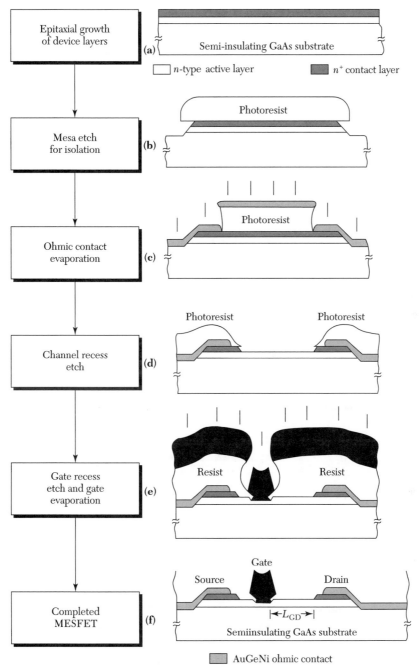

Figure 9.33 Fabrication sequence of a GaAs MESFET.[14]

be produced at low cost by using batch fabrication, just as for ICs. In addition to the IC fabrication process, some specialized techniques have been developed for MEMS. This section considers three specialized etching techniques: bulk micromachining, surface micromachining, and the LIGA process.

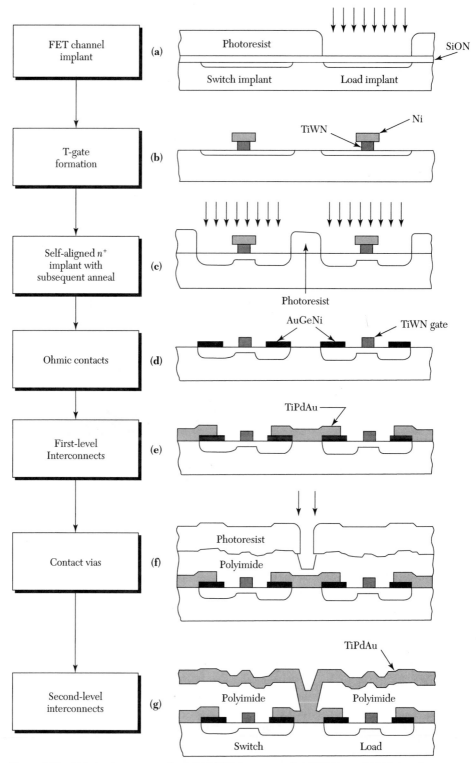

Figure 9.34 Fabrication process for MESFET direct-coupled FET logic (DCFL) with active loads. Note that the n^+ source and drain regions are self-aligned to the gate.[15]

9.5.1 Bulk Micromachining

In bulk micromachining, the device (e.g., a sensor or an actuator) is shaped by etching a large single-crystal substrate. The films are patterned on the bulk substrate to define the isolation and transducer functions. Orientation-dependent wet chemical etching techniques provide high-resolution etch and tight dimensional control. Often, a bulk-micromachined device uses two-sided processing, creating a self-isolated structure with one side exposed to the measured variables, such as mechanical or chemical signals, while the other side is enclosed in a clean package. Two-sided structures are very robust for operation in environments hostile to microelectronic devices. Simple mechanical devices such as diaphragm pressure sensors, membranes, and cantilever-beam piezoresistive acceleration sensors are fabricated commercially by this technique. Figure 9.35 illustrates a fabrication process of a simple silicone rubber membrane.[18]

9.5.2 Surface Micromachining

Surface-micromachined devices are constructed entirely from thin films. There are several differences and trade-offs between structures made from bulk and thin film materials. Typical dimensions for bulk-micromachined sensors are in the millimeter range, but surface-micromachined devices are of micrometer dimensions. Surface micromachining permits the fabrication of structurally complex devices by stacking and patterning layers or "building blocks" of thin films, whereas multilayered bulk devices are difficult to construct. Free-standing and movable parts can be fabricated using sacrificial layers. Figure 9.36 illustrates how sacrificial etching techniques can be used to create an electrostatic micromotor with well-defined, submicron tolerance between the rotor and the center hub.[17]

9.5.3 LIGA Process

LIGA is a German acronym for *lithographic, galvanoformung, abformung*.[19] It consists of three basic processing steps: lithography, electroplating, and molding. The LIGA process is based on x-ray radiation from a synchrotron. The process can produce microstructures with lateral dimensions in the micrometer range and structural heights of several hundred micrometers from a variety of materials. Its potential applications cover microelectronics, sensors, microoptics, micromechanics, and biotechnology.

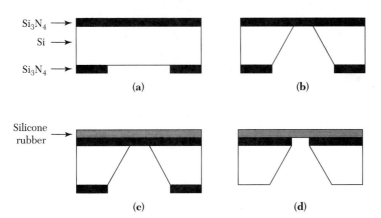

Figure 9.35 Fabrication process of simple silicone rubber membrane. (*a*) Nitride deposition and patterning. (*b*) KOH etching. (*c*) Silicone rubber spin coating. (*d*) Nitride removal on back side.[18]

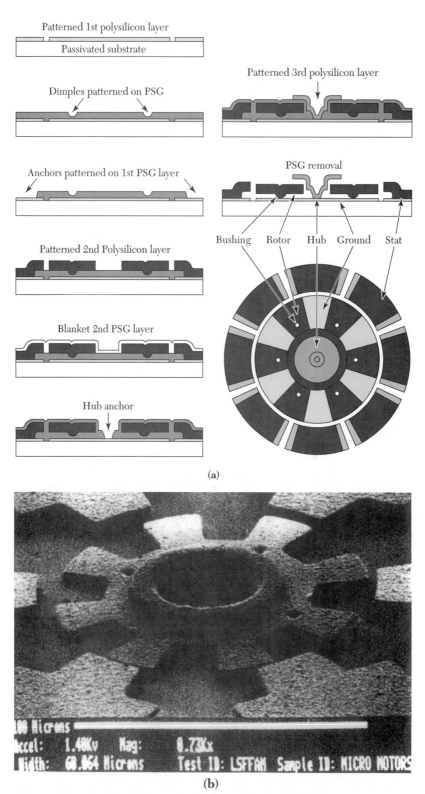

Patterned 1st polysilicon layer

Passivated substrate

Dimples patterned on PSG

Anchors patterned on 1st PSG layer

Patterned 2nd Polysilicon layer

Blanket 2nd PSG layer

Hub anchor

Patterned 3rd polysilicon layer

PSG removal

Bushing   Rotor   Hub   Ground   Stat

(a)

(b)

Figure 9.36 (*a*) Sacrificial process flow for an electrostatic micromotor. PSG, phosphosilicate glass. (*b*) Photograph of a micromotor.[17]

216

An example of the LIGA process is shown in Figure 9.37. An x-ray resist, ranging from 300 μm to more than 500 μm in thickness, is deposited on a substrate with an electrically conductive surface. Lithographic patterning is done with extended exposure from highly collimated x-ray radiation through an x-ray mask, as shown Figure 9.37a. A flower-shaped trench structure is formed in the thick resist after developer treatment (Fig. 9.37b).

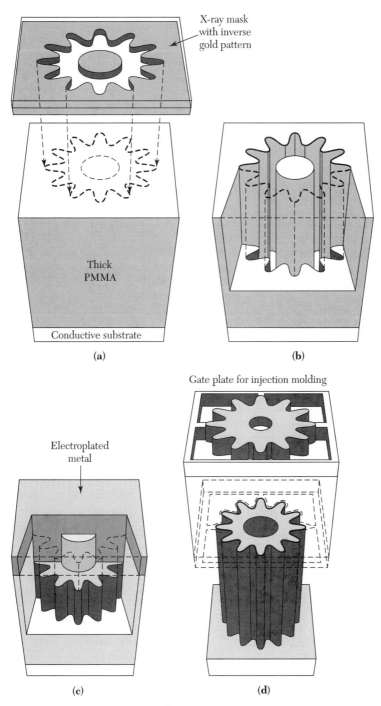

Figure 9.37 The LIGA process.[19]

(*continued*)

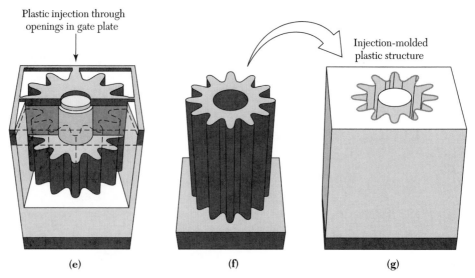

Plastic injection through
openings in gate plate

Injection-molded
plastic structure

(e)   (f)   (g)

Figure 9.37   *(continued)*

Metal is then electroplated on the exposed bottom conductive surface, filling the trench space and covering the top surface of the resist (Fig. 9.37*c*). The metal structure is formed after removing the resist (Fig. 9.37*d*). This structure can be used repeatedly as a mold insert for injection molding to form multiple plastic replicas of the original plating base (Fig. 9.37*e*). The plating base replicas, in turn, can be used to electroplate many metal structures as the final products, as shown in Figures 9.37*f* and *g*.

The distinct advantage of the LIGA process is the ability to create three-dimensional structures as thick as bulk-micromachined devices, while retaining the same degree of design flexibility as surface micromachining. However, the initial synchrotron radiation process is a very costly step, and the mold separation steps may result in degradation of the original mold insert.

▶ 9.6 PROCESS SIMULATION

SUPREM can be extremely useful for the simulation of a complete IC fabrication sequence. As an example,[20] consider a simulation of the NMOS polysilicon gate process described in Section 9.3.1. The cross section of the device to be simulated was shown initially in Figure 9.17*c* and is repeated here in Figure 9.38. Three vertical regions of the device, denoted by cut lines *A–A′*, *B–B′*, and *C–C′*, are simulated. These three simulations represent the center of the device, the source/drain region, and the field region, respectively.

The structure is simulated using a total of five SUPREM input decks. The first deck simulates processing in the active region of the device, up to the point at which the process sequence diverges for the gate and source/drain regions. The second and third decks start with the results from the first deck and complete processing for the gate and source/drain regions, respectively. This is accomplished by using the SAVEFILE statement at the end of the first deck to save the structure and subsequently using the saved structure in the INITIALIZE statements in the second and third decks. The fourth deck is similar to the first, except that it simulates the processing in the field region. The fifth deck completes field region processing.

The complete process sequence is as follows:

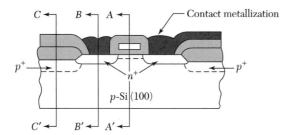

Figure 9.38   NMOS device for SUPREM simulation.

1. Begin with a high-resistivity <100> *p*-type silicon substrate.
2. Grow a 400-Å SiO₂ pad layer.
3. Deposit an 800-Å silicon nitride layer on top of the pad oxide.
4. Strip the nitride from areas outside of the active region.
5. Implant boron into the field regions.
6. Oxidize the field regions for 3 hours at 1000°C in wet O₂.
7. Etch down to the silicon in the active region.
8. Implant boron to set the threshold voltage of the MOSFET.
9. Grow a 400-Å gate oxide.
10. Deposit 0.5 µm of polysilicon.
11. Dope the polysilicon with phosphorus using POCl₃.
12. Etch the polysilicon in the areas outside the gate region.
13. Implant arsenic to form the source/drain regions.
14. Drive in the source/drain As implant for 10 minutes at 1000°C in dry O₂.
15. Open contact holes in the gate, source, and drain regions.
16. Deposit phosphorus-doped SiO₂ (P-glass) over the wafer surface.
17. Reflow the P-glass at 1000°C for 30 minutes.
18. Reopen the contact holes, and deposit aluminum.

Plots of the doping profiles in the gate (section *A–A′*), source/drain (section *B–B′*), and field (section *C–C′*) regions are shown in Figures 9.39, 9.40, and 9.41, respectively. The SUPREM input listing is as follows:

```
TITLE         NMOS Polysilicon Gate-Deck 1
COMMENT       Active device region initial processing
COMMENT       Initialize silicon substrate
INITIALIZE    <100> Silicon Boron Concentration=1e15
COMMENT       Grow 400A pad oxide
DIFFUSION     Time=40 Temperature=1000 DryO2
COMMENT       Deposit 800A CVD nitride
DEPOSITION    Nitride Thickness=0.08
COMMENT       Grow field oxide
DIFFUSION     Time=180 Temperature=1000 WetO2
COMMENT       Etch to silicon surface
ETCH          Oxide all
ETCH          Nitride all
```

```
ETCH            Oxide all
COMMENT         Implant boron to shift threshold voltage
IMPLANT         Boron Dose=4e11 Energy=50
COMMENT         Grow gate oxide
DIFFUSION       Time=30 Temperature=1050 DryO2 HCl%=3
COMMENT         Deposit polysilicon
DEPOSITION      Polysilicon Thickness=0.5 Temperature=600
COMMENT         Dope the polysilicon using POCl3
DIFFUSION       Time=25 Temperature=1000 Phosphorus solidsol
PRINT           Layers
PLOT            Chemical Boron Phosphor Net
SAVEFILE        Structur Filename=nmosactiveinit.str
STOP            End Deck 1

TITLE           NMOS Polysilicon Gate-Deck 2
COMMENT         Gate region
COMMENT         Initialize silicon substrate
INITIALIZE      Structur=nmosactiveinit.str
COMMENT         Implant arsenic for source/drain regions
IMPLANT         Arsenic Dose=5e15 Energy=150
COMMENT         Drive-in arsenic and re-oxidize source/drain regions
DIFFUSION       Time=30 Temperature=1000 DryO2
COMMENT         Etch contact holes to gate, source, and drain regions
ETCH            Oxide
COMMENT         Deposit phosphorus-doped SiO2 using CVD
DEPOSITION      Oxide Thickness=0.75 C.phosphor=1e21
COMMENT         Reopen contact holes
ETCH            Oxide
COMMENT         Deposit Aluminum
DEPOSITION      Aluminum Thickness=1.2
PRINT           Layers
PLOT            Chemical Boron Arsenic Phosphor Net
STOP            End Deck 2

TITLE           NMOS Polysilicon Gate-Deck 3
COMMENT         Source/drain regions
COMMENT         Initialize silicon substrate
INITIALIZE      Structur=nmosactiveinit.str
COMMENT         Etch polysilicon and oxide over source/drain regions
ETCH            Polysilicon
ETCH            Oxide
COMMENT         Implant arsenic for source/drain regions
IMPLANT         Arsenic Dose=5e15 Energy=150
COMMENT         Drive-in arsenic and re-oxidize source/drain regions
DIFFUSION       Time=30 Temperature=1000 DryO2
COMMENT         Etch contact holes to gate, source, and drain regions
ETCH            Oxide
COMMENT         Deposit phosphorus-doped SiO2 using CVD
DEPOSITION      Oxide Thickness=0.75 C.phosphor=1e21
COMMENT         Reflow glass to smooth surface and dope contact holes
DIFFUSION       Time=30 Temperature=1000
```

```
COMMENT          Reopen contact holes
ETCH             Oxide
COMMENT          Deposit Aluminum
DEPOSITION       Aluminum Thickness=1.2
PRINT            Layers
PLOT             Chemical Boron Arsenic Phosphor Net
STOP             End Deck 3

TITLE            NMOS Polysilicon Gate-Deck 4
COMMENT          Isolation region initial processing
COMMENT          Initialize silicon substrate
INITIALIZE       <100> Silicon Boron Concentration=1e15
COMMENT          Grow 400A pad oxide
DIFFUSION        Time=40 Temperature=1000 DryO2
COMMENT          Implant boron to increase field doping
IMPLANT          Boron Dose=1e13 Energy=150
COMMENT          Grow field oxide
DIFFUSION        Time=180 Temperature=1000 WetO2
COMMENT          Implant boron to shift threshold voltage
IMPLANT          Boron Dose=4e11 Energy=50
COMMENT          Grow gate oxide
DIFFUSION        Time=30 Temperature=1050 DryO2 HCl%=3
COMMENT          Deposit polysilicon
DEPOSITION       Polysilicon Thickness=0.5 Temperature=600
COMMENT          Dope the polysilicon using POCl3
DIFFUSION        Time=25 Temperature=1000 Phosphorus solidsol
PRINT            Layers
PLOT             Chemical Boron Phosphor Net
SAVEFILE         Structur Filename=nmosfieldinit.str
STOP             End Deck 4

TITLE            NMOS Polysilicon Gate-Deck 5
COMMENT          Isolation region final processing
COMMENT          Initialize silicon substrate
INITIALIZE       Structur=nmosfieldinit.str
COMMENT          Etch polysilicon and oxide over source/drain regions
ETCH             Polysilicon
ETCH             Oxide Thickness=0.07
COMMENT          Implant arsenic for source/drain regions
IMPLANT          Arsenic Dose=5e15 Energy=150
COMMENT          Drive-in arsenic and re-oxidize source/drain regions
DIFFUSION        Time=30 Temperature=1000 DryO2
COMMENT          Deposit phosphorus-doped SiO2 using CVD
DEPOSITION       Oxide Thickness=0.75 C.phosphor=1e21
COMMENT          Reflow glass to smooth surface and dope contact holes
DIFFUSION        Time=30 Temperature=1000
COMMENT          Deposit Aluminum
DEPOSITION       Aluminum Thickness=1.2
PRINT            Layers
PLOT             Chemical Boron Arsenic Phosphor Net
STOP             End Deck 5
```

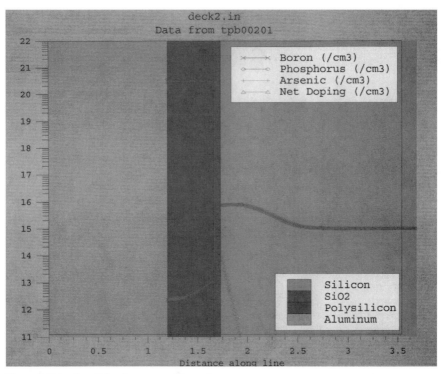

Figure 9.39 Plot of the doping profile in the gate region.

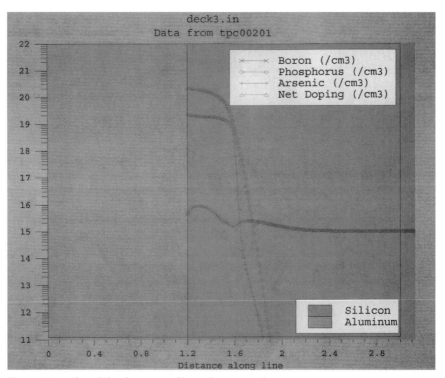

Figure 9.40 Plot of the doping profile in the source/drain region.

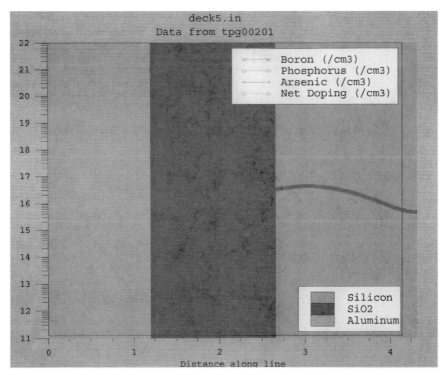

Figure 9.41 Plot of the doping profile in the field region.

▶ 9.7 SUMMARY

This chapter considered processing technologies for passive components, active devices, ICs, and MEMS. Three major IC technologies based on the bipolar transistor, the MOS-FET, and the MESFET were discussed in detail. It appears that the MOSFET will be the dominant technology at least until 2014 because of its superior performance compared with the bipolar transistor. For 100-nm CMOS technology, a good candidate is the combination of an SOI substrate with interconnections using Cu and low-k materials.

MEMS is still an emerging field. MEMS has adopted the lithographic and etching technologies from IC fabrication. Specialized etching techniques have also been developed for MEMS: bulk micromachining using an orientation-dependent etching process, surface micromachining using sacrifical layers, and the LIGA process using x-ray lithography with highly collimated radiation.

▶ REFERENCES

1. For a detailed discussion on IC process integration, see C. Y. Liu and W. Y. Lee, "Process Integration," in C. Y. Chang and S. M. Sze, Eds., *ULSI Technology,* McGraw-Hill, New York, 1996.

2. T. Tachikawa, "Assembly and Packaging," in C. Y. Chang and S. M. Sze, Eds., *ULSI Technology,* McGraw-Hill, New York, 1996.

3. T. H. Lee, *The Design of CMOS Radio-Frequency Integrated Circuits,* Cambridge University Press, Cambridge, U.K., 1998, Ch. 2.

4. D. Rise, "Isoplanar-S Scales Down for New Heights in Performance," *Electronics,* **53**, 137 (1979).

5. T. C. Chen, et al., "A Submicrometer High-Performance Bipolar Technology," *IEEE Electron. Device Lett.,* **10**(8), 364 (1989).

6. G. P. Li et al., "An Advanced High-Performance Trench-Isolated Self-Aligned Bipolar Technology," *IEEE Trans. Electron Devices,* **34**(10), 2246 (1987).

7. W. E. Beasle, J. C. C. Tsai, and R. D. Plummer, Eds., *Quick Reference Manual for Semiconductor Engineering,* Wiley, New York, 1985.

8. R. W. Hunt, "Memory Design and Technology," in M. J. Howes and D. V. Morgan, Eds., *Large Scale Integration,* Wiley, New York, 1981.

9. A. K. Sharma, *Semiconductor Memories—Technology, Testing, and Reliability,* IEEE, New York, 1997.

10. U. Hamann, "Chip Cards—The Application Revolution," *IEEE Tech. Dig. Int. Electron Devices Meet.,* p. 15 (1997).

11. R. D. Rung, H. Momose, and Y. Nagakubo, "Deep Trench Isolation CMOS Devices," *IEEE Tech. Dig. Int. Electron. Devices Meet.,* p. 237 (1982).

12. D. M. Bron, M. Ghezzo, and J. M. Primbley, "Trends in Advanced CMOS Process Technology," *Proc. IEEE,* p. 1646 (1986).

13. H. Higuchi, et al., "Performance and Structure of Scaled-Down Bipolar Devices Merge with CMOSFETs," *IEEE Tech. Dig. Int. Electron. Devices Meet.,* p. 694 (1984).

14. M. A. Hollis and R. A. Murphy, "Homogeneous Field-Effect Transistors," in S. M. Sze, Ed., *High-Speed Semiconductor Devices,* Wiley, New York, 1990.

15. H. P. Singh, et al., "GaAs Low Power Integrated Circuits for a High Speed Digital Signal Processor," *IEEE Trans. Electron Devices,* **36**, 240 (1989).

16. C. H. Mastrangelo and W. C. Tang, "Semiconductor Sensor Technology," in S. M. Sze, Ed., *Semiconductor Sensors,* Wiley, New York, 1994.

17. L. S. Fan, Y. C. Tai, and R. S. Muller, "IC-Processed Electrostatic Micromotors," in *IEEE Int. Electron Devices Meet,* p. 666 (1988).

18. X. Yang, et al., "A MEMS Thermopneumatic Silicone Rubber Membrane Valve," *Sens. Actuators,* **A64**, 101 (1998).

19. W. Ehrfeld, et al. "Fabrication of Microstructures Using the LIGA Process," *Proc. IEEE Micro Robots and Teleoperators Workshop,* Hyannis, MA, Nov. 1987.

20. C. P. Ho and S. E. Hansen, *SUPREM III User's Manual,* Stanford University, 1983.

▶ PROBLEMS

Asterisks denote difficult problems

SECTION 9.1: PASSIVE COMPONENTS

1. For a sheet resistance of 1 k$\Omega/\square$, find the maximum resistance that can be fabricated on a 2.5 × 2.5 mm chip using 2-µm lines with a 4-µm pitch (i.e., distance between the centers of the parallel lines).

2. Design a mask set for a 5-pF MOS capacitor. The oxide thickness is 30 nm. Assume that the minimum window size is 2 × 10 µm and the maximum registration errors are 2 µm.

3. Draw a complete step-by-step set of masks for the spiral inductor with three turns on a substrate.

4. Design a 10-nH square spiral inductor in which the total length of the interconnect is 350 µm; the spacing between turns is 2 µm.

SECTION 9.2: BIPOLAR TECHNOLOGY

5. Draw the circuit diagram and device cross section of a clamped transistor.

6. Identify the purpose of the following steps in self-aligned double-polysilicon bipolar structure: (a) undoped polysilicon in trench in Figure 9.13*a*, (b) the poly 1 in Figure 9.13*b*, and (c) the poly 2 in Figure 9.13*d*.

SECTION 9.3: MOSFET TECHNOLOGY

°**7.** In NMOS processing, the starting material is a p-type 10 Ω-cm <100>-oriented silicon wafer. The source and drain are formed by arsenic implantation of 10^{16} ions/cm² at 30 keV through a gate oxide of 25 nm. (a) Estimate the threshold voltage change of the device. (b) Draw the doping profile along a coordinate perpendicular to the surface and passing through the channel region or the source region.

8. (a) Why is <100> orientation preferred in NMOS fabrication? (b) What are the disadvantages if too thin a field oxide is used in NMOS devices? (c) What problems occur if a polysilicon gate is used for gate lengths less than 3 μm? Can another material be substituted for polysilicon? (d) How is a self-aligned gate obtained, and what are its advantages? (e) What purpose does P-glass serve?

°**9.** For a floating-gate nonvolatile memory, the lower insulator has a dielectric constant of 4 and is 10 nm thick. The insulator above the floating gate has a dielectric constant of 10 and is 100 nm thick. If the current density J in the lower insulators is given by $J = \sigma E$, where $\sigma = 10^{-7}$ S/cm, and the current in the other insulator is negligibly small, find the threshold voltage shift of the device caused by a voltage of 10 V applied to the control gate for (a) 0.25 μs, and (b) a sufficiently long time that J in the lower insulator becomes negligibly small.

10. Draw a complete step-by-step set of masks for the CMOS inverter shown in Figure 9.23. Pay particular attention to the cross section shown in Figure 9.23*c* for your scale.

°**11.** A 0.5-μm digital CMOS technology has 5-μm-wide transistors. The minimum wire width is 1 μm, and the metallization layer consists of 1-μm-thick aluminum. Assume that μ_n is 400 cm²/V-s, d is 10 nm, V_{DD} is 3.3 V, and the threshold voltage is 0.6 V. Finally, assume that the maximum voltage drop that can be tolerated is 0.1 V when a 1 μm² cross section aluminum wire is carrying the maximum current that can be supplied by the NMOS transistor. How long a wire can be allowed? Use a simple square-law, long-channel model to predict the MOS current drive (resistivity of aluminum is 2.7×10^{-8} Ω-cm).

12. Plot the cross-sectional views of a twin-tub CMOS structure of the following stages of processing: (a) n-tub implant, (b) p-tub implant, (c) twin-tub drive-in, (d) nonselective p^+ source/drain implant, (e) selective n^+ source/drain implant using photoresist as mask, and (f) P-glass deposition.

13. Why do we use a p^+ polysilicon gate for PMOS?

14. What is the boron penetration problem in p^+ polysilicon PMOS? How would you eliminate it?

15. To obtain a good interfacial property, a buffered layer is usually deposited between the high-k material and substrate. Calculate the effective oxide thickness if the stacked gate dielectric structure is (a) a buffered nitride of 0.5 nm and (b) a Ta₂O₅ of 10 nm.

16. Describe the disadvantages of LOCOS technology and the advantages of shallow-trench isolation technology.

SECTION 9.4: MESFET TECHNOLOGY

17. What is the purpose of the polyimide used in Figure 9.34*f*?

18. What is the reason that it is difficult to make bipolar transistors and MOSFETs in GaAs?

SECTION 9.6: PROCESS SIMULATION

°**19.** Use SUPREM to simulate the bipolar process described in Section 9.2.1. Plot the doping profiles along the following vertical cross sections: (a) from the top of the base contact, (b) from the top of the emitter contact, and (c) from the top of the collector contact.

°**20.** Use SUPREM to simulate the CMOS process described by Figure 9.23. Plot the doping profiles along the following vertical cross sections: (a) through the PMOS source/drain regions, (b) through the PMOS gate region, (c) through the NMOS source/drain regions, and (d) through the NMOS gate region.

10

IC Manufacturing

Manufacturing is defined as the process by which raw materials are converted into finished products. As illustrated in Figure 10.1, a manufacturing operation can be viewed graphically as a system with raw materials and supplies serving as its inputs and finished commercial products serving as outputs. In integrated circuit manufacturing, input materials include semiconductor wafers, insulators, dopants, and metals. The outputs are the ICs themselves. The types of processes that arise in IC manufacturing, which have been the subjects of previous chapters in this text, include oxidation and deposition processes, photolithography, etching, and doping (implantation and/or diffusion).

However, before finished ICs can be put to their intended use in various commercial electronic systems and products (such as computers, cellular phones, and digital cameras), several other key processes must take place. These include electrical testing and packaging. Testing is necessary to yield high-quality products. Quality requires conformance of all products to a set of specifications and the reduction of any variability in the manufacturing process. Maintaining quality often involves the use of statistical process control. A designed experiment is an extremely useful tool for discovering key variables that influence quality characteristics. Statistical experimental design is a powerful approach for systematically varying controllable process conditions and determining their impact on output parameters that measure quality.

A key metric that can be used to evaluate any manufacturing process is cost, and cost is directly impacted by yield. *Yield* refers to the proportion of manufactured products that perform as required by a set of specifications. Yield is inversely proportional to the total manufacturing cost: The higher the yield, the lower the cost. Finally, computer-integrated manufacturing is aimed at optimizing the cost-effectiveness of electronics manufacturing by using the latest developments in computer hardware and software technology to enhance expensive manufacturing methods.

This chapter describes each of these concepts. Specifically, it discusses the following topics:

- Electrical testing and test structures
- Electronics packaging processes
- Statistical process control and experimental design in the context of IC manufacturing

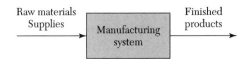

Figure 10.1 Block diagram of a manufacturing system.

• IC yield and various yield models
• Computer-integrated manufacturing systems

Viewed from a systems-level perspective, IC manufacturing intersects with nearly all parts of the production process, including design, fabrication, integration, assembly, testing, and packaging. The end result is an electronic system that meets all specified performance, quality, cost, and reliability requirements.

▶ 10.1 ELECTRICAL TESTING

Electrical measurements performed on test structures are a major mechanism for assessing integrated circuit yield (see Section 10.5), as well as other indicators of product performance. Such measurements are performed both during and at the conclusion of the fabrication process. In addition, electrical testing of the final product is crucial to ensure quality. These concepts are discussed in more detail in the following subsections.

10.1.1 Test Structures

To assess the impact of the presence of defects on semiconductor wafers caused by particles, contamination, or other sources, specially designed test structures are used. These structures, also known as *process control monitors* (PCMs), include single transistors, single lines of conducting material, MOS capacitors, and interconnect monitors. Product wafers typically contain several PCMs distributed across the surface, either in die sites or in the scribe lines between die (see Fig. 10.2).[1]

Process quality can be checked at various stages of manufacturing through in-line measurements on PCM structures. Three typical interconnect test structures are shown in Figure 10.3.[1] Using such test structures, measurements are performed to assess the presence of defects, which can be inferred by the presence of short circuits or open circuits using simple resistance measurements. For example, the meander structure facilitates the detection of open circuits through increased end-to-end resistance of the

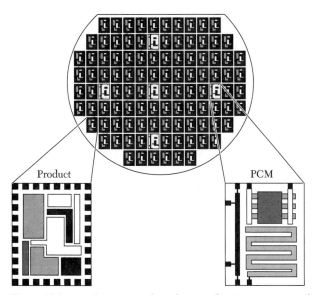

Figure 10.2 Configuration of products and PCMs on a typical semiconductor wafer.[1]

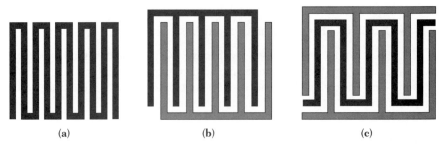

Figure 10.3   Basic test structures for interconnect layers.[1] (*a*) Meander structure. (*b*) Double-comb structure. (*c*) Comb-meander-comb structure.

meander. The double-comb structure can likewise be used to detect shorts, since any extra conducting material bridging the two combs will reduce the resistance between combs significantly. The comb-meander-comb structure combines the capabilities of the other two structures and permits the detection of both shorts and opens. Various combinations of widths of lines and spaces in these test structures allow the collection of statistics on defects of various sizes.

10.1.2   Final Test

Functional testing at the completion of manufacturing is the final arbiter of process quality and yield. The purpose of final testing is to ensure that all products perform to the specifications for which they were designed. For integrated circuits, the test process depends a great deal on whether the chip tested is a logic or memory device. In either case, automated test equipment (ATE) is used to apply a measurement stimulus to the chip and record the results. The major functions of the ATE are input pattern generation, pattern application, and output response detection.[2]

During each functional test cycle, input vectors are sent through the chip by the ATE in a timed sequence. Output responses are read and compared with expected results. This sequence is repeated for each input pattern. It is often necessary to perform such tests at various supply voltages and operating temperatures to ensure device operation at all potential regimes. The number and sequence of failures in the output signature are indicative of manufacturing process faults.

Test results may be expressed in a variety of ways.[2] Two examples are shown in Figures 10.4 and 10.5. Figure 10.4 shows a two-dimensional plot called a *shmoo plot* for a hypothetical bipolar product. In a shmoo plot, the outlined shaded region is where the device is intended to operate, while the blank area outside represents the failure region. Another typical test output is the cell map shown in Figure 10.5. Cell maps are very useful in identifying and isolating device failures, particularly in memory arrays. In addition, the patterns generated in the cell map may be compiled, catalogued, and later compared with a library of existing defect types, thereby aiding in the diagnosis of faults.

▶ 10.2 PACKAGING

Loosely defined, the term *packaging* refers to the set of technologies and processes that connect ICs with electronic systems. A useful analogy is to consider an electronic product as the human body. Like the body, these products have "brains," which are analogous to ICs. Electronic packaging provides the "nervous system" as well as the "skeletal system." The package is responsible for interconnecting, powering, cooling, and protecting the IC.[3] This concept is illustrated in Figure 10.6.

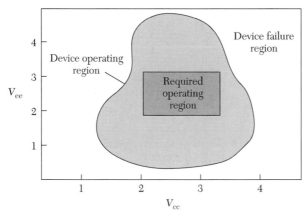

Figure 10.4 Example of two-dimensional voltage shmoo plot for a bipolar IC.[2]

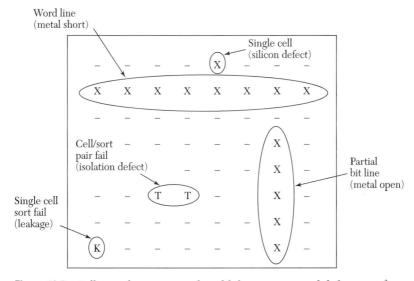

Figure 10.5 Cell map showing examples of failure patterns and defect types.[2]

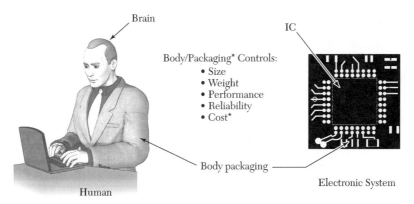

Figure 10.6 Analogy between electronic packaging and the human body.[3]

Overall, electronic systems consist of several levels of packaging, each with distinctive types of interconnection devices. Figure 10.7 depicts this packaging hierarchy.[4] Level 0 consists of on-chip interconnections. Chip-to-printed circuit board or chip-to-module connections comprise Level 2, and board-to-board interconnections comprise Level 3. Levels 4 and 5 consist of connections between subassemblies and between systems (such as computer to printer), respectively.

10.2.1 Die Separation

Following functional testing, individual ICs (or dice) must be separated from the substrate. This is essentially the first step in the packaging process. In a common method that has been used for many years, the substrate wafer is mounted on a holder and scribed in both the x and y directions using a diamond scribe. This is done along scribe borders 75 to 250 μm in width that are formed around the periphery of the dice during fabrication. These borders are aligned with the crystal planes of the substrate if possible. After scribing, the wafer is removed from the holder and placed upside-down on a soft support. A roller is then used to apply pressure, fracturing the wafer along the scribe lines. This must be accomplished with minimal damage to the individual die.

More modern die separation processes use a diamond saw rather than a diamond scribe. In this procedure, the wafer is attached to an adhesive sheet of mylar film. The saw is then used to either scribe the wafer or to cut completely through it. After separation, the dice are removed from the mylar. The separated dice are then ready to be placed into packages.

10.2.2 Package Types

There are a number of approaches to the packaging of single ICs. The *dual in-line package*, or DIP (Fig. 10.8), is the package most people envision when they think of integrated circuits. The DIP was developed in the 1960s, quickly became the primary package for ICs, and has long dominated the electronics packaging market. The DIP can be made

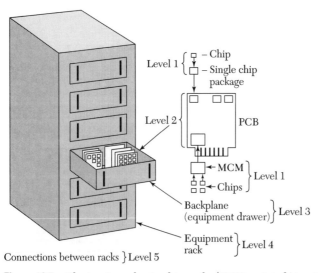

Figure 10.7 Electronic packaging hierarchy.[4] PCB, printed circuit board; MCM, multichip module.

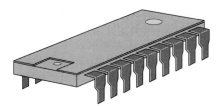

Figure 10.8   Dual in-line package.[4]

of plastic or ceramic, with the latter called the *CerDIP*. The CerDIP consists of a DIP constructed of two pieces of sandwiched ceramic with leads protruding from between the ceramic plates.

In the 1970s and 1980s, *surface mount packages* were developed in response to a need for higher-density interconnect than the DIP approach could provide. In contrast to DIPs, the leads of a surface-mounted package do not penetrate the printed circuit board (PCB) upon which it is mounted. This means that the package can be mounted on both sides of the board, thereby allowing higher density. One example of such a package is the quad flatpack, or QFP (Fig. 10.9), which has leads on all four sides to further increase the number of input/output (I/O) connections.

More recently, the need for greater and greater numbers of I/O connections has led to the development of pin grid array (PGA) and ball grid array (BGA) packages (Figs. 10.10 and 10.11, respectively). PGAs have an I/O density of about 600, and BGAs can have densities greater than 1000, as compared with about 200 for QFPs. BGAs can be identified by the solder bumps on the bottom of the package. With QFPs, as the spacing between leads becomes tighter, the manufacturing yield decreases rapidly. The BGA allows higher density and takes up less space than the QFP, but its manufacturing process is inherently more expensive.

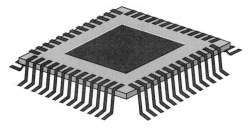

Figure 10.9   Quad flatpack.[4]

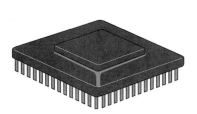

Figure 10.10   Pin grid array package.[4]

Figure 10.11   Ball grid array package.[4]

The most recent development in packaging is the chip scale package (CSP), which is shown in Figure 10.12. CSPs, defined as packages no larger than 20% greater than the size of the IC die itself, often take the form of miniaturized ball grid arrays. They are designed to be flip-chip mounted (see Section 10.2.3) using conventional equipment and solder reflow. CSPs are typically manufactured in a process that creates external power and signal I/O contacts and encapsulates the finished silicon die prior to dicing the wafer. Essentially, CSPs provide an interconnection framework for ICs so that before dicing, each die has all the functions (e.g., external electrical contacts, encapsulation of the finished silicon) of a conventional fully packaged IC. Two essential features of this approach are that the leads and interposer layer (an added layer on the IC used to provide electrical functionality and mechanical stability) are flexible enough so that the packaged device is compliant to the test fixture for full testing and burn-in, and the package can accommodate the vertical nonplanarity and thermal expansion and contraction of the underlying printed circuit board during assembly and operation.

10.2.3 Attachment Methodologies

An IC must be mounted and bonded to a package, and that package must be attached to a printed circuit board before the IC can be used in an electronic system. Methods of attaching ICs are referred to as *Level 1 packaging*. The technique used to bond a bare die to a package has a significant effect on the ultimate electrical, mechanical, and thermal properties of the electronic system being manufactured. Chip-to-package interconnection is generally accomplished by either wire bonding, tape-automated bonding, or flip-chip bonding (see Fig. 10.13).

Wire Bonding
Wire bonding is the oldest attachment method and is still the dominant technique for chips with fewer than 200 I/O connections. Wire bonding requires connecting gold or aluminum wires between chip bonding pads and contact points on the package. ICs are first attached to the substrate using a thermally conductive adhesive, with their bonding

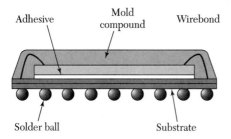

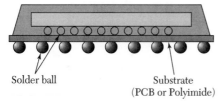

Figure 10.12   Examples of two typical CSPs.[3]

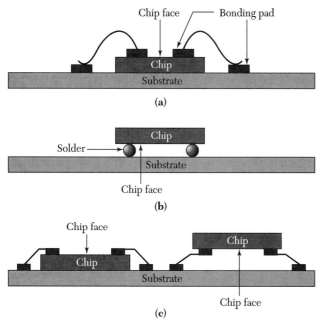

Figure 10.13 Illustration of (*a*) wire, (*b*) flip-chip, and (*c*) tape-automated bonding.[4]

pads facing upward. The Au or Al wires are then attached between the pads and substrate using ultrasonic, thermosonic, or thermocompression bonding.[5] Although automated, this process is still time-consuming since each wire must be attached individually.

In the *thermocompression* technique (Fig. 10.14), a fine wire (15–75 μm diameter) is fed from a spool through a heated capillary. A small hydrogen torch or electric spark then melts the end of the wire, forming a ball. The ball is then positioned over the chip bonding pad, the capillary is lowered, and the ball deforms into a "nail head" due to pressure and heat from the capillary. (The substrate is maintained at a temperature of 150°C to 200°C, and the bonding interface temperature ranges from 280°C to 350°C). Next, the capillary is raised, and wire is fed from the spool and positioned over the package substrate. The bond to the package is a wedge bond produced by deforming the wire with the edge of the capillary. The capillary is then raised and the wire is broken near the edge of the bond.

Oxidation of Al at high temperatures makes it difficult to form a good ball at the end of the wire; in addition, many epoxies cannot withstand the temperatures needed in thermocompression bonding. *Ultrasonic bonding* represents a lower-temperature alternative that relies on a combination of pressure and rapid mechanical vibration to form bonds (Fig. 10.15). In this approach, the wire is fed from a spool through a hole in the bonding tool, which is then lowered into position as an ultrasonic vibration at 20 to 60 kHz causes the metal to deform and flow (even at room temperature). As the tool is raised after the bond to the package is formed, a clamp pulls and breaks the wire.

Thermosonic bonding is a combination of the other two techniques. The substrate temperature is maintained at about 150°C, and ultrasonic vibration and pressure are used to cause the metal to flow under pressure to form a weld. Thermosonic bonders are quite fast—they are capable of producing 5 to 10 bonds per second.

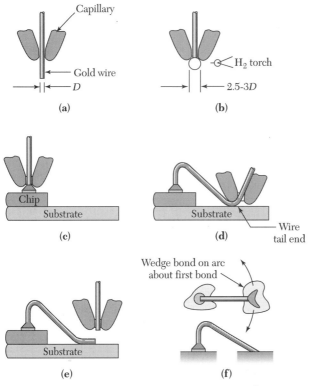

Figure 10.14 Thermocompression bonding process.[5] (*a*) Gold wire in a capillary. (*b*) Ball formation. (*c*) Bonding. (*d*) Wire loop and edge bonding. (*e*) Wire broken at edge. (*f*) Geometry of ball–wedge bond.

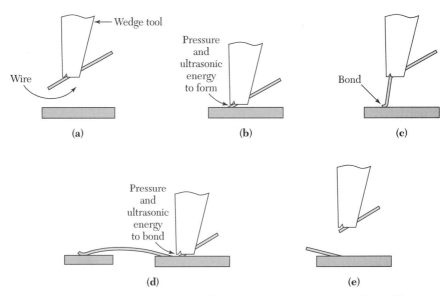

Figure 10.15 Ultrasonic bonding process.[5] (*a*) Tool guides wire to the package. (*b*) Pressure and ultrasonic energy form bond. (*c* and *d*) Tool feeds wire and repositions above the IC. (*e*) Wire broken at bond.

Tape-Automated Bonding

Tape-automated bonding (TAB) was developed in the early 1970s and is often used to bond packages to printed circuit boards. In TAB, chips are first mounted on a flexible polymer tape (usually polyimide) containing repeated copper interconnection patterns (Fig. 10.16). The copper leads are defined by lithography and etching, and the lead pattern can contain hundreds of connections. After aligning the IC pads to metal interconnection stripes on the tape, attachment takes place by thermocompression (Fig. 10.17). Gold bumps are formed on either side of the die or tape and are used to bond the die to the leads on the tape.

A benefit of TAB is that all bonds are formed simultaneously, which significantly improves manufacturing throughput. However, unless all the leads are coplanar, reliability problems can result. TAB also requires multilayer solder bumps with complex metallurgy. Generally, these bumps use gold or copper as the primary constituent, with titanium or tungsten serving as a diffusion barrier to prevent alloying. In addition, a particular tape can only be used for a chip and package that matches its interconnect pattern, thereby rendering TAB an extremely customized process for which the bonding equipment is relatively expensive.

Flip-Chip Bonding

Flip-chip bonding is a direct interconnection approach in which the IC is mounted upside-down onto a module or printed circuit board. Electrical connections are made via solder bumps (or solderless materials such as epoxies or conductive adhesives) located over the surface of the chip. Since bumps can be located anywhere on the chip, flip-chip bonding ensures that the interconnect distance between the chip and package is minimized. The I/O density is limited only by the minimum distance between adjacent bond pads.

In flip-chip processing, chips are placed face down on the module substrate so that I/O pads on the chip are aligned with those on the substrate (Fig. 10.18). A solder reflow process is then used to simultaneously form all the required connections, thereby drastically improving throughput compared with wire bonding. However, the bump fabrication process itself is fairly complex and capital intensive.

Solderless flip-chip technology involves stencil printing of an organic polymer onto an IC, leaving the bond pads uncoated. A high-conductivity organic polymer paste is then stenciled onto the bond pads to form the solderless bumps, which are then cured. The same organic polymer is stenciled onto the bond pads of the substrate. Alignment is then accomplished, and the final bond is formed by applying pressure and heat to the bumps.

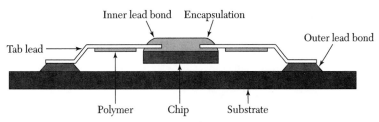

Figure 10.16 Tape-automated bonding.[4]

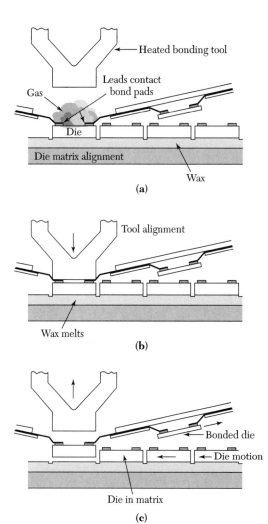

Figure 10.17 TAB procedure.[5] (*a*) Leads lowered into position and aligned above bonding pads. (*b*) Tool descends and performs bonds. (*c*) Tool and film are raised so a new die can be moved into position.

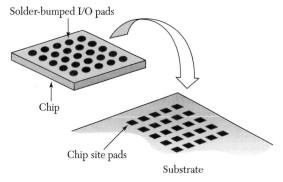

Figure 10.18 Flip-chip bonding.[4]

▶ 10.3 STATISTICAL PROCESS CONTROL

IC manufacturing processes must be stable, repeatable, and of high quality to yield products with acceptable performance. This implies that all individuals involved in manufacturing an IC (including operators, engineers, and management) must continuously seek to improve manufacturing process output and reduce variability. Variability reduction is accomplished in large part by strict process control. This section focuses on *statistical process control* (SPC) techniques as a means to achieve high-quality products.

SPC refers to a powerful collection of problem-solving tools used to achieve process stability and reduce variability. Perhaps the primary and most technically sophisticated of these tools is the control chart. The control chart was developed by Dr. Walter Shewhart of Bell Telephone Laboratories in the 1920s.[6] For this reason, control charts are also often referred to as *Shewhart control charts*.

A control chart is an online SPC technique that is used to detect the occurrence of shifts in process performance so that investigation and corrective action may be undertaken to bring an incorrectly behaving manufacturing process back under control. A typical control chart is shown in Figure 10.19. This chart is a graphical display of a quality characteristic that has been measured from a sample versus the sample number or time. The chart consists of (a) a *center line*, which represents the mean value of the characteristic corresponding to an in-control state; (b) an *upper control limit* (UCL); and (c) a *lower control limit* (LCL). The control limits are selected such that if the process is under statistical control, nearly all the sample points will plot between them. If the variance of the quality characteristic is σ^2 and the standard deviation of the characteristic is σ, then the control limits are typically set at $\pm 3\sigma$ from the center line. Points that plot outside of the control limits are interpreted as evidence that the process is out of control.

10.3.1 Control Charts for Attributes

Some quality characteristics cannot be easily represented numerically. For example, we may be concerned with whether or not a wire bond is defective. In this case, the bond is classified as either defective or nondefective (or equivalently, conforming or nonconforming), and there is no numerical value associated with the quality of the bond. Quality characteristics of this type are referred to as *attributes*.

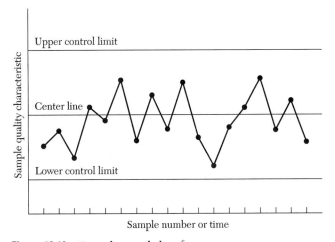

Figure 10.19 Typical control chart.[6]

Two commonly used control charts for attributes are the *defect chart* (or c-*chart*) and the *defect density chart* (or u-*chart*). When a specification is not satisfied in a product, a defect or nonconformity may result. In such cases, it is possible to develop control charts for either the total number of defects or the defect density. These charts assume that the presence of defects in samples of constant size is appropriately modeled by the *Poisson distribution*,[6] in which the probability of a defect occurring is given by

$$P(x) = \frac{e^{-c}c^x}{x!} \tag{1}$$

where x is the number of defects and c is a constant greater than zero. For the Poisson distribution, c is both the mean and variance. Therefore, the c-chart with $\pm 3\sigma$ control limits is given by

$$
\begin{aligned}
\text{UCL} &= c + 3\sqrt{c} \\
\text{Centerline} &= c \\
\text{LCL} &= c - 3\sqrt{c}
\end{aligned} \tag{2}
$$

assuming that c is known. (*Note*: if these calculations yield a negative value for the LCL, the standard practice is to set the LCL equal to 0). If c is not known, it may be estimated from an observed average number of defects in a sample ($\bar{c}$). In this case, the control chart becomes

$$
\begin{aligned}
\text{UCL} &= \bar{c} + 3\sqrt{\bar{c}} \\
\text{Center line} &= \bar{c} \\
\text{LCL} &= \bar{c} - 3\sqrt{\bar{c}}
\end{aligned} \tag{3}
$$

EXAMPLE 1

Suppose the inspection of 25 silicon wafers yields 37 defects. Set up a c-chart for this situation.

SOLUTION We estimate c using

$$\bar{c} = \frac{37}{25} = 1.48$$

This is the center line for the c-chart. The upper and lower control limits can be found from Eq. 3 as follows:

$$
\begin{aligned}
\text{UCL} &= \bar{c} + 3\sqrt{\bar{c}} = 5.13 \\
\text{LCL} &= \bar{c} - 3\sqrt{\bar{c}} = -2.17
\end{aligned}
$$

Since $-2.17 < 0$, we set the LCL equal to 0 in this case.   ◀

Suppose we would like to set up a control chart for the *average* number of defects over a sample size of n products. If there were c total defects among the n samples, then the average number of defects per sample is

$$u = \frac{c}{n} \tag{4}$$

The parameters of a 3σ defect density chart (u-chart) are then given by

$$\text{UCL} = \bar{u} + 3\sqrt{\frac{\bar{u}}{n}}$$

$$\text{Center line} = \bar{u} \qquad (5)$$

$$\text{LCL} = \bar{u} - 3\sqrt{\frac{\bar{u}}{n}}$$

where u is the average number of defects over m groups of sample size n.

EXAMPLE 2

Suppose an IC manufacturer wants to establish a defect density chart. Twenty different samples of size $n = 5$ wafers are inspected, and a total of 183 defects are found. Set up the u-chart for this situation.

SOLUTION We estimate u using

$$\bar{u} = \frac{u}{m} = \frac{c}{mn} = \frac{183}{(20)(5)} = 1.83$$

This is the center line for the u-chart. The upper and lower control limits can be found from Eq. 5 as follows:

$$\text{UCL} = \bar{u} + 3\sqrt{\frac{\bar{u}}{n}} = 3.64$$

$$\text{LCL} = \bar{u} - 3\sqrt{\frac{\bar{u}}{n}} = 0.02 \qquad ◀$$

10.3.2 Control Charts for Variables

In many cases, quality characteristics are expressed as specific numeric measurements, rather than assessing the probability of presence of defects. For example, the thickness of a film is an important characteristic to be measured and controlled. Control charts for continuous variables such as this can provide more information regarding manufacturing process performance than attribute control charts like the c- and u-charts.

When attempting to control continuous variables, it is important to control both the mean and variance of the quality characteristic. This is true because shifts or drifts in either of these parameters can result in significant misprocessing. Control of the mean is achieved using an $\bar{x}$-chart, and variance can be monitored using the standard deviation as in an s-chart. The names of these two charts originate from the *sample mean* ($\bar{x}$) and *sample variance* (s^2), which, respectively, are given by

$$\bar{x} = \frac{x_1 + x_2 + \ldots + x_n}{n} = \frac{1}{n}\sum_{i=1}^{n} x_i \qquad (6)$$

$$s^2 = \frac{1}{n-1}\sum_{i=1}^{n}(x_i - \bar{x})^2 \qquad (7)$$

where $x_1, x_2, \ldots, x_n$ are observations in a sample of size n. The square root of the sample variance is known as the *sample standard deviation* (s).

Suppose m samples of size n are collected. If $\bar{x}_1, \bar{x}_2, \ldots, \bar{x}_m$ are the sample means, the best estimator for the true mean (μ) is the *grand average* ($\bar{\bar{x}}$), which is given by

$$\bar{\bar{x}} = \frac{\bar{x}_1 + \bar{x}_2 + \ldots + \bar{x}_m}{m} \tag{8}$$

Since $\bar{\bar{x}}$ estimates μ, it is used as the center line of the $\bar{x}$-chart. It can also be shown[6] that if a quality characteristic is normally distributed with a known mean μ and standard deviation σ, then $\bar{x}$ is also normally distributed with mean μ and standard deviation $\sigma/\sqrt{n}$. Thus, the center line and control limits for the $\bar{x}$-chart are

$$\text{UCL} = \bar{\bar{x}} + 3\sqrt{\frac{\sigma}{n}}$$
$$\text{Center line} = \bar{\bar{x}} \tag{9}$$
$$\text{LCL} = \bar{\bar{x}} - 3\sqrt{\frac{\sigma}{n}}$$

Since σ is unknown, it must also be estimated by analyzing past data. Caution must be applied in doing so since s itself cannot be used directly as the estimate because s is not an *unbiased* estimator of σ. (The term *unbiased* refers to the situation in which the expected value of the estimator is equal to the parameter being estimated.) Instead s actually estimates $c_4 s$, where c_4 is a statistical parameter that is dependent on the sample size (see Table 10.1). For m samples of size n, the average sample standard deviation is

$$\bar{s} = \frac{1}{m} \sum_{i=1}^{m} s_i \tag{10}$$

It turns out that the statistic $\bar{s} / c_4$ is, in fact, an unbiased estimator of σ.

In addition, the standard deviation of s is $\sigma\sqrt{1 - c_4^2}$. Using this information, the control limits for the s-chart can be set up as follows:

$$\text{UCL} = \bar{s} + 3\frac{\bar{s}}{c_4}\sqrt{1 - c_4^2}$$
$$\text{Center line} = \bar{s} \tag{11}$$
$$\text{LCL} = \bar{s} - 3\frac{\bar{s}}{c_4}\sqrt{1 - c_4^2}$$

When $\bar{s}/c_4$ is used to estimate σ, the limits on the corresponding $\bar{x}$-chart may be defined as

$$\text{UCL} = \bar{\bar{x}} + \frac{3\bar{s}}{c_4\sqrt{n}}$$
$$\text{Center line} = \bar{\bar{x}} \tag{12}$$
$$\text{LCL} = \bar{\bar{x}} - \frac{3\bar{s}}{c_4\sqrt{n}}$$

EXAMPLE 3

Suppose $\bar{x}$- and s-charts are to be established to control linewidth for a lithography process. Twenty-five different samples of size $n = 5$ linewidths are measured. Suppose the grand average for the 125 lines is 4.01 μm. If $\bar{s} = 0.09$ mm, what are the control limits for the s-chart?

TABLE 10.1 c_4 **Parameter for** s**-Chart**

Sample Size (n)	c_4
2	0.7979
3	0.8862
4	0.9213
5	0.9400
6	0.9515
7	0.9594
8	0.9650
9	0.9693
10	0.9727
11	0.9754
12	0.9776
13	0.9794
14	0.9810
15	0.9823
16	0.9835
17	0.9845
18	0.9854
19	0.9862
20	0.9869
21	0.9876
22	0.9882
23	0.9887
24	0.9892
25	0.9896
$n > 25$	$c_4 \cong \dfrac{4(n-1)}{4n-3}$

SOLUTION   The value for c_4 for $n = 5$ (found in Table 10.1) is 0.94. The upper and lower control limits for $\bar{x}$ can be found from Eq. 12 as follows:

$$\text{UCL} = \bar{\bar{x}} + \frac{3\bar{s}}{c_4\sqrt{n}} = 4.14 \ \mu\text{m}$$

$$\text{LCL} = \bar{\bar{x}} - \frac{3\bar{s}}{c_4\sqrt{n}} = 3.88 \ \mu\text{m}$$

The upper and lower control limits for s can be found from Eq. 11 as follows:

$$\text{UCL} = \bar{s} + 3\frac{\bar{s}}{c_4}\sqrt{1-c_4^2} = 0.19 \ \mu\text{m}$$

$$\text{LCL} = \bar{s} - 3\frac{\bar{s}}{c_4}\sqrt{1-c_4^2} \approx 0$$

Note: Since the LCL is (slightly) negative, we automatically set it to zero.   ◄

▶ 10.4 STATISTICAL EXPERIMENTAL DESIGN

Experiments allow investigators to determine the effects of several variables on a given process or product. A *designed experiment* is a test or series of tests that involve purposeful changes to these variables in order to observe the effect of the changes on that process or product. *Statistical experimental design* is an efficient approach for systematically varying these controllable process variables and ultimately determining their impact on process or product quality, or both. This approach is useful for comparing methods, deducing dependencies, and creating models to predict effects.

Statistical process control and experimental design are closely interrelated. Both techniques can be used to reduce variability. However, SPC is a passive approach in which a process is monitored and data collected, whereas experimental design requires active intervention in performing tests on the process under different conditions. Experimental design can also be beneficial in implementing SPC, since designed experiments may help to identify the most influential process variables, as well as their optimum settings.

Overall, experimental design is a powerful engineering tool for improving a manufacturing process. Application of experimental design techniques can lead to improved yield, reduced variability, reduced development time, and reduced cost. Ultimately, the result is enhanced manufacturability, performance, and product reliability. The following sections illustrate the use of experimental design methods in IC fabrication.

10.4.1 Comparing Distributions

Consider the yield data in Table 10.2 obtained from an IC manufacturing process in which two batches of 10 wafers were fabricated using a standard method (Method A) and a modified method (Method B). The question to be answered from the experiment is what evidence (if any) does the data collected provide that Method B is really better than Method A?

To answer this question, we examine the average yields for each process. The modified method (Method B) gave an average yield that was 1.30% higher than the standard method. However, due to the considerable variability in the individual test results, it might not be correct to immediately conclude that Method B is superior to Method A. In fact,

TABLE 10.2 Yield Data from a Hypothetical IC Manufacturing Process

Wafer	Method A Yield (%)	Method B Yield (%)
1	89.7	84.7
2	81.4	86.1
3	84.5	83.2
4	84.8	91.9
5	87.3	86.3
6	79.7	79.3
7	85.1	86.2
8	81.7	89.1
9	83.7	83.7
10	84.5	88.5
Average	84.24	85.54

it is conceivable that the difference observed could be due to experimental error, operator error, or even pure chance.

The proper approach to determine whether the difference between the two manufacturing processes is significant is a *hypothesis test*. A statistical hypothesis is a statement about the values of the parameters of a probability distribution. A hypothesis test is an evaluation of the validity of the hypothesis according to some criterion. Hypotheses are expressed in the following manner:

$$H_0: \mu = \mu_0$$
$$H_1: \mu \neq \mu_0 \tag{13}$$

where the statement $H_0: \mu = \mu_0$ is called the *null hypothesis*, and $H_1: \mu \neq \mu_0$ is called the *alternative hypothesis*. To perform a hypothesis test, we select a random sample from a population, compute an appropriate test statistic, and then either accept or reject the null hypothesis. For the yield experiment, the hypothesis test can be represented as

$$H_0: \mu_A = \mu_B$$
$$H_1: \mu_A \neq \mu_B \tag{14}$$

where μ_A and μ_B represent the mean yields for the two methods.

To evaluate this hypothesis, a *test statistic* is required. The appropriate test statistic in this case is[6]

$$t_0 = \frac{(\bar{y}_A - \bar{y}_B)}{s_p \sqrt{\dfrac{1}{n_A} + \dfrac{1}{n_B}}} \tag{15}$$

where $\bar{y}_A$ and $\bar{y}_B$ are the sample means of the yields for each method, n_A and n_B are the number of trials in each sample (10 each in this case), and

$$s_p^2 = \frac{(n_A - 1)s_A^2 + (n_B - 1)s_B^2}{n_A + n_B - 2} \tag{16}$$

which is referred to as the *pooled estimate of the common variance* of the two processes. The demoninator of Eq. 16 is called the number of *degrees of freedom* for the hypothesis test. The values of the sample variances are calculated using Eq. 7: $s_A = 2.90$ and $s_B = 3.65$. Using Eqs. 16 and 15 then gives values of $s_p = 3.30$ and $t_0 = 0.88$, respectively.

We can use Appendix K to determine the probability of computing a given t statistic with a certain number of degrees of freedom. This probability is represented by the shaded region in the figure in the appendix. Interpolating from Appendix K, we find that the likelihood of computing a t statistic with $n_A + n_B - 2 = 18$ degrees of freedom equal to 0.88 is 0.195. The value 0.195 is the *statistical significance* of the hypothesis test. This means that there is only a 19.5% chance that the observed difference between the mean yields is due to pure chance. In other words, we can be 80.5% confident that Method B is really superior to Method A.

10.4.2 Analysis of Variance

The previous example shows how we might use hypothesis testing to compare two distributions. However, it is often important in IC manufacturing applications to be able to compare several distributions. Moreover, we might also be interested in determining which process conditions in particular have a significant impact on process quality. *Analysis of variance* (ANOVA) is an excellent technique for accomplishing these objectives. ANOVA builds on the idea of hypothesis testing and allows us to compare different sets of process conditions (i.e., treatments), as well as to determine whether a given treatment results in a statistically significant variation in quality.

The ANOVA procedure is best illustrated by example. In the following discussion, consider the data in Table 10.3, which represents hypothetical defect densities measured on wafers fabricated using four different sets of process recipes (labeled 1 through 4). Through the use of ANOVA, we will determine whether the discrepancies *between* recipes (i.e, treatments) is truly greater than the variation of the via diameters *within* the individual groups of vias processed with the same recipe.

Let k be the number of treatments ($k = 4$ in this case). Note that the sample size (n) for each treatment varies ($n_1 = 4$, $n_2 = n_3 = 6$, and $n_4 = 8$). The treatment means (in cm^{-2}) are as follows: $\bar{y}_1 = 61$, $\bar{y}_2 = 66$, $\bar{y}_3 = 68$, and $\bar{y}_4 = 61$. The total number of samples (N) is 24, and the grand average of all 24 samples is $\bar{\bar{y}} = 64$ cm^{-2}.

Sums of Squares

To perform ANOVA, several key parameters must be computed. These parameters, called *sums of squares*, serve to quantify deviations within and between different treatments. Let y_{ti} represent the ith observation for the tth treatment. The sum of squares within the tth treatment is given by

$$S_t = \sum_{1}^{n_t} \left(y_{ti} - \bar{y}_t \right)^2 \tag{17}$$

where n_t is the sample size for the treatment in question and y_t is the treatment mean. The *within-treatment sum of squares* for all treatments is

$$S_R = S_1 + S_2 + \ldots + S_k = \sum_{t=1}^{k} \sum_{t=1}^{n_t} \left(y_{ti} - \bar{y}_t \right)^2 \tag{18}$$

To quantify the deviations of the treatment averages from the grand average, we use the *between-treatment sum of squares*, which is given by

$$S_T = \sum_{t=1}^{k} n_t \left(\bar{y}_t - \bar{\bar{y}} \right)^2 \tag{19}$$

Finally, the total sum of squares for all the data about the grand average is

$$S_D = \sum_{t=1}^{k} \sum_{t=1}^{n_t} \left(\bar{y}_{ti} - \bar{\bar{y}} \right)^2 \tag{20}$$

TABLE 10.3   Hypothetical Defect Densities (in cm^{-2}) for Four Different Process Recipes

Recipe 1	Recipe 2	Recipe 3	Recipe 4
62	63	68	56
60	67	66	62
63	71	71	60
59	64	67	61
	65	68	63
	66	68	64
			63
			59

Each sum of squares has an associated number of degrees of freedom required for its computation. The degrees of freedom for the within-treatment, between-treatment, and total sums of squares, respectively, are

$$
\begin{aligned}
v_R &= N - k \\
v_T &= k - 1 \\
v_D &= N - 1
\end{aligned}
\tag{21}
$$

The final quantity needed to carry out analysis of variance is the pooled estimate of the variance quantified by each sum of squares. This quantity, known as the *mean square*, is the ratio of the sum of squares to its associated number of degrees of freedom. The within-treatment, between-treatment, and total mean squares are therefore

$$
s_R^2 = \frac{S_R}{v_R} = \frac{\sum_{t=1}^{k} \sum_{t=1}^{n_t} (y_{ti} - \bar{y}_t)^2}{N - k}
$$

$$
s_T^2 = \frac{S_T}{v_T} = \frac{\sum_{t=1}^{k} n_t (\bar{y}_t - \bar{\bar{y}})^2}{k - 1}
\tag{22}
$$

$$
s_D^2 = \frac{S_D}{v_D} = \frac{\sum_{t=1}^{k} \sum_{t=1}^{n_t} (y_{ti} - \bar{\bar{y}})^2}{N - 1}
$$

ANOVA Table

Once the parameters just described have been computed, it is customary to arrange them in a tabular format called an *ANOVA table.* The general form of the ANOVA table is depicted in Table 10.4. The ANOVA table that corresponds to the defect density data in Table 10.3 is shown in Table 10.5. Note that in both the Sum of Squares and Degrees of Freedom columns, the values for between and within treatments add up to give the corresponding total value. This additive property of the sum of squares arises from the algebraic identity

$$
\sum_{t=1}^{k} \sum_{i=1}^{n_t} (y_{ti} - \bar{\bar{y}})^2 = \sum_{t=1}^{k} n_t (\bar{y}_t - \bar{\bar{y}})^2 + \sum_{t=1}^{k} \sum_{i=1}^{n_t} (y_{ti} - \bar{y}_i)^2
\tag{23}
$$

or equivalently,

$$
S_D = S_T + S_R
$$

The complete ANOVA table provides a mechanism for testing the hypothesis that all of the treatment means are equal. The null hypothesis in this case is thus

$$
H_0: \mu_1 = \mu_2 = \mu_3 = \mu_4
$$

TABLE 10.4   General Format of the ANOVA Table

Source of Variation	Sum of Squares	Degrees of Freedom	Mean Square	F Ratio
Between treatments	S_T	$v_T = k - 1$	s_T^2	s_T^2 / s_R^2
Within treatments	S_R	$v_R = N - k$	s_R^2	
Total	S_D	$v_D = N - 1$	s_D^2	

TABLE 10.5   ANOVA Table for Via Diameter Data

Source of Variation	Sum of Squares	Degrees of Freedom	Mean Square	F Ratio
Between treatments	$S_T = 228$	$\nu_T = 3$	$s_T^2 = 76.0$	$s_T^2/s_R^2 = 13.6$
Within treatments	$S_R = 112$	$\nu_R = 20$	$s_R^2 = 5.6$	
Total	$S_D = 340$	$\nu_D = 23$	$s_D^2 = 14.8$	

If the null hypothesis were true, the ratio s_T^2/s_R^2 would follow the F distribution with ν_T and ν_R degrees of freedom. Interpolating from Appendix L, the *significance level* (i.e., the first subscript of the F values in the tables) for the observed F ratio of 13.6 with 3 and 30 degrees of freedom is 0.000046. This means that there is only a 0.0046% chance that the means are in fact equal, and the null hypothesis is discredited. In other words, we can be 99.9954% sure that real differences exist among the four different processes in our example.

10.4.3   Factorial Designs

Experimental design is an organized method of conducting experiments to extract the maximum amount of information from a limited number of experiments. Experimental design techniques are employed in manufacturing to systematically and efficiently explore the effects of a set of input variables, or *factors* (such as processing temperature), on responses (such as yield). The unifying feature in statistically designed experiments is that all factors are varied simultaneously, as opposed to the more traditional "one-variable-at-a-time" technique. A properly designed experiment can minimize the number of experimental runs that would otherwise be required if this approach or random sampling were used.

Factorial experimental designs are of great practical importance for IC manufacturing applications. To perform a factorial experiment, an investigator selects a fixed number of *levels* for each of a number of variables (factors) and runs experiments at all possible combinations of the levels. Two of the most important issues in factorial experimental designs are choosing the set of factors to be varied in the experiment and specifying the ranges over which variation will take place. The choice of the number of factors directly affects the number of experimental runs (and therefore the cost of the experiment).

Two-Level Factorials

The ranges of the process variables investigated in factorial experiments can be discretized into minimum, maximum, and "center" levels. In a *two-level factorial design*, the minimum and maximum levels of each factor (normalized to take on values −1 and +1, respectively) are used together in every possible combination. Thus, a full two-level factorial experiment with n factors requires 2^n experimental runs. The various factor level combinations of a three-factor experiment can be represented pictorially as the vertices of a cube, as shown in Figure 10.20.

Table 10.6 shows a 2^3 factorial experiment for a CVD process. The three factors are temperature (T), pressure (P), and gas flow rate (F). The response being measured is the deposition rate (D) in angstroms per minute. The highest and lowest levels of each factor are represented by the + and − signs, respectively. The display of levels depicted in the first three columns of this table is called a *design matrix*.

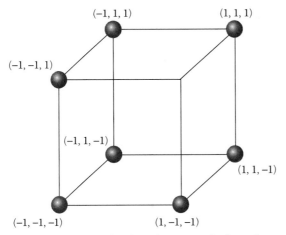

Figure 10.20 Factor level combinations of a three-factor experiment represented as the vertices of a cube.

TABLE 10.6 Two-Level Factorial Experiment

Run	P	T	F	D (Å/min)
1	−	−	−	$d_1 = 94.8$
2	+	−	−	$d_2 = 110.96$
3	−	+	−	$d_3 = 214.12$
4	+	+	−	$d_4 = 255.82$
5	−	−	+	$d_5 = 94.14$
6	+	−	+	$d_6 = 145.92$
7	−	+	+	$d_7 = 286.71$
8	+	+	+	$d_8 = 340.52$

What can we determine from this factorial design? Furthermore, what does the data collected tell us about the effect of pressure on deposition rate? The effect of any single variable on the response is called a *main effect*. The method used to compute such a main effect is to find the difference between the average deposition rate when the pressure is high (i.e., runs 2, 4, 6, and 8) and the average deposition rate when the pressure is low (runs 1, 3, 5, and 7). Mathematically, this is expressed as

$$P = d_{p+} - d_{p-} = 1/4[(d_2 + d_4 + d_6 + d_8) - (d_1 + d_3 + d_5 + d_7)] = 40.86 \qquad (24)$$

where P is the main effect for pressure, d_{p+} is the average deposition rate when the pressure is high, and d_{p-} is the average deposition rate when the pressure is low. We interpret this result as showing that the average effect of increasing pressure from its lowest to its highest level is to increase the deposition rate by 40.86 Å/min. The other main effects for temperature and flow rate are computed in a similar manner. In general, the main effect for each variable in a two-level factorial experiment is the difference between the two averages of the response (y), or

$$\text{Main effect} = y_+ - y_- \qquad (25)$$

We are also interested in quantifying how two or more factors *interact*. For example, suppose that the pressure effect is much greater at high temperatures than it is at low temperatures. A measure of this interaction is provided by the difference between the average pressure effect with temperature high and the average pressure effect with temperature low. By convention, *half* of this difference is called the *pressure-by-temperature interaction*, or symbolically, the $P \times T$ interaction. This interaction may also be thought of as one-half the difference in the average temperature effects at the two levels of pressure. Mathematically, this is

$$P \times T = d_{PT+} - d_{PT-} = 1/4[(d_1 + d_4 + d_5 + d_8) - (d_2 + d_3 + d_6 + d_7)] = 6.89 \qquad (26)$$

The $P \times F$ and $T \times F$ interactions are obtained in a similar fashion. Finally, we might also be interested in the interaction of all three factors, denoted as the *pressure-by-temperature-by flow rate* or the $P \times T \times F$ interaction. This interaction defines the average difference between any two-factor interaction at the high and low levels of the third factor. It is given by

$$P \times T \times F = d_{PTF+} - d_{PTF-} = -5.88 \qquad (27)$$

It is important to note that the main effect of any factor can be individually interpreted only if there is no evidence that the factor interacts with other factors.

The Yates Algorithm

It can be tedious to calculate the effects and interactions for two-level factorial experiments using the method just described, particularly if there are more than three factors involved. Fortunately, the *Yates algorithm* provides a quicker method of computation that is also relatively easily programmed via computer. To implement this algorithm, the experimental design matrix is first arranged in what is called *standard order*. A 2^n factorial design is in standard order when the first column of the design matrix consists of alternating minus and plus signs, the second column consists of successive pairs of minus and plus signs, the third column consists of four minus signs followed by four plus signs, and so on. In general, the kth column consists of 2^{k-1} minus signs followed by 2^{k-1} plus signs.

The Yates calculations for the deposition rate data are shown in Table 10.7. Column y contains the deposition rates for each run. These are considered in successive pairs. The first four entries in column (1) are obtained by adding the pairs together, and the next four are obtained by subtracting the top number from the bottom number of each pair. Column (2) is obtained from column (1) in the same way, and column (3) is obtained from column (2). To obtain the experimental effects, one need only divide the column

TABLE 10.7   Illustration of the Yates Algorithm

P	T	F	y	(1)	(2)	(3)	Divisor	Effect	ID
−	−	−	94.8	205.76	675.70	1543.0	8	192.87	Avg
+	−	−	110.96	469.94	867.29	163.45	4	40.86	P
−	+	−	214.12	240.06	57.86	651.35	4	162.84	T
+	+	−	255.82	627.23	105.59	27.57	4	6.89	PT
−	−	+	94.14	16.16	264.18	191.59	4	47.90	F
+	−	+	145.92	41.70	387.17	47.73	4	11.93	PF
−	+	+	286.71	51.78	25.54	122.99	4	30.75	TF
+	+	+	340.52	53.81	2.03	−23.51	4	−5.88	PTF

(3) entries by the Divisor column entries. In general, the first divisor will be 2^n, and the remaining divisors will be 2^{n-1}. The first element in the Identification (ID) column is the grand average of all of the observations, and the remaining identifications are derived by locating the plus signs in the design matrix.

Although the Yates algorithm provides a relatively straightforward methodology for computing experimental effects, it should be pointed out that modern analysis of statistical experiments is accomplished almost exclusively by commercially available statistical software packages. A few of the more common packages include RS/1, SAS, and Minitab. These packages completely alleviate the necessity of performing any tedious hand calculations.

Fractional Factorial Designs

A disadvantage of the two-level factorial design is that the number of experimental runs increases exponentially with the number of factors. To alleviate this concern, *fractional factorial* designs are constructed by systematically eliminating some of the runs in a full factorial design. For example, a half fractional design with n factors requires only 2^{n-1} runs. Full or fractional two-level factorial designs can be used to estimate the main effects of individual factors as well as the interaction effects between factors. However, they cannot be used to estimate quadratic or higher-order effects. This is not a serious shortcoming, since higher-order effects and interactions tend to be smaller than low-order effects (i.e., main effects tend to be larger than two-factor interactions, which tend to be larger than three-factor interactions, etc.). Ignoring high-order effects is conceptually similar to ignoring higher-order terms in a Taylor series expansion.

To illustrate the use of fractional factorial designs, let $n = 5$ and consider a 2^5 factorial design. The full factorial implementation of this design would require 32 experimental runs. However, a 2^{5-1} fractional factorial design only requires 16 runs. This 2^{5-1} design is generated by first writing the design matrix for a 2^4 full factorial design in standard order. Then plus and minus signs in the four columns of the 2^4 design matrix are each "multiplied" together to form a fifth column.

For example, let's take another look at our CVD experiment. Suppose we only have the time or resources available to perform four deposition experiments, rather than the eight required for a 2^3 full factorial design. This calls for a 2^{3-1} fractional factorial alternative. This new design could be generated by writing the full 2^2 design for the pressure and temperature variables, and then multiplying those columns to obtain a third column for flow rate. This procedure is illustrated in Table 10.8. The only drawback in using this procedure is that since we have used the PT relation to define column F, we can no longer distinguish between the effects of the $P \times T$ interaction and the F main effect. When this occurs, the two effects are said to be *confounded*.

TABLE 10.8 Illustration of 2^{3-1} Fractional Factorial Design for CVD Example

Run	P	T	F
1	−	−	+
2	+	−	−
3	−	+	−
4	+	+	+

▶ 10.5 YIELD

Variability in IC manufacturing processes can lead to deformations or nonconformities in finished products. Such process disturbances often result in *faults*, or unintentional changes in the performance or conformance of electronic products. The presence of such faults is quantified by the manufacturing yield. *Yield* is defined as the percentage of devices or circuits that meet a nominal performance specification.

Yield can be categorized as either *functional* or *parametric*. Functional yield is determined by the proportion of fully functional products. Often referred to as hard yield, the functional yield of ICs is usually characterized by open circuits or short circuits caused by physical defects (such as particles). In some cases, however, a fully functional product still fails to meet performance specifications for one or more parameters (such as speed, noise level, or power consumption). These situations are described by parametric yield (or soft yield).

10.5.1 Functional Yield

The development of models to estimate the functional yield of ICs is fundamental to manufacturing. A model that provides accurate estimates of manufacturing yield can help predict product cost, determine optimum equipment utilization, or be used as a metric against which actual measured manufacturing yields can be evaluated. Yield models are also critical to support decisions involving new technologies and the identification of problematic products or processes.

As previously mentioned, functional yield is significantly affected by the presence of defects. Defects can result from many random sources, including contamination from equipment, processes or handling, mask imperfections, and airborne particles. Physically, these defects include shorts, opens, misalignment, photoresist splatters and flakes, pinholes, scratches, and crystallographic flaws. This is illustrated by Figure 10.21.

Yield models are usually presented as a function of the average number of defects per unit area (D_0) and the *critical area* (A_c) of the electronic system. In other words,

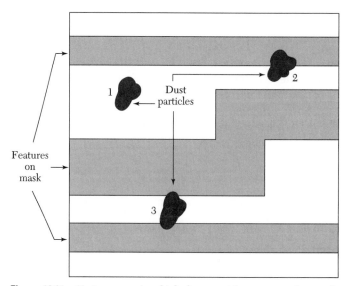

Figure 10.21   Various ways in which dust particles can interfere with interconnect mask patterns.

$$Y = f(A_c, D_0) \tag{28}$$

where Y is the functional yield. The critical area is the area in which a defect occurring has a high probability of resulting in a fault. For example, if particle 3 in Figure 10.21 is large enough and conductive, it has fallen into an area in which it causes a short between the two metal lines it bridges. The relationship between the yield, defect density, and critical area is complex. It depends on the circuit geometry, the density of photolithographic patterns, the number of photolithography steps used in the manufacturing process, and other factors. A few of the more common models that attempt to quantify this relationship are described next.

Poisson Model

The Poisson yield model assumes that defects are uniformly distributed across a substrate, and that each defect results in a fault. Pineda de Gyvez provides an excellent derviation of this model.[7] Let C be the number of circuits on a substrate (i.e., the number of ICs), and let M be the number of possible defect types. Under these conditions, there are C^M unique ways in which the M defects can be distributed on the C circuits. For example, if there are three circuits (C1, C2, and C3) and three defect types (such as M1 = metal open, M2 = metal short, and M3 = metal 1 to metal 2 short, for example), then there are

$$C^M = 3^3 = 27 \tag{29}$$

possible ways in which these three defects can be distributed over three chips. These combinations are illustrated in Table 10.9.

If one circuit is removed (i.e., is found to contain no defects), the number of ways to distribute the M defects among the remaining circuits is

$$(C - 1)^M \tag{30}$$

Thus, the probability that a circuit will contain zero defects of any type is

$$\frac{(C-1)^M}{C^M} = \left(1 - \frac{1}{C}\right)^M \tag{31}$$

TABLE 10.9 Truth Table for Unique Fault Combinations

Combination	C1	C2	C3	Combination	C1	C2	C3
1	M1M2M3			15	M3		M2M1
2		M1M2M3		16		M1M2	M3
3			M1M2M3	17		M1M3	M2
4	M1M2	M3		18		M2M3	M1
5	M1M3	M2		19		M1	M2M3
6	M2M3	M1		20		M2	M1M3
7	M1M2		M3	21		M3	M2M1
8	M1M3		M2	22	M1	M2	M3
9	M2M3		M1	23	M1	M3	M2
10	M1	M2M3		24	M2	M1	M3
11	M2	M1M3		25	M2	M3	M1
12	M3	M2M1		26	M3	M1	M2
13	M1		M2M3	27	M3	M2	M1
14	M2		M1M3				

Substituting $M = CA_cD_0$, the yield is the number of circuits with zero defects, or

$$Y = \lim_{C \to \infty}\left(1 - \frac{1}{C}\right)^{CA_cD} = \exp(-A_cD_0) \tag{32}$$

For N circuits to have zero defects, this becomes

$$Y = \exp(-A_cD_0)^N = \exp(-NA_cD_0) \tag{33}$$

The Poisson model is simple and relatively easy to derive. It provides a reasonably good estimate of yield when the critical area is small. However, if D_0 is calculated based on small-area circuits, using the same D_0 for large-area yield computations results in a yield estimate that is overly pessimistic compared with actual measured data.

Murphy's Yield Integral

B. T. Murphy first proposed that the value of the defect density (D) should not be constant.[8] Instead, he reasoned that D must be summed over all circuits and substrates using a normalized probability density function, $f(D)$. The yield can then be calculated using the integral

$$Y = \int_0^\infty e^{-A_cD}f(D)\,dD \tag{34}$$

Various forms of $f(D)$ form the basis for the differences between many analytical yield models. The Poisson model assumes that $f(D)$ is a delta function, that is,

$$f(D) = \delta(D - D_0) \tag{35}$$

where D_0 is the average defect density as before (see Fig. 10.22a). Using this density function, the yield is determined from Eq. 34 to be

$$Y_{\text{Poisson}} = \int_0^\infty e^{-A_cD}f(D)\,dD = \exp(-A_cD_0) \tag{36}$$

as shown before.

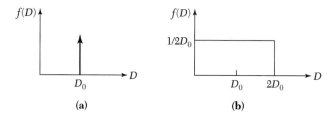

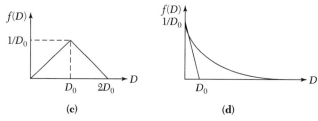

Figure 10.22 Probability density functions for (a) the Poisson model, (b) the uniform Murphy model, (c) the triangular Murphy model, and (d) the exponential Seeds model.[7]

Murphy initially investigated a uniform density function, as shown in Figure 10.22b. The evaluation of the yield integral for the uniform density function gives

$$Y_{\text{uniform}} = \frac{1 - e^{-2D_0 A_c}}{2D_0 A_c} \tag{37}$$

Murphy later believed that a Gaussian distribution would be a better reflection of the true defect density distribution than the delta function. However, since he was unable to integrate the yield integral with a Gaussian function substituted for $f(D)$, he approximated it using the triangular function shown in Figure 10.22c. This function results in the yield expression

$$Y_{\text{triangular}} = \left(\frac{1 - e^{D_0 A_c}}{2D_0 A_c} \right)^2 \tag{38}$$

The triangular Murphy yield model is widely used today in industry to determine the effect of manufacturing process defect density.

R. B. Seeds was the first to verify Murphy's predictions.[9] However, Seeds theorized that high yields were caused by a large population of low defect densities (which are not high enough to cause faults) and a small proportion of high defect densities (i.e., high enough to cause faults). He therefore proposed the exponential density function given by

$$f(D) = \frac{1}{D_0} \exp\left(\frac{-D}{D_0} \right) \tag{39}$$

and shown in Figure 10.22d. This function implies that the probability of observing a low defect density is significantly higher than that of observing a high defect density. Substituting this exponential function in the Murphy integral and integrating yields

$$Y_{\text{exponential}} = \frac{1}{1 + D_0 A_c} \tag{40}$$

Although the Seeds model is simple, its yield predictions for large-area substrates are too optimistic. Therefore, this model has not been widely used.

Okabe, Nagata, and Shimada recognized the physical nature of defect distributions and proposed the gamma probability density function.[10] Stapper has likewise developed and applied yield models using the gamma density function.[11] The gamma distribution is given by

$$f(D) = \left[\Gamma(\alpha)\beta^\alpha \right]^{-1} D^{\alpha-1} e^{-D/\beta} \tag{41}$$

where α and β are two parameters of the distribution, and $\Gamma(\alpha)$ is the gamma function. The shape of $\Gamma(\alpha)$ is shown for several values of α in Figure 10.23. In this distribution, the average defect density is $D_0 = \alpha\beta$.

The yield model derived by substituting Eq. 41 into Murphy's integral is

$$Y_{\text{gamma}} = \left(1 + \frac{A_c D_0}{\alpha} \right)^{-\alpha} \tag{42}$$

This model is commonly referred to as the *negative binomial* model. The parameter α must be empirically determined. It is generally called the cluster parameter because it increases with decreasing variance in the distribution of defects. If α is high, then the variability of defects is low (little clustering). Under these conditions, the gamma density function approaches a delta function, and the negative binomial model reduces to the Poisson model. Mathematically, this means

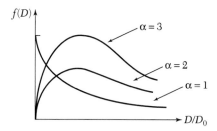

Figure 10.23 Probability density function for the gamma distribution.

$$Y = \lim_{\alpha \to \infty}\left(1 + \frac{A_c D_0}{\alpha}\right)^{-\alpha} = \exp(-A_c D_0) \tag{43}$$

If α is low, on the other hand, the variability of defects across the wafer is significant (much clustering), and the gamma model reduces to the Seeds exponential model, or

$$Y = \lim_{\alpha \to 0}\left(1 + \frac{A_c D_0}{\alpha}\right)^{-\alpha} = \frac{1}{1 + A_c D_0} \tag{44}$$

If the critical area and defect density are known (or can be accurately measured), the negative binomial model is an excellent general-purpose yield predictor that can be used for a variety of IC manufacturing processes.

10.5.2 Parametric Yield

Even in a defect-free manufacturing environment, random processing variations can lead to varying levels of system performance. These variations result from the fluctuation of numerous physical and environmental parameters (linewidths, film thicknesses, ambient humidity, etc.), which in turn manifest themselves as variations in final system performance (such as speed or noise level). These performance variations lead to "soft" faults and are characterized by the *parametric yield* of the manufacturing process. Parametric yield is a measure of the quality of functioning systems, whereas functional yield measures the proportion of functioning units produced by the manufacturing process.

A common method used to evaluate parametric yield is *Monte Carlo simulation.* In the Monte Carlo approach, a large number of pseudo-random sets of values for circuit or system parameters are generated according to an assumed probability distribution (usually the normal distribution) based on sample means and standard deviations extracted from measured data. For each set of parameters, a simulation is performed to obtain information about the predicted behavior of a circuit or system. The overall performance distribution is then extracted from the set of simulation results.

To illustrate the Monte Carlo technique, consider as a performance metric the drive current of an n-channel MOSFET in saturation (I_{Dsat}). It can be shown that[12]

$$I_{Dsat} \cong \left(\frac{Z\mu_n C_o}{2L}\right)(V_G - V_T)^2 \tag{45}$$

where Z is the width of the device, L is its length, μ_n is the electron mobility in the channel, C_o is the oxide capacitance per unit area, V_G is the applied gate voltage, and V_T is the threshold voltage of the transistor. In this equation, C_o is a function of the thickness of the oxide (d), and V_T is a function of the oxide thickness as well as the doping in the channel, or $I_{Dsat} = f(C_o, V_T)$. Both of these dimensions are subject to manufacturing process variations. They can thus be characterized as varying according to normal distributions with means μ_c and μ_v and standard deviations σ_c and σ_v, respectively (see Fig. 10.24).

Using the Monte Carlo approach, we can estimate the parametric yield of MOSFETs produced by a given manufacturing process within a certain range of saturation drain currents by computing the value of I_{Dsat} for every possible combination of C_o and V_T. The result of these computations is a final performance distribution like the one shown in Figure 10.24b. This probability density function can then be used to compute the proportion of transistors having a given range of drive currents. For example, if we wanted to compute the percentage of MOSFETs manufactured that would have a value of I_{Dsat} between two limits a and b, we would evaluate the integral

$$Y\left(\text{MOSFETs with } a < I_{Dsat} < b\right) = \int_a^b f(x)dx \tag{46}$$

Thus, once the overall distribution of a given output metric is known, it is possible to estimate the fraction of manufactured parts with any range of performance. Estimation of parametric yield is useful for circuit designers because it helps identify the limits of the manufacturing process to facilitate and encourage design for manufacturability.

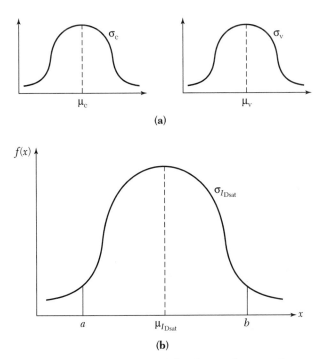

Figure 10.24 (a) Normal probability density functions for C_o and V_T. (b) Overall probability density function for I_{Dsat}.

▶ 10.6 COMPUTER-INTEGRATED MANUFACTURING

The vast majority of quantitative evaluation of IC manufacturing processes is accomplished via computers rather than by hand calculations. Not only is this more efficient, but in today's economy, it is essential. The fabrication of integrated circuits can be quite expensive. In fact, the last decade has seen electronics manufacturing become so capital intensive that small companies often find it too expensive to support their own manufacturing operations. A typical state-of-the-art high-volume manufacturing facility today costs several orders of magnitude more than a comparable facility 20 years ago. This has led to the rise of a large contract manufacturing industry.

As a result of rising costs, the challenge before manufacturers today is to offset such large capital investment with a greater amount of technological innovation in the fabrication process. In other words, the objective now is to make use of the latest developments in computer hardware and software technology to enhance manufacturing methods that have become prohibitively expensive. In effect, this effort in *computer-integrated manufacturing* of integrated circuits (IC-CIM) is aimed at optimizing the cost-effectiveness of electronics manufacturing in the same manner in which *computer-aided design* (CAD) has dramatically affected the economics of circuit design.

Under the overall heading of reducing manufacturing cost, several subtasks have been identified. These include increasing fabrication yield, reducing product cycle time, maintaining consistent levels of product quality and performance, and improving the reliability of processing equipment. Since fabrication processes often consist of hundreds of sequential steps, yield loss may potentially occur at every step. Consequently, maintaining product quality in an electronics manufacturing facility requires the strict control of literally hundreds or even thousands of process variables. The interdependent issues of high yield, high quality, and low cycle time can be addressed by the development of several critical capabilities in a state-of-the-art IC-CIM system: work-in-process (WIP) monitoring, equipment communication, data acquisition and storage, process/equipment modeling, and real-time process control, to name a few. The emphasis of each of these activities is to increase throughput and reduce yield loss by preventing potential misprocessing, but each presents significant engineering challenges in its effective implementation and deployment.

A block diagram of a typical modern IC-CIM system is shown in Figure 10.25. This diagram outlines many of the key features required for efficient manufacturing operations.[13]

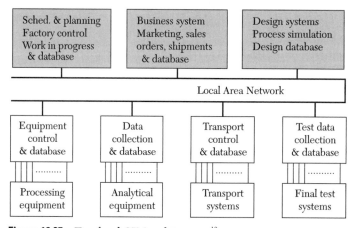

Figure 10.25 Two-level CIM architecture.[13]

The lower level of this two-level architecture includes embedded controllers that provide real-time control and analysis of fabrication equipment. These controllers often consist of personal computers and the associated control software dedicated to each individual piece of equipment. The second level of this IC-CIM architecture is composed of a distributed local area network of computer workstations and file servers linked by a common distributed database. Equipment communication with host computers is facilitated by an electronics manufacturing standard called the *generic equipment model* (GEM). The GEM standard is used in both semiconductor manufacturing and printed circuit board assembly. This standard is based on the *semiconductor equipment communications standard* (SECS) protocol.

This type of IC-CIM architecture has great flexibility, allowing extension and adaptation to meet constantly changing requirements. Over the past several years, powerful, flexible, and cost-effective information systems based on models such as this have become an integral part of the IC manufacturing enterprise.

▶ 10.7 SUMMARY

This chapter provided an overview of the relevant issues in IC manufacturing. This included a description of electrical testing and basic packaging processes, as well as a presentation of statistical process control, statistical experimental design, and yield modeling. The chapter concluded with a brief introduction to IC-CIM systems. In IC manufacturing, process and equipment reliability directly influence throughput, yield, and ultimately cost. Over the next several years, significant enhancement of manufacturing operations will be required to reach projected targets for future generations of microelectronic devices, packages, and systems.

▶ REFERENCES

1. Pineda de Gyvez and D. Pradhan, *Integrated Circuit Manufacturability*, IEEE Press, Piscataway, NJ, 1999.

2. A. Landzberg, *Microelectronics Manufacturing Diagnostics Handbook*, Van Nostrand Reinhold, New York, 1993.

3. R. Tummala, Ed., *Fundamentals of Microsystems Packaging*, McGraw-Hill, New York, 2001.

4. W. Brown, Ed., *Advanced Electronic Packaging*, IEEE Press, New York, 1999.

5. R. Jaeger, *Introduction to Microelectronic Fabrication*, 2nd Ed., Prentice-Hall, Upper Saddle River, NJ, 2002.

6. D. Montgomery, *Introduction to Statistical Quality Control*, Wiley, New York, 1985.

7. J. Pineda de Gyvez and D. Pradhan, *Integrated Circuit Manufacturability*, IEEE Press, New York, 1999.

8. B. Murphy, "Cost-Size Optima of Monolithic Integrated Circuits," *Proc. IEEE*, **52**, (12), 1537–1545 (1964).

9. R. Seeds, "Yield and Cost Analysis of Bipolar LSI," *IEEE Int. Electron Devices Meet.*, Washington, DC, October 1967.

10. T. Okabe, M. Nagata, and S. Shimada, "Analysis of Yield of Integrated Circuits and a New Expression for the Yield," in C. Strapper, Ed., *Defect and Fault Tolerance in VLSI Systems*, Vol. 2, Plenum Press, New York, pp. 47–61, 1990.

11. C. Stapper, "Fact and Fiction in Yield Modeling," *Microelectronics J.*, **210**, 129–151 (1989).

12. S. Sze, *Semiconductor Devices: Physics and Technology*, 2nd Ed., Wiley, New York, 2002.

13. D. Hodges, L. Rowe, and C. Spanos, "Computer Integrated Manufacturing of VLSI," Proceedings of the 11th IEEE/CHMT International Electronics Manufacturing Technology Symposium, September 1989, pp. 1–3.

▶ PROBLEMS

Asterisks denote difficult problems.

SECTION 10.3: STATISTICAL PROCESS CONTROL

1. Control charts for $\bar{x}$ and s are to be maintained for the threshold voltage of short-channel MOSFETs using sample sizes of $n = 10$. It is known that the process is normally distributed with $\mu = 0.75$ V and $\sigma = 0.10$ V. Find the center line and control limits for each of these control charts.

2. Repeat Problem 1 assuming that μ and σ are unknown and that we have collected 50 observations of sample size 10. These samples yielded a grand average of 0.734 V and an average s_i of 0.125 V.

SECTION 10.4: STATISTICAL EXPERIMENTAL DESIGN

3. The following 2^3 factorial experiment was used to analyze a photolithography process. Analyze the experimental results using the Yates algorithm.

Run	Exposure Dose	Develop Time	Bake Temperature	Yield (%)
1	−	−	−	60
2	+	−	−	77
3	−	+	−	59
4	+	+	−	68
5	−	−	+	57
6	+	−	+	83
7	−	+	+	45
8	+	+	+	85

°4. Consider the throughputs (i.e., wafers processed per hour) of five different manufacturing processes (labelled A to E in the following table). For each process, data was collected on three different dates. Perform an analysis of variance to determine whether the processes and processing dates are significantly different.

DAY	A	B	C	D	E
1	509	512	532	506	509
2	505	507	542	520	519
3	465	472	498	483	475

SECTION 10.5: YIELD

5. Assuming a Poisson model, calculate the maximum defect density allowable on 100,000 NMOS transistors in order to achieve a functional yield of 95%. Assume the gate of each device is 10 μm wide and 1 μm long.

6. Use Murphy's yield integral to derive Eqs. 37, 38, and 40.

7. Suppose the probability density function of the defect density for a given interconnect manufacturing process is given by

$$f(D) = -100D + 10 \quad 0 \le D \le 0.1$$

If the critical area for this interconnect is 100 cm², calculate the functional yield we can expect for the process over the range of defect densities from 0.05 to 0.1 cm⁻².

11

Future Trends and Challenges

Since the beginning of the integrated circuit era in 1959, the minimum device dimension, also called the *minimum feature length*, has been reduced at an annual rate of about 13% (i.e., a reduction of 30% every 3 years). According to the prediction by the *International Technology Roadmap for Semiconductors*,[1] the minimum feature length will shrink from 130 nm (0.13 μm) in the year 2002 to 35 nm (0.035 μm) around 2014, as shown in Table 11.1. Also shown in Table 11.1 is DRAM size. The DRAM has increased its memory cell capacity four times every 3 years, and 64-Gbit DRAM is expected to be available in 2011 using 50-nm design rules. The table also shows that wafer size will increase to 450 mm (18 in. diameter) in 2014. In addition to the feature size reduction, challenges come from the device level, material level, and system level, as discussed in the following sections.

▶ 11.1 CHALLENGES FOR INTEGRATION

Figure 11.1 shows the trends of power supply voltage (V_{DD}), threshold voltage (V_T), and gate oxide thickness (d) versus channel length for CMOS logic technology.[2] From this figure, one can see that the gate oxide thickness will soon approach the tunneling-current limit of 2 nm. V_{DD} scaling will slow down because of the nonscalable V_T (i.e., to a minimum V_T of about 0.3 V due to subthreshold leakage and circuit noise immunity). Some challenges of the 180-nm technology and beyond are shown in Figure 11.2.[3] The most stringent requirements are detailed in the following subsections.

TABLE 11.1   The Technology Generation from 1997 to 2014

Year of the first product shipment	1997	1999	2002	2005	2008	2011	2014
Feature size (nm)	250	180	130	100	70	50	35
DRAM size (bit)	256M	1 G	—	8 G	—	64 G	—
Wafer size (mm)	200	300	300	300	300	300	450
Gate oxide (nm)	3–4	1.9–2.5	1.3–1.7	0.9–1.1	<1.0	—	—
Junction depth (nm)	50–100	42–70	25–43	20–33	15–30	—	—

DRAM, dynamic random access memory.

Source: *International Technology Roadmap for Semiconductors*, Semiconductor Industry Association, San Jose, CA, 1999.

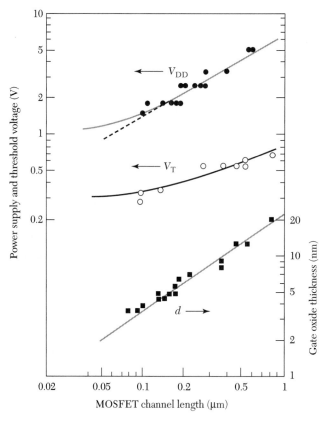

Figure 11.1 Trends of power supply voltage V_{DD}, threshold voltage V_T, and gate oxide thickness d versus channel length for CMOS logic technologies. Points are collected from data published over recent years.[2]

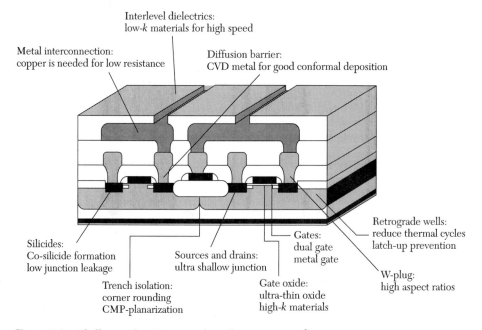

Figure 11.2 Challenges for 180-nm and smaller MOSFETs.[3]

11.1.1 Ultrashallow Junction Formation

So-called short-channel effects happen as the channel length is reduced. This problem becomes critical as the device dimension is scaled down to 100 nm. To achieve an ultrashallow junction with low sheet resistance, low-energy (i.e., less than 1 keV) implantation technology with high dosage must be employed. Table 11.1 shows the required junction depth versus the technology generation. The requirements of the junction for 100 nm are depths around 20 to 33 nm with a doping concentration of $1 \times 10^{20}/\text{cm}^3$.

11.1.2 Ultrathin Oxide

As the gate length shrinks below 130 nm, the oxide equivalent thickness of the gate dielectric must be reduced around 2 nm to maintain performance. However, if only SiO_2 (with a dielectric constant of 3.9) is used, the leakage through the gate becomes very high because of direct tunneling. For this reason, thicker high-k dielectric materials that have lower leakage current are needed. Candidates for the short term are silicon nitride (with a dielectric constant of 7), Ta_2O_5 (25), and TiO_2 (60–100).

11.1.3 Silicide Formation

Silicide-related technology has become an integral part of submicron devices for reducing the parasitic resistance to improve device and circuit performance. The conventional Ti–silicide process has been widely used in 350- to 250-nm technology. However, the sheet resistance of a $TiSi_2$ line increases with decreasing linewidth, which limits the use of $TiSi_2$ in 100-nm CMOS applications and beyond. $CoSi_2$ or $NiSi$ processes will replace $TiSi_2$ for technology beyond 100 nm.

11.1.4 New Materials for Interconnection

To achieve high-speed operation, the RC time delay of the interconnection must be reduced. Figure 8.14 showed the delay as a function of feature size.[4] It is obvious that gate delay decreases as the channel length decreases. Meanwhile, the delay resulting from interconnect increases significantly as the size decreases. This causes the total delay time to increase as the dimension of the device size scales down to 250 nm. Consequently, both high-conductivity metals, such as Cu, and low-dielectric-constant (low-k) insulators, such as organic (polyimide) or inorganic (F-doped oxide) materials, offer major performance gains. Cu exhibits superior performance because of its high conductivity ($1.7~\mu\Omega$-cm compared with $2.7~\mu\Omega$-cm for Al) and is 10 to 100 times more resistant to electromigration. The delay using Cu and the low-k material shows a significant decrease compared with that of conventional Al and oxide. Hence, Cu with the low-k material is essential in multilevel interconnection for future deep-submicron technology.

11.1.5 Power Limitations

The power required merely to charge and discharge circuit nodes in an IC is proportional to the number of gates and the frequency at which they are switched (clock frequency). The power can be expressed as $P \cong 1/2CV^2nf$, where C is the capacitance per device, V is the applied voltage, n is the number of devices per chip, and f is the clock frequency. The temperature rise caused by this power dissipation in an IC package is limited by the thermal conductivity of the package material, unless auxiliary liquid or gas cooling is used. The maximum allowable temperature rise is limited by the bandgap of the semiconductor

(~100°C for Si with a bandgap of 1.1 eV). For such a temperature rise, the maximum power dissipation of a typical high-performance package is about 10 W. As a result, we must limit either the maximum clock rate or the number of gates on a chip. As an example, in an IC containing 100-nm MOS devices with $C = 5 \times 10^{-2}$ fF, running at a 20 GHz clock rate, the maximum number of gates we can have is about 10^7 if we assume a 10% duty cycle. This is a design constraint fixed by basic material parameters.

11.1.6 SOI Integration

Section 9.2.2 mentioned the isolation of SOI wafers. Recently, SOI technology has received a great deal of attention. The advantages of SOI integration become significant as the minimum feature length approaches 100 nm. From a process point of view, SOI does not need the complex well structure and isolation processes. In addition, shallow junctions are directly obtained through the SOI film thickness. There is no risk of nonuniform interdiffusion of silicon and Al in the contact regions because of oxide isolation at the bottom of the junction. Hence, the contact barrier is not necessary. From a device point of view, the modern bulk silicon device needs high doping at the drain and substrate to eliminate short-channel effects and punchthrough. This high doping results in high capacitance when the junction is reverse biased. On the other hand, in SOI, the maximum capacitance between the junction and substrate is the capacitance of the buried insulator, whose dielectric constant is three times smaller than that of silicon (3.9 versus 11.9). Based on ring oscillator performance, 130-nm SOI CMOS technology can achieve 25% faster speeds or require 50% less power compared with a similar bulk technology.[5] SRAM, DRAM, CPU, and rf CMOS have all been successfully fabricated using SOI technology. Therefore, SOI is a key candidate for the future system-on-a-chip technology, considered in the following section.

EXAMPLE 1

For an equivalent oxide thickness of 1.5 nm, what will be the physical thickness when the high-k materials nitride ($\varepsilon_i/\varepsilon_0 = 7$), Ta_2O_5 (25), or TiO_2 (80) are used?

SOLUTION For nitride,

$$\left(\frac{\varepsilon_{ox}}{1.5}\right) = \left(\frac{\varepsilon_{nitride}}{d_{nitride}}\right)$$

$$d_{nitride} = 1.5\left(\frac{7}{3.9}\right) = 2.69 \, nm$$

Using the same calculation, we obtain 9.62 nm for Ta_2O_5 and 10.77 nm for TiO_2. ◀

▶ 11.2 SYSTEM-ON-A-CHIP

Increased component density and improved fabrication technology have helped the realization of the *system-on-a-chip* (SOC), that is, an IC chip that contains a complete electronic system. Designers can build all the circuitry needed for a complete electronic system, such as a camera, radio, television, or personal computer (PC), on a single chip. Figure 11.3 shows an SOC application for a PC's motherboard. Components (11 chips in this case) once found on boards become virtual components on the chip at the right.[6]

There are two obstacles in the realization of the SOC. The first is the huge complexity of the design. Since the component board is presently designed by different companies

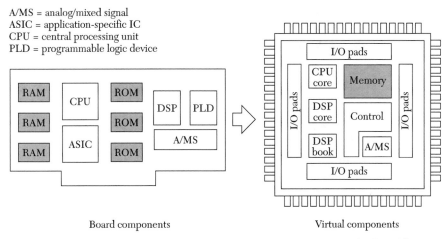

Board components   Virtual components

Figure 11.3   System-on-a-chip of a conventional personal computer motherboard.[6]

using different design tools, it is difficult to integrate the components into one chip. Another difficulty lies in fabrication. In general, the fabrication processes of a DRAM are significantly different from those of logic ICs (e.g., CPUs). Speed is the first priority for the logic, whereas leakage of the stored charge is the priority for memory. Therefore, multilevel interconnection schemes using five to six levels of metal are essential for logic ICs to improve the speed. However, DRAM circuits need only two to three levels. In addition, to increase the speed, a silicide process must be used to reduce the series resistance, and ultrathin gate oxide is needed to increase the drive current in logic circuits. These requirements are not critical for memory.

To achieve the SOC goal, an embedded DRAM technology has been introduced to merge logic and DRAM into a single chip with compatible processes. Figure 11.4 shows a schematic cross section of an embedded DRAM, including the DRAM cells and the logic CMOS devices.[7] Some processing steps are modified as a compromise. The trench-type

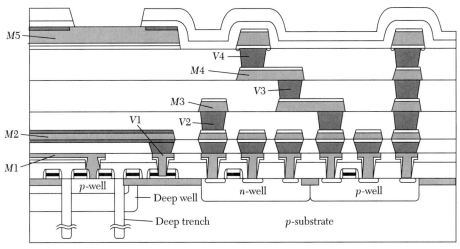

Figure 11.4   Schematic cross section of the embedded DRAM including DRAM cells and logic MOSFETs. There is no height difference in the trench capacitor cell because of the DRAM cell structure. $M1$ to $M5$ are metal interconnections, and $V1$ to $V4$ are via holes.[7]

capacitor, instead of the stacked type, is used so that there is no height difference in the DRAM cell structure. In addition, multiple gate oxide thicknesses exist on the same wafer to accommodate multiple supply voltages or to combine memory and logic circuits on one chip, or both.

▶ 11.3 SUMMARY

Because of the rapid reduction in feature length, IC technology will soon reach its practical limit as the channel length is reduced to about 20 nm. What ICs will be beyond CMOS is a key question being asked by research scientists. Major candidates include innovative devices based on quantum mechanical effects, because when the lateral dimension is reduced to below 100 nm, electronic structures will exhibit nonclassical behavior depending on the materials and the temperature of operation. The operation of such devices will be on the scale of single-electron transport. This approach has been demonstrated by the single-electron memory cell. The realization of such systems with trillions of components will be a major challenge beyond CMOS.[8]

▶ REFERENCES

1. *International Technology Roadmap for Semiconductors,* Semiconductor Industry Association, San Jose, 1999.

2. Y. Taur and E. J. Nowak, "CMOS Devices below 0.1 μm: How High Will Performance Go?" *IEEE Tech. Dig. Int. Electron Devices Meet.,* p. 215 (1997).

3. L. Peters, "Is the 0.18 μm Node Just a Roadside Attraction?" *Semicond. Int.,* **22**, 46 (1999).

4. M. T. Bohr, "Interconnect Scaling—The Real Limiter to High Performance ULSI," *IEEE Tech. Dig. Int. Electron Devices Meet.,* p. 241 (1995).

5. E. Leobandung, et al., "Scalability of SOI Technology into 0.13 μm 1.2 V CMOS Generation," *IEEE Int. Electron Devices Meet.,* p. 403 (1998).

6. B. Martin, "Electronic Design Automation," *IEEE Spectr.,* **36**, 61 (1999).

7. H. Ishiuchi, et al., "Embedded DRAM Technologies," *IEEE Tech. Dig. Int. Electron Devices Meet.,* p. 33 (1997).

8. S. Luryi, J. Xu, and A. Zaslavsky, Eds., *Future Trends in Microelectronics,* Wiley, New York, 1999.

▶ PROBLEMS

1. (a) Calculate the RC time constant of an aluminum runner 0.5 μm thick formed on a thermally grown SiO_2 0.5 μm thick. The length and width of the runner are 1 cm and 1 μm, respectively. The resistivity of the runner is 10^{-5} Ω-cm. (b) What will be the RC time constant for a polysilicon runner (R_h = 30 Ω/h) of identical dimension?

2. Why do we need multiple oxide thicknesses for a system-on-a-chip?

3. Normally we need a buffered layer placed between a high-k Ta_2O_5 and the silicon substrate. Calculate the effective oxide thickness (EOT) when the stacked gate dielectric is Ta_2O_5 (k = 25) with a thickness of 75 Å on a buffered nitride layer (k = 7 and a thickness of 10 Å). Also calculate EOT for a buffered oxide layer (k = 3.9 and a thickness of 5 Å).

Appendix A

List of Symbols

Symbol	Description	Unit
a	Lattice constant	Å
c	Speed of light in vacuum	cm/s
C	Capacitance	F
D	Diffusion coefficient	cm^2/s
E	Energy	eV
$\mathscr{E}$	Electric field	V/cm
f	Frequency	Hz(cps)
h	Planck constant	J·s
I	Current	A
J	Current density	A/cm^2
k	Boltzmann constant	J/K
L	Length	cm or μm
m_0	Electron rest mass	kg
$\bar{n}$	Refractive index	
n	Density of free electrons	cm^{-3}
n_i	Intrinsic carrier concentration	cm^{-3}
p	Density of free holes	cm^{-3}
P	Pressure	Pa
q	Magnitude of electronic charge	C
Q_{it}	Interface trapped charge	charges/cm^2
R	Resistance	Ω
t	Time	s
T	Absolute temperature	K
v	Carrier velocity	cm/s
V	Voltage	V
ε_0	Permittivity in vacuum	F/cm
ε_s	Semiconductor permittivity	F/cm
ε_{ox}	Insulator permittivity	F/cm
$\varepsilon_s/\varepsilon_0$ or $\varepsilon_{ox}/\varepsilon_0$	Dielectric constant	
λ	Wavelength	μm or nm
ν	Frequency of light	Hz
μ_0	Permeability in vacuum	H/cm
μ_n	Electron mobility	cm^2/V·s
μ_p	Hole mobility	cm^2/V·s
ρ	Resistivity	Ω-cm
Ω	Ohm	Ω

Appendix B

International System of Units (SI Units)

Quantity	Unit	Symbol	Dimensions
Length[a]	Meter	m	
Mass	Kilogram	kg	
Time	Second	s	
Temperature	Kelvin	K	
Current	Ampere	A	
Light intensity	Candela	Cd	
Angle	Radian	rad	
Frequency	Hertz	Hz	1/s
Force	Newton	N	$kg\text{-}m/s^2$
Pressure	Pascal	Pa	N/m^2
Energy[a]	Joule	J	N-m
Power	Watt	W	J/s
Electric charge	Coulomb	C	A·s
Potential	Volt	V	J/C
Conductance	Siemens	S	A/V
Resistance	Ohm	Ω	V/A
Capacitance	Farad	F	C/V
Magnetic flux	Weber	Wb	V·s
Magnetic induction	Tesla	T	Wb/m^2
Inductance	Henry	H	Wb/A
Light flux	Lumen	Lm	Cd-rad

[a] It is more common in the semiconductor field to use cm for length and eV for energy (1 cm = 10^{-2} m, 1 eV = 1.6×10^{-19} J).

Appendix C

Unit Prefixes*

Multiple	Prefix	Symbol
10^{18}	exa	E
10^{15}	peta	P
10^{12}	tera	T
10^{9}	giga	G
10^{6}	mega	M
10^{3}	kilo	k
10^{2}	hecto	h
10	deka	da
10^{-1}	deci	d
10^{-2}	centi	c
10^{-3}	milli	m
10^{-6}	micro	μ
10^{-9}	nano	n
10^{-12}	pico	p
10^{-15}	femto	f
10^{-18}	atto	a

* Adopted by International Committee on Weights and Measures. (Compound prefixes should not be used, e.g., not μμ but p.)

Appendix D

Greek Alphabet

Letter	Lowercase	Uppercase
Alpha	α	A
Beta	β	B
Gamma	γ	Γ
Delta	δ	Δ
Epsilon	ε	E
Zeta	ζ	Z
Eta	η	H
Theta	θ	Θ
Iota	ι	I
Kappa	κ	K
Lambda	λ	Λ
Mu	μ	M
Nu	ν	N
Xi	ξ	Ξ
Omicron	o	O
Pi	π	Π
Rho	ρ	P
Sigma	σ	Σ
Tau	τ	T
Upsilon	υ	Y
Phi	ϕ	Φ
Chi	χ	X
Psi	ψ	Ψ
Omega	ω	Ω

Appendix E

Physical Constants

Quantity	Symbol	Value
Angstrom unit	Å	$10 \text{ Å} = 1 \text{ nm} = 10^{-3} \text{ μm} = 10^{-7} \text{ cm} = 10^{-9} \text{ m}$
Avogadro constant	N_{av}	6.02214×10^{23}
Bohr radius	a_B	0.52917 Å
Boltzmann constant	k	$1.38066 \times 10^{-23} \text{ J/K } (R/N_{av})$
Elementary charge	q	$1.60218 \times 10^{-19} \text{ C}$
Electron rest mass	m_0	$0.91094 \times 10^{-30} \text{ kg}$
Electron volt	eV	$1 \text{ eV} = 1.60218 \times 10^{-19} \text{ J}$ $= 23.053 \text{ kcal/mol}$
Gas constant	R	$1.98719 \text{ cal/mol-K}$
Permeability in vacuum	μ_0	$1.25664 \times 10^{-8} \text{ H/cm } (4\pi \times 10^{-9})$
Permittivity in vacuum	ε_0	$8.85418 \times 10^{-14} \text{ F/cm } (1/\mu_0 c^2)$
Planck constant	h	$6.62607 \times 10^{-34} \text{ J·s}$
Reduced Planck constant	$\hbar$	$1.05457 \times 10^{-34} \text{ J·s } (h/2\pi)$
Proton rest mass	M_p	$1.67262 \times 10^{-27} \text{ kg}$
Speed of light in vacuum	c	$2.99792 \times 10^{10} \text{ cm/s}$
Standard atmosphere		$1.01325 \times 10^5 \text{ Pa}$
Thermal voltage at 300 K	kT/q	0.025852 V
Wavelength of 1 eV quantum	λ	1.23984 μm

Appendix F

Properties of Si and GaAs at 300 K

Properties	Si	GaAs
Atoms/cm^3	5.02×10^{22}	4.42×10^{22}
Atomic weight	28.09	144.63
Breakdown field (V/cm)	$\sim 3 \times 10^5$	$\sim 4 \times 10^5$
Crystal structure	Diamond	Zincblende
Density (g/cm^3)	2.329	5.317
Dielectric constant	11.9	12.4
Effective density of states in conduction band, N_C (cm^{-3})	2.86×10^{19}	4.7×10^{17}
Effective density of states in valence band, N_V (cm^{-3})	2.66×10^{19}	7.0×10^{18}
Effective mass (conductivity)		
Electrons (m_n/m_0)	0.26	0.063
Holes (m_p/m_0)	0.69	0.57
Electron affinity, χ (V)	4.05	4.07
Energy gap (eV)	1.12	1.42
Index of refraction	3.42	3.3
Intrinsic carrier concentration (cm^{-3})	9.65×10^9	2.25×10^6
Intrinsic resistivity (Ω-cm)	3.3×10^5	2.9×10^8
Lattice constant (Å)	5.43102	5.65325
Linear coefficient of thermal expansion, $\Delta L/L \times T$ (°C^{-1})	2.59×10^{-6}	5.75×10^{-6}
Melting point (°C)	1412	1240
Minority-carrier lifetime (s)	3×10^{-2}	$\sim 10^{-8}$
Mobility (cm^2/V·s)		
μ_n (electrons)	1450	9200
μ_p (holes)	505	320
Specific heat (J/g -°C)	0.7	0.35
Thermal conductivity (W/cm-K)	1.31	0.46
Vapor pressure (Pa)	1 at 1650°C	100 at 1050°C
	10^{-6} at 900°C	1 at 900°C

Appendix G

Some Properties of the Error Function

w	$\text{erf}(w)$	w	$\text{erf}(w)$	w	$\text{erf}(w)$	w	$\text{erf}(w)$
0.00	0.000 000	0.32	0.349 126	0.64	0.634 586	0.96	0.825 424
0.01	0.011 283	0.33	0.359 279	0.65	0.642 029	0.97	0.829 870
0.02	0.022 565	0.34	0.369 365	0.66	0.649 377	0.98	0.834 232
0.03	0.033 841	0.35	0.379 382	0.67	0.656 628	0.99	0.838 508
0.04	0.045 111	0.36	0.389 330	0.68	0.663 782	1.00	0.842 701
0.05	0.056 372	0.37	0.399 206	0.69	0.670 840	1.01	0.846 810
0.06	0.067 622	0.38	0.409 009	0.70	0.677 801	1.02	0.850 838
0.07	0.078 858	0.39	0.418 739	0.71	0.684 666	1.03	0.854 784
0.08	0.090 078	0.40	0.428 392	0.72	0.691 433	1.04	0.858 650
0.09	0.101 281	0.41	0.437 969	0.73	0.698 104	1.05	0.862 436
0.10	0.112 463	0.42	0.447 468	0.74	0.704 678	1.06	0.866 144
0.11	0.123 623	0.43	0.456 887	0.75	0.711 156	1.07	0.869 773
0.12	0.134 758	0.44	0.466 225	0.76	0.717 537	1.08	0.873 326
0.13	0.145 867	0.45	0.475 482	0.77	0.723 822	1.09	0.876 803
0.14	0.156 947	0.46	0.484 655	0.78	0.730 010	1.10	0.880 205
0.15	0.167 996	0.47	0.493 745	0.79	0.736 103	1.11	0.883 533
0.16	0.179 012	0.48	0.502 750	0.80	0.742 101	1.12	0.886 788
0.17	0.189 992	0.49	0.511 668	0.81	0.748 003	1.13	0.889 971
0.18	0.200 936	0.50	0.520 500	0.82	0.753 811	1.14	0.893 082
0.19	0.211 840	0.51	0.529 244	0.83	0.759 524	1.15	0.896 124
0.20	0.222 703	0.52	0.537 899	0.84	0.765 143	1.16	0.899 096
0.21	0.233 522	0.53	0.546 464	0.85	0.770 668	1.17	0.902 000
0.22	0.244 296	0.54	0.554 939	0.86	0.776 110	1.18	0.904 837
0.23	0.255 023	0.55	0.563 323	0.87	0.781 440	1.19	0.907 608
0.24	0.265 700	0.56	0.571 616	0.88	0.786 687	1.20	0.910 314
0.25	0.276 326	0.57	0.579 816	0.89	0.719 843	1.21	0.912 956
0.26	0.286 900	0.58	0.587 923	0.90	0.796 908	1.22	0.915 534
0.27	0.297 418	0.59	0.595 936	0.91	0.801 883	1.23	0.918 050
0.28	0.307 880	0.60	0.603 856	0.92	0.806 768	1.24	0.920 505
0.29	0.318 283	0.61	0.611 681	0.93	0.811 564	1.25	0.922 900
0.30	0.328 627	0.62	0.619 411	0.94	0.816 271	1.26	0.925 236
0.31	0.338 908	0.63	0.627 046	0.95	0.820 891	1.27	0.927 514

(continued)

(continued)

w	$\mathrm{erf}(w)$	w	$\mathrm{erf}(w)$	w	$\mathrm{erf}(w)$	w	$\mathrm{erf}(w)$
1.28	0.929 734	1.74	0.986 135	2.21	0.998 224	2.67	0.999 841
1.29	0.931 899	1.75	0.986 672	2.22	0.998 308	2.68	0.999 849
1.30	0.934 008	1.76	0.987 190	2.23	0.998 388	2.69	0.999 858
1.31	0.936 063	1.77	0.987 691	2.24	0.998 464	2.70	0.999 866
1.32	0.938 065	1.79	0.988 641	2.25	0.998 537	2.71	0.999 873
1.33	0.940 015	1.80	0.989 091	2.26	0.998 607	2.72	0.999 880
1.34	0.941 914	1.81	0.989 525	2.27	0.998 674	2.73	0.999 887
1.35	0.943 762	1.82	0.989 943	2.28	0.998 738	2.74	0.999 893
1.36	0.945 561	1.83	0.990 347	2.29	0.998 799	2.75	0.999 899
1.37	0.947 312	1.84	0.990 736	2.30	0.998 857	2.76	0.999 905
1.38	0.949 016	1.85	0.991 111	2.31	0.998 912	2.77	0.999 910
1.39	0.950 673	1.86	0.991 472	2.32	0.998 966	2.78	0.999 916
1.40	0.952 285	1.87	0.991 821	2.33	0.999 016	2.79	0.999 920
1.41	0.953 852	1.88	0.992 156	2.34	0.999 065	2.80	0.999 925
1.42	0.955 376	1.89	0.992 479	2.35	0.999 111	2.81	0.999 929
1.43	0.956 857	1.90	0.992 790	2.36	0.999 155	2.82	0.999 933
1.44	0.958 297	1.91	0.993 090	2.37	0.999 197	2.83	0.999 937
1.45	0.959 695	1.92	0.993 378	2.38	0.999 237	2.85	0.999 944
1.46	0.961 054	1.93	0.993 656	2.39	0.999 275	2.86	0.999 948
1.47	0.962 373	1.94	0.993 923	2.40	0.999 311	2.87	0.999 951
1.48	0.963 654	1.95	0.994 179	2.41	0.999 346	2.88	0.999 954
1.49	0.964 898	1.96	0.994 426	2.42	0.999 379	2.89	0.999 956
1.50	0.966 105	1.97	0.994 664	2.43	0.999 411	2.90	0.999 959
1.51	0.967 277	1.98	0.994 892	2.44	0.999 441	2.91	0.999 961
1.52	0.968 413	1.99	0.995 111	2.45	0.999 469	2.92	0.999 964
1.53	0.969 516	2.00	0.995 322	2.46	0.999 497	2.93	0.999 966
1.54	0.970 586	2.01	0.995 525	2.47	0.999 523	2.94	0.999 968
1.55	0.971 623	2.02	0.995 719	2.48	0.999 547	2.95	0.999 970
1.56	0.972 628	2.03	0.995 906	2.49	0.999 571	2.96	0.999 972
1.57	0.973 603	2.04	0.996 086	2.50	0.999 593	2.97	0.999 973
1.58	0.974 547	2.05	0.996 258	2.51	0.999 614	2.98	0.999 975
1.59	0.975 462	2.06	0.996 423	2.52	0.999 634	2.99	0.999 976
1.60	0.976 348	2.07	0.996 582	2.53	0.999 654	3.00	0.999 977 91
1.61	0.977 207	2.08	0.996 734	2.54	0.999 672	3.01	0.999 979 26
1.62	0.978 038	2.09	0.996 880	2.55	0.999 689	3.02	0.999 980 53
1.63	0.978 843	2.10	0.997 021	2.56	0.999 706	3.03	0.999 981 73
1.64	0.979 622	2.11	0.997 155	2.57	0.999 722	3.04	0.999 982 86
1.65	0.980 376	2.12	0.997 284	2.58	0.999 736	3.05	0.999 983 92
1.66	0.981 105	2.13	0.997 407	2.59	0.999 751	3.06	0.999 984 92
1.67	0.981 810	2.14	0.997 525	2.60	0.999 764	3.07	0.999 985 86
1.68	0.982 493	2.15	0.997 639	2.61	0.999 777	3.08	0.999 986 74
1.69	0.983 153	2.16	0.997 747	2.62	0.999 789	3.09	0.999 987 57
1.70	0.983 790	2.17	0.997 851	2.63	0.999 800	3.10	0.999 988 35
1.71	0.984 407	2.18	0.997 951	2.64	0.999 811	3.11	0.999 989 08
1.72	0.985 003	2.19	0.998 046	2.65	0.999 822	3.12	0.999 989 77
1.73	0.985 578	2.20	0.998 137	2.66	0.999 831	3.13	0.999 990 42

(continued)

(continued)

w	erf(w)	w	erf(w)	w	erf(w)	w	erf(w)
3.14	0.999 991 03	3.36	0.999 997 983	3.58	0.999 999 587	3.80	0.999 999 923
3.15	0.999 991 60	3.37	0.999 998 120	3.59	0.999 999 617	3.81	0.999 999 929
3.16	0.999 992 14	3.38	0.999 998 247	3.60	0.999 999 644	3.82	0.999 999 934
3.17	0.999 992 64	3.39	0.999 998 367	3.61	0.999 999 670	3.83	0.999 999 939
3.18	0.999 993 11	3.40	0.999 998 478	3.62	0.999 999 694	3.84	0.999 999 944
3.19	0.999 993 56	3.41	0.999 998 582	3.63	0.999 999 716	3.85	0.999 999 948
3.20	0.999 993 97	3.42	0.999 998 679	3.64	0.999 999 736	3.86	0.999 999 952
3.21	0.999 994 36	3.43	0.999 998 770	3.65	0.999 999 756	3.87	0.999 999 956
3.22	0.999 994 73	3.44	0.999 998 855	3.66	0.999 999 773	3.88	0.999 999 959
3.23	0.999 995 07	3.45	0.999 998 934	3.67	0.999 999 790	3.89	0.999 999 962
3.24	0.999 995 40	3.46	0.999 999 008	3.68	0.999 999 805	3.90	0.999 999 965
3.25	0.999 995 70	3.47	0.999 999 077	3.69	0.999 999 820	3.91	0.999 999 968
3.26	0.999 995 98	3.48	0.999 999 141	3.70	0.999 999 833	3.92	0.999 999 970
3.27	0.999 996 24	3.49	0.999 999 201	3.71	0.999 999 845	3.93	0.999 999 973
3.28	0.999 996 49	3.50	0.999 999 257	3.72	0.999 999 857	3.94	0.999 999 975
3.29	0.999 996 72	3.51	0.999 999 309	3.73	0.999 999 867	3.95	0.999 999 977
3.30	0.999 996 94	3.52	0.999 999 358	3.74	0.999 999 877	3.96	0.999 999 979
3.31	0.999 997 15	3.53	0.999 999 403	3.75	0.999 999 886	3.97	0.999 999 980
3.32	0.999 997 34	3.54	0.999 999 445	3.76	0.999 999 895	3.98	0.999 999 982
3.33	0.999 997 51	3.55	0.999 999 485	3.77	0.999 999 903	3.99	0.999 999 983
3.34	0.999 997 68	3.56	0.999 999 521	3.78	0.999 999 910		
3.35	0.999 997 838	3.57	0.999 999 555	3.79	0.999 999 917		

Appendix H

Basic Kinetic Theory of Gases

The ideal gas law states that

$$PV = RT = N_{av}kT \qquad (1)$$

where P is the pressure, V is the volume of 1 mole of gas, R is the gas constant (1.98 cal/mol-K, or 82 atm-cm^3/mol-K), T is the absolute temperature in K, N_{av} is the Avogadro constant (6.02×10^{23} molecules/mole), and k is the Boltzmann constant (1.38×10^{-23} J/K, or 1.37×10^{-22} atm-cm^2/K). Since real gases behave more and more like the ideal gas as the pressure is lowered, Eq. 1 is valid for most vacuum processes. We can use Eq. 1 to calculate the molecular concentration n (the number of molecules per unit volume):

$$n = \frac{N_{av}}{V} = \frac{P}{kT} \qquad (2)$$

$$= 7.25 \times 10^{16} \frac{P}{T} \text{ molecules/cm}^3 \qquad (2a)$$

where P is in Pa. The density ρ_d of a gas is given by the product of its molecular weight and its concentration:

$$\rho_d = \text{Molecular weight} \times \left(\frac{P}{kT} \right) \qquad (3)$$

The gas molecules are in constant motion and their velocities are temperature dependent. The distribution of velocities is described by the Maxwell–Boltzmann distribution law, which states that for a given speed v,

$$\frac{1}{n}\frac{dn}{dv} \equiv f_v = \frac{4}{\sqrt{\pi}} \left(\frac{m}{2kT} \right)^{3/2} v^2 \exp\left(-\frac{mv^2}{2kT} \right) \qquad (4)$$

where m is the mass of a molecule. This equation states that if there are n molecules in the volume, there will be dn molecules having a speed between v and $v + dv$. The average speed is obtainable from Eq. 4:

$$v_{au} = \frac{\int_0^\infty v f_v dv}{\int_0^\infty f_v dv} = \frac{2}{\sqrt{\pi}} \sqrt{\frac{2kT}{m}} \qquad (5)$$

An important parameter for vacuum technology is the molecular *impingement rate*, that is, how many molecules impinge on a unit area per unit time. To obtain this parameter, first consider the distribution function f_{vx} for the velocities of molecules in the x direction. This function can be expressed by an equation similar to Eq. 4:

$$\frac{1}{n}\frac{dn_x}{dv_x} \equiv f_{v_x} = \left(\frac{m}{2\pi kT}\right)^{1/2} v_x^2 \exp\left(\frac{-mv_x^2}{2kT}\right) \tag{6}$$

The molecular impingement rate ϕ is given by

$$\phi = \int_0^\infty v_x dn_x \tag{7}$$

Substituting dn_x from Eq. 6 and integrating gives

$$\phi = n\sqrt{\frac{kT}{2\pi m}} \tag{8}$$

The relationship between the impingement rate and the gas pressure is obtained by using Eq. 2:

$$\phi = P(2\pi mkT)^{-1/2} \tag{9}$$

$$= 2.64 \times 10^{20}\left(\frac{P}{\sqrt{MT}}\right) \tag{9a}$$

where P is the pressure in Pa and M is the molecular weight.

Appendix I

SUPREM Commands

The Stanford University Process Engineering Modeling (SUPREM) program is a simulation package that allows a user to model various process steps used in the fabrication of integrated circuits. SUPREM can predict the results of oxidation, deposition, etching, diffusion, epitaxial growth, and ion implantation processes. SUPREM III models the changes to the semiconductor structure that result from these processes in one dimension. The primary results are the thicknesses of various layers and the distribution of impurities within those layers. The program can also determine certain material properties, such as the sheet resistance of diffused regions in silicon layers.

To run SUPREM, an input deck must be provided. This file contains a series of statements and comments. The deck begins with a TITLE statement, which is merely a comment repeated on each page of the program output. The next command, INITIALIZE, is a control statement that sets the substrate type, orientation, and doping. This command can also be used to specify the thickness of the region to be simulated and establish a grid. After the substrate and materials are established, a series of statements is used to specify the sequence of process steps as they occur. Finally, the output of the simulation can be printed or plotted using PRINT or PLOT statements, respectively. Simulation ends with a STOP statement. Several COMMENT statements will typically appear throughout the deck.

A description of a few commonly used SUPREM statements is provided in Table I.1. *This table is by no means complete.* To obtain the complete SUPREM software package and its associated documentation, contact

Silvaco Data Systems, Inc.

4701 Patrick Henry Drive

Building 2

Santa Clara, CA 95054

Phone: 408-654-4372

Fax: 408-727-5297

www.silvaco.com

SUPREM is a trademark of the Board of Trustees of Stanford University.

TABLE I.1   Common SUPREM Commands

Name	Description	Basic Syntax	Typical Flags and Parameters
COMMENT	Outouts character string to label an input sequence	COMMENT <text>	None
DEPOSITION	Deposits specified material on top of current structure	DEPOSITION <Material> Thickness=<n> Temperature=<n>	Aluminum Nitride Oxide Polysilicon Silicon C.Phosphor (cm^{-3}) C.Arsenic (cm^{-3}) C.Boron (cm^{-3}) Thickness (μm) Temperature $(^\circ\text{C})$
DIFFUSION	Models high-temperature diffusion in oxidizing and nonoxidizing ambients	DIFFUSION Time=<n> Temperature=<n> <Dopant> <Ambient>	Arsenic Boron DryO2 Nitrogen Phosphorus WetO2 Solidsol HCI% $(\%)$ T.Rate $(^\circ\text{C/min})$ Temperature $(^\circ\text{C})$ Time (minutes)
ETCH	Etches specified material from the top of current structure	ETCH <Material> Thickness=<n>	All Aluminum Nitride Oxide Polysilicon Silicon Thickness (μm)
IMPLANT	Simulates ion implantation of impurities	IMPLANT <Dopant> Dose=<n> Energy=<n>	Arsenic Boron Phosphorus Dose (cm^{-2}) Energy (keV)
INITIALIZE	Sets up initial coefficients and structure to be used in the simulation	INITIALIZE <Structure> <Substrate> <Dopant> Concentration=<n> INITIALIZE Structure=<filename>	<100> <110> <111> Silicon Arsenic Boron Phosphorus Concentration (cm^{-3})

(continued)

(continued)

Name	Description	Basic Syntax	Typical Flags and Parameters
PLOT	Specifies that impurity concentrations or results of electrical calculations versus depth into the substrate are to be plotted	PLOT *<Parameters>* Cmin=*<n>* Cmax=*<n>*	Active Arsenic Boron Chemical Net Phosphorus Cmin (cm^{-3}) Cmax (cm^{-3})
PRINT	Outputs information about the structure being simulated and coefficients used	PRINT *<Parameters>*	Arsenic Boron Chemical Concentration Layers Net Phosphorus
SAVEFILE	Saves the current structure being processed, the coefficients being used, or both	SAVEFILE *<Feature>* Filename=*<Text>*	Structure
STOP	Terminates simulation	STOP *<Text>*	None
TITLE	Inputs a character string to label the following input	TITLE *<Text>*	None

Appendix J

Running PROLITH

PROLITH is a Windows-based photolithography simulation program marketed by FINLE Technologies in Austin, Texas. PROLITH simulates the complete one- and two-dimensional optical lithography process from aerial image formation through resist exposure and development. The output of the program is an accurate prediction of the final resist profile, which is presented in a wide variety of images, plots, graphs, and calculations.

PROLITH accepts lithography information in the form of data files and input parameters and uses this information to simulate standard and advanced lithography processes. After the software has been installed, a user can run PROLITH by simply clicking on the PROLITH icon from the Windows Start menu. After a successful license search, the Imaging Tool parameters window appears (see Fig. 4.20). Before running a simulation, the user must choose the simulation options and enter a set of input parameters. This is accomplished by selecting Options from the File menu to open the Options dialog. The settings in this dialog box are used to establish basic simulation options, such as the Image Calculation Mode, Physical Model, and Speed Factor.

Once the options have been established, the user may enter simulation input parameters from the appropriate parameters window. These windows are opened by selecting various options from the View menu or by clicking the corresponding toolbar button for the window. In each window, parameters may be entered by marking or clearing checkboxes or option buttons, entering values in text boxes, selecting files or other values from lists, and so forth. Many parameter windows, such as the Resist parameters window, for example, provide instant graphical views of the information entered.

After input parameters have been entered, PROLITH displays simulation results from the Graphs menu. PROLITH can produce graphs simulating formation of a mask feature by an optical projection system, exposure of photoresist using this image, or development of the exposed photoresist. The following options from the Graphs menu (or corresponding toolbar buttons) are available to display such simulations:

- *Aerial Image*: The relative intensity of the image as a function of position
- *Image in Resist*: The image projected into the photoresist at the start of exposure
- *Exposed Latent Image*: The latent image before post-exposure bake (PEB)
- *PEB Latent Image*: The latent image after post-exposure bake
- *Develop Time Contours*: Contours of constant develop time as a function of position in the resist
- *Resist Profile*: Two-dimensional photoresist profile after development

This overview of PROLITH capabilities is by no means complete. To obtain the complete PROLITH software package and its associated documentation, contact

FINLE Technologies, Inc.

P.O. Box 162712

Austin, TX 78716

Phone: 512-327-3781

Fax: 512-327-1510

www.finle.com

Appendix K

Percentage Points of the *t* Distribution

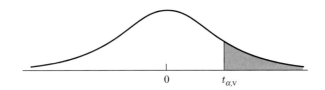

α ν	0.40	0.25	0.10	0.05	0.025	0.01	0.005	0.0025	0.001	0.0005
1	0.325	1.000	3.078	6.314	12.706	31.821	63.657	127.32	318.31	636.62
2	0.289	0.816	1.886	2.920	4.303	6.965	9.925	14.089	23.326	31.598
3	0.277	0.765	1.638	2.353	3.182	4.541	5.841	7.453	10.213	12.924
4	0.271	0.741	1.533	2.132	2.776	3.747	4.604	5.598	7.173	8.610
5	0.267	0.727	1.476	2.015	2.571	3.365	4.032	4.773	5.893	6.869
6	0.265	0.727	1.440	1.943	2.447	3.143	3.707	4.317	5.208	5.959
7	0.263	0.711	1.415	1.895	2.365	2.998	3.499	4.019	4.785	5.408
8	0.262	0.706	1.397	1.860	2.306	2.896	3.355	3.833	4.501	5.041
9	0.261	0.703	1.383	1.833	2.262	2.821	3.250	3.690	4.297	4.781
10	0.260	0.700	1.372	1.812	2.228	2.764	3.169	3.581	4.144	4.587
11	0.260	0.697	1.363	1.796	2.201	2.718	3.106	3.497	4.025	4.437
12	0.259	0.695	1.356	1.782	2.179	2.681	3.055	3.428	3.930	4.318
13	0.259	0.694	1.350	1.771	2.160	2.650	3.012	3.372	3.852	4.221
14	0.258	0.692	1.345	1.761	2.145	2.624	2.977	3.326	3.787	4.140
15	0.258	0.691	1.341	1.753	2.131	2.602	2.947	3.286	3.733	4.073
16	0.258	0.690	1.337	1.746	2.120	2.583	2.921	3.252	3.686	4.015
17	0.257	0.689	1.333	1.740	2.110	2.567	2.898	3.222	3.646	3.965
18	0.257	0.688	1.330	1.734	2.101	2.552	2.878	3.197	3.610	3.922
19	0.257	0.688	1.328	1.729	2.093	2.539	2.861	3.174	3.579	3.883
20	0.257	0.687	1.325	1.725	2.086	2.528	2.845	3.153	3.552	3.850
21	0.257	0.686	1.323	1.721	2.080	2.518	2.831	3.135	3.527	3.819
22	0.256	0.686	1.321	1.717	2.074	2.508	2.819	3.119	3.505	3.792

(continued)

(continued)

α ν	0.40	0.25	0.10	0.05	0.025	0.01	0.005	0.0025	0.001	0.0005
23	0.256	0.685	1.319	1.714	2.069	2.500	2.807	3.104	3.485	3.767
24	0.256	0.685	1.318	1.711	2.064	2.492	2.797	3.091	3.467	3.745
25	0.256	0.684	1.316	1.708	2.060	2.485	2.787	3.078	3.450	3.725
26	0.256	0.684	1.315	1.706	2.056	2.479	2.779	3.067	3.435	3.707
27	0.256	0.684	1.314	1.703	2.052	2.473	2.771	3.057	3.421	3.690
28	0.256	0.683	1.313	1.701	2.048	2.467	2.763	3.047	3.408	3.674
29	0.256	0.683	1.311	1.699	2.045	2.462	2.756	3.038	3.396	3.659
30	0.256	0.683	1.310	1.697	2.042	2.457	2.750	3.030	3.385	3.646
40	0.255	0.681	1.303	1.684	2.021	2.423	2.704	2.971	3.307	3.551
60	0.254	0.679	1.296	1.671	2.000	2.390	2.660	2.915	3.232	3.460
120	0.254	0.677	1.289	1.658	1.980	2.358	2.617	2.860	3.160	3.373
∞	0.253	0.674	1.282	1.645	1.960	2.326	2.576	2.807	3.090	3.291

ν, degrees of freedom.

Source: Adapted with permission from *Biometrika Tables for Statisticians*, Vol. 1, 3rd ed., by E. S. Pearson and H. O. Hartley, Cambridge University Press, Cambridge, 1966.

Appendix L

Percentage Points of the *F* Distribution

$F_{0.25, v_1, v_2}$

v_2 \ v_1	1	2	3	4	5	6	7	8	9	10	12	15	20	24	30	40	60	120	∞
1	5.83	7.50	8.20	8.58	8.82	8.98	9.10	9.19	9.26	9.32	9.41	9.49	9.58	9.63	9.67	9.71	9.76	9.80	9.85
2	2.57	3.00	3.15	3.23	3.28	3.31	3.34	3.35	3.37	3.38	3.39	3.41	3.43	3.43	3.44	3.45	3.46	3.47	3.48
3	2.02	2.28	2.36	2.39	2.41	2.42	2.43	2.44	2.44	2.44	2.45	2.46	2.46	2.46	2.47	2.47	2.47	2.47	2.47
4	1.81	2.00	2.05	2.06	2.07	2.08	2.08	2.08	2.08	2.08	2.08	2.08	2.08	2.08	2.08	2.08	2.08	2.08	2.08
5	1.69	1.85	1.88	1.89	1.89	1.89	1.89	1.89	1.89	1.89	1.89	1.89	1.88	1.88	1.88	1.88	1.87	1.87	1.87
6	1.62	1.76	1.78	1.79	1.79	1.78	1.78	1.78	1.77	1.77	1.77	1.76	1.76	1.75	1.75	1.75	1.74	1.74	1.74
7	1.57	1.70	1.72	1.72	1.71	1.71	1.70	1.70	1.70	1.69	1.68	1.68	1.67	1.67	1.66	1.66	1.65	1.65	1.65
8	1.54	1.66	1.67	1.66	1.66	1.65	1.64	1.64	1.63	1.63	1.62	1.62	1.61	1.60	1.60	1.59	1.59	1.58	1.58
9	1.51	1.62	1.63	1.63	1.62	1.61	1.60	1.60	1.59	1.59	1.58	1.57	1.56	1.56	1.55	1.54	1.54	1.53	1.53
10	1.49	1.60	1.60	1.59	1.59	1.58	1.57	1.56	1.56	1.55	1.54	1.53	1.52	1.52	1.51	1.51	1.50	1.49	1.48
11	1.47	1.58	1.58	1.57	1.56	1.55	1.54	1.53	1.53	1.52	1.51	1.50	1.49	1.49	1.48	1.47	1.47	1.46	1.45
12	1.46	1.56	1.56	1.55	1.54	1.53	1.52	1.51	1.51	1.50	1.49	1.48	1.47	1.46	1.45	1.45	1.44	1.43	1.42
13	1.45	1.55	1.55	1.53	1.52	1.51	1.50	1.49	1.49	1.48	1.47	1.46	1.45	1.44	1.43	1.42	1.42	1.41	1.40
14	1.44	1.53	1.53	1.52	1.51	1.50	1.49	1.48	1.47	1.46	1.45	1.44	1.43	1.42	1.41	1.41	1.40	1.39	1.38
15	1.43	1.52	1.52	1.51	1.49	1.48	1.47	1.46	1.46	1.45	1.44	1.43	1.41	1.41	1.40	1.39	1.38	1.37	1.36
16	1.42	1.51	1.51	1.50	1.48	1.47	1.46	1.45	1.44	1.44	1.43	1.41	1.40	1.39	1.38	1.37	1.36	1.35	1.34
17	1.42	1.51	1.50	1.49	1.47	1.46	1.45	1.44	1.43	1.43	1.41	1.40	1.39	1.38	1.37	1.36	1.35	1.34	1.33
18	1.41	1.50	1.49	1.48	1.46	1.45	1.44	1.43	1.42	1.42	1.40	1.39	1.38	1.37	1.36	1.35	1.34	1.33	1.32
19	1.41	1.49	1.49	1.47	1.46	1.44	1.43	1.42	1.41	1.41	1.40	1.38	1.37	1.36	1.35	1.34	1.33	1.32	1.30
20	1.40	1.49	1.48	1.47	1.45	1.44	1.43	1.42	1.41	1.40	1.39	1.37	1.36	1.35	1.34	1.33	1.32	1.31	1.29
21	1.40	1.48	1.48	1.46	1.44	1.43	1.42	1.41	1.40	1.39	1.38	1.37	1.35	1.34	1.33	1.32	1.31	1.30	1.28
22	1.40	1.48	1.47	1.45	1.44	1.42	1.41	1.40	1.39	1.39	1.37	1.36	1.34	1.33	1.32	1.31	1.30	1.29	1.28
23	1.39	1.47	1.47	1.45	1.43	1.42	1.41	1.40	1.39	1.38	1.37	1.35	1.34	1.33	1.32	1.31	1.30	1.28	1.27
24	1.39	1.47	1.46	1.44	1.43	1.41	1.40	1.39	1.38	1.38	1.36	1.35	1.33	1.32	1.31	1.30	1.29	1.28	1.26
25	1.39	1.47	1.46	1.44	1.42	1.41	1.40	1.39	1.38	1.37	1.36	1.34	1.33	1.32	1.31	1.29	1.28	1.27	1.25
26	1.38	1.46	1.45	1.44	1.42	1.41	1.39	1.38	1.37	1.37	1.35	1.34	1.32	1.31	1.30	1.29	1.28	1.26	1.25
27	1.38	1.46	1.45	1.43	1.42	1.40	1.39	1.38	1.37	1.36	1.35	1.33	1.32	1.31	1.30	1.28	1.27	1.26	1.24
28	1.38	1.46	1.45	1.43	1.41	1.40	1.39	1.38	1.37	1.36	1.34	1.33	1.31	1.30	1.29	1.28	1.27	1.25	1.24
29	1.38	1.45	1.45	1.43	1.41	1.40	1.38	1.37	1.36	1.35	1.34	1.32	1.31	1.30	1.29	1.27	1.26	1.25	1.23
30	1.38	1.45	1.44	1.42	1.41	1.39	1.38	1.37	1.36	1.35	1.34	1.32	1.30	1.29	1.28	1.27	1.26	1.24	1.23
40	1.36	1.44	1.42	1.40	1.39	1.37	1.36	1.35	1.34	1.33	1.31	1.30	1.28	1.26	1.25	1.24	1.22	1.21	1.19
60	1.35	1.42	1.41	1.38	1.37	1.35	1.33	1.32	1.31	1.30	1.29	1.27	1.25	1.24	1.22	1.21	1.19	1.17	1.15
120	1.34	1.40	1.39	1.37	1.35	1.33	1.31	1.30	1.29	1.28	1.26	1.24	1.22	1.21	1.19	1.18	1.16	1.13	1.10
∞	1.32	1.39	1.37	1.35	1.33	1.31	1.29	1.28	1.27	1.25	1.24	1.22	1.19	1.18	1.16	1.14	1.12	1.08	1.00

Degrees of Freedom for the Numerator (v_1)

Degrees of Freedom for the Denominator (v_2)

$F_{0.75, v_1, v_2} = 1/F_{0.25, v_2, v_1}$;

Source: Adapted with permission from Biometrika Tables for Statisticians, Vol. 1, 3rd ed., by E. S. Pearson and H. O. Hartley, Cambridge University Press, Cambridge, 1966.

(continued)

$$F_{0.10,\nu_1,\nu_2}$$

Degrees of Freedom for the Numerator (ν_1)

ν_2	1	2	3	4	5	6	7	8	9	10	12	15	20	24	30	40	60	120	∞
1	39.86	49.50	53.59	55.83	57.24	58.20	58.91	59.44	59.86	60.19	60.71	61.22	61.74	62.00	62.26	62.53	62.79	63.06	63.33
2	8.53	9.00	9.16	9.24	9.29	9.33	9.35	9.37	9.38	9.39	9.41	9.42	9.44	9.45	9.46	9.47	9.47	9.48	9.49
3	5.54	5.46	5.39	5.34	5.31	5.28	5.27	5.25	5.24	5.23	5.22	5.20	5.18	5.18	5.17	5.16	5.15	5.14	5.13
4	4.54	4.32	4.19	4.11	4.05	4.01	3.98	3.95	3.94	3.92	3.90	3.87	3.84	3.83	3.82	3.80	3.79	3.78	3.76
5	4.06	3.78	3.62	3.52	3.45	3.40	3.37	3.34	3.32	3.30	3.27	3.24	3.21	3.19	3.17	3.16	3.14	3.12	3.10
6	3.78	3.46	3.29	3.18	3.11	3.05	3.01	2.98	2.96	2.94	2.90	2.87	2.84	2.82	2.80	2.78	2.76	2.74	2.72
7	3.59	3.26	3.07	2.96	2.88	2.83	2.78	2.75	2.72	2.70	2.67	2.63	2.59	2.58	2.56	2.54	2.51	2.49	2.47
8	3.46	3.11	2.92	2.81	2.73	2.67	2.62	2.59	2.56	2.54	2.50	2.46	2.42	2.40	2.38	2.36	2.34	2.32	2.29
9	3.36	3.01	2.81	2.69	2.61	2.55	2.51	2.47	2.44	2.42	2.38	2.34	2.30	2.28	2.25	2.23	2.21	2.18	2.16
10	3.29	2.92	2.73	2.61	2.52	2.46	2.41	2.38	2.35	2.32	2.28	2.24	2.20	2.18	2.16	2.13	2.11	2.08	2.06
11	3.23	2.86	2.66	2.54	2.45	2.39	2.34	2.30	2.27	2.25	2.21	2.17	2.12	2.10	2.08	2.05	2.03	2.00	1.97
12	3.18	2.81	2.61	2.48	2.39	2.33	2.28	2.24	2.21	2.19	2.15	2.10	2.06	2.04	2.01	1.99	1.96	1.93	1.90
13	3.14	2.76	2.56	2.43	2.35	2.28	2.23	2.20	2.16	2.14	2.10	2.05	2.01	1.98	1.96	1.93	1.90	1.88	1.85
14	3.10	2.73	2.52	2.39	2.31	2.24	2.19	2.15	2.12	2.10	2.05	2.01	1.96	1.94	1.91	1.89	1.86	1.83	1.80
15	3.07	2.70	2.49	2.36	2.27	2.21	2.16	2.12	2.09	2.06	2.02	1.97	1.92	1.90	1.87	1.85	1.82	1.79	1.76
16	3.05	2.67	2.46	2.33	2.24	2.18	2.13	2.09	2.06	2.03	1.99	1.94	1.89	1.86	1.84	1.81	1.78	1.75	1.72
17	3.03	2.64	2.44	2.31	2.22	2.15	2.10	2.06	2.03	2.00	1.96	1.91	1.86	1.84	1.81	1.78	1.75	1.72	1.69
18	3.01	2.62	2.42	2.29	2.20	2.13	2.08	2.04	2.00	1.98	1.93	1.89	1.84	1.81	1.78	1.75	1.72	1.69	1.66
19	2.99	2.61	2.40	2.27	2.18	2.11	2.06	2.02	1.98	1.96	1.91	1.86	1.81	1.79	1.76	1.73	1.70	1.67	1.63
20	2.97	2.59	2.38	2.25	2.16	2.09	2.04	2.00	1.96	1.94	1.89	1.84	1.79	1.77	1.74	1.71	1.68	1.64	1.61
21	2.96	2.57	2.36	2.23	2.14	2.08	2.02	1.98	1.95	1.92	1.87	1.83	1.78	1.75	1.72	1.69	1.66	1.62	1.59
22	2.95	2.56	2.35	2.22	2.13	2.06	2.01	1.97	1.93	1.90	1.86	1.81	1.76	1.73	1.70	1.67	1.64	1.60	1.57
23	2.94	2.55	2.34	2.21	2.11	2.05	1.99	1.95	1.92	1.89	1.84	1.80	1.74	1.72	1.69	1.66	1.62	1.59	1.55
24	2.93	2.54	2.33	2.19	2.10	2.04	1.98	1.94	1.91	1.88	1.83	1.78	1.73	1.70	1.67	1.64	1.61	1.57	1.53
25	2.92	2.53	2.32	2.18	2.09	2.02	1.97	1.93	1.89	1.87	1.82	1.77	1.72	1.69	1.66	1.63	1.59	1.56	1.52
26	2.91	2.52	2.31	2.17	2.08	2.01	1.96	1.92	1.88	1.86	1.81	1.76	1.71	1.68	1.65	1.61	1.58	1.54	1.50
27	2.90	2.51	2.30	2.17	2.07	2.00	1.95	1.91	1.87	1.85	1.80	1.75	1.70	1.67	1.64	1.60	1.57	1.53	1.49
28	2.89	2.50	2.29	2.16	2.06	2.00	1.94	1.90	1.87	1.84	1.79	1.74	1.69	1.66	1.63	1.59	1.56	1.52	1.48
29	2.89	2.50	2.28	2.15	2.06	1.99	1.93	1.89	1.86	1.83	1.78	1.73	1.68	1.65	1.62	1.58	1.55	1.51	1.47
30	2.88	2.49	2.28	2.14	2.03	1.98	1.93	1.88	1.85	1.82	1.77	1.72	1.67	1.64	1.61	1.57	1.54	1.50	1.46
40	2.84	2.44	2.23	2.09	2.00	1.93	1.87	1.83	1.79	1.76	1.71	1.66	1.61	1.57	1.54	1.51	1.47	1.42	1.38
60	2.79	2.39	2.18	2.04	1.95	1.87	1.82	1.77	1.74	1.71	1.66	1.60	1.54	1.51	1.48	1.44	1.40	1.35	1.29
120	2.75	2.35	2.13	1.99	1.90	1.82	1.77	1.72	1.68	1.65	1.60	1.55	1.48	1.45	1.41	1.37	1.32	1.26	1.19
∞	2.71	2.30	2.08	1.94	1.85	1.77	1.72	1.67	1.63	1.60	1.55	1.49	1.42	1.38	1.34	1.30	1.24	1.17	1.00

Degrees of Freedom for the Denominator (ν_2)

(continued)

$F_{0.90,\nu_1,\nu_2} = 1/F_{0.10,\nu_2,\nu_1}.$

$F_{0.05, v_1, v_2}$

Degrees of Freedom for the Numerator (v_1)

v_2 \ v_1	1	2	3	4	5	6	7	8	9	10	12	15	20	24	30	40	60	120	∞
1	161.4	199.5	215.7	224.6	230.2	234.0	236.8	238.9	240.5	241.9	243.9	245.9	248.0	249.1	250.1	251.1	252.2	253.3	254.3
2	18.51	19.00	19.16	19.25	19.30	19.33	19.35	19.37	19.38	19.40	19.41	19.43	19.45	19.45	19.46	19.47	19.48	19.49	19.50
3	10.13	9.55	9.28	9.12	9.01	8.94	8.89	8.85	8.81	8.79	8.74	8.70	8.66	8.64	8.62	8.59	8.57	8.55	8.53
4	7.71	6.94	6.59	6.39	6.26	6.16	6.09	6.04	6.00	5.96	5.91	5.86	5.80	5.77	5.75	5.72	5.69	5.66	5.63
5	6.61	5.79	5.41	5.19	5.05	4.95	4.88	4.82	4.77	4.74	4.68	4.62	4.56	4.53	4.50	4.46	4.43	4.40	4.36
6	5.99	5.14	4.76	4.53	4.39	4.28	4.21	4.15	4.10	4.06	4.00	3.94	3.87	3.84	3.81	3.77	3.74	3.70	3.67
7	5.59	4.74	4.35	4.12	3.97	3.87	3.79	3.73	3.68	3.64	3.57	3.51	3.44	3.41	3.38	3.34	3.30	3.27	3.23
8	5.32	4.46	4.07	3.84	3.69	3.58	3.50	3.44	3.39	3.35	3.28	3.22	3.15	3.12	3.08	3.04	3.01	2.97	2.93
9	5.12	4.26	3.86	3.63	3.48	3.37	3.29	3.23	3.18	3.14	3.07	3.01	2.94	2.90	2.86	2.83	2.79	2.75	2.71
10	4.96	4.10	3.71	3.48	3.33	3.22	3.14	3.07	3.02	2.98	2.91	2.85	2.77	2.74	2.70	2.66	2.62	2.58	2.54
11	4.84	3.98	3.59	3.36	3.20	3.09	3.01	2.95	2.90	2.85	2.79	2.72	2.65	2.61	2.57	2.53	2.49	2.45	2.40
12	4.75	3.89	3.49	3.26	3.11	3.00	2.91	2.85	2.80	2.75	2.69	2.62	2.54	2.51	2.47	2.43	2.38	2.34	2.30
13	4.67	3.81	3.41	3.18	3.03	2.92	2.83	2.77	2.71	2.67	2.60	2.53	2.46	2.42	2.38	2.34	2.30	2.25	2.21
14	4.60	3.74	3.34	3.11	2.96	2.85	2.76	2.70	2.65	2.60	2.53	2.46	2.39	2.35	2.31	2.27	2.22	2.18	2.13
15	4.54	3.68	3.29	3.06	2.90	2.79	2.71	2.64	2.59	2.54	2.48	2.40	2.33	2.29	2.25	2.20	2.16	2.11	2.07
16	4.49	3.63	3.24	3.01	2.85	2.74	2.66	2.59	2.54	2.49	2.42	2.35	2.28	2.24	2.19	2.15	2.11	2.06	2.01
17	4.45	3.59	3.20	2.96	2.81	2.70	2.61	2.55	2.49	2.45	2.38	2.31	2.23	2.19	2.15	2.10	2.06	2.01	1.96
18	4.41	3.55	3.16	2.93	2.77	2.66	2.58	2.51	2.46	2.41	2.34	2.27	2.19	2.15	2.11	2.06	2.02	1.97	1.92
19	4.38	3.52	3.13	2.90	2.74	2.63	2.54	2.48	2.42	2.38	2.31	2.23	2.16	2.11	2.07	2.03	1.98	1.93	1.88
20	4.35	3.49	3.10	2.87	2.71	2.60	2.51	2.45	2.39	2.35	2.28	2.20	2.12	2.08	2.04	1.99	1.95	1.90	1.84
21	4.32	3.47	3.07	2.84	2.68	2.57	2.49	2.42	2.37	2.32	2.25	2.18	2.10	2.05	2.01	1.96	1.92	1.87	1.81
22	4.30	3.44	3.05	2.82	2.66	2.55	2.46	2.40	2.34	2.30	2.23	2.15	2.07	2.03	1.98	1.94	1.89	1.84	1.78
23	4.28	3.42	3.03	2.80	2.64	2.53	2.44	2.37	2.32	2.27	2.20	2.13	2.05	2.01	1.96	1.91	1.86	1.81	1.76
24	4.26	3.40	3.01	2.78	2.62	2.51	2.42	2.36	2.30	2.25	2.18	2.11	2.03	1.98	1.94	1.89	1.84	1.79	1.73
25	4.24	3.39	2.99	2.76	2.60	2.49	2.40	2.34	2.28	2.24	2.16	2.09	2.01	1.96	1.92	1.87	1.82	1.77	1.71
26	4.23	3.37	2.98	2.74	2.59	2.47	2.39	2.32	2.27	2.22	2.15	2.07	1.99	1.95	1.90	1.85	1.80	1.75	1.69
27	4.21	3.35	2.96	2.73	2.57	2.46	2.37	2.31	2.25	2.20	2.13	2.06	1.97	1.93	1.88	1.84	1.79	1.73	1.67
28	4.20	3.34	2.95	2.71	2.56	2.45	2.36	2.29	2.24	2.19	2.12	2.04	1.96	1.91	1.87	1.82	1.77	1.71	1.65
29	4.18	3.33	2.93	2.70	2.55	2.43	2.35	2.28	2.22	2.18	2.10	2.03	1.94	1.90	1.85	1.81	1.75	1.70	1.64
30	4.17	3.32	2.92	2.69	2.53	2.42	2.33	2.27	2.21	2.16	2.09	2.01	1.93	1.89	1.84	1.79	1.74	1.68	1.62
40	4.08	3.23	2.84	2.61	2.45	2.34	2.25	2.18	2.12	2.08	2.00	1.92	1.84	1.79	1.74	1.69	1.64	1.58	1.51
60	4.00	3.15	2.76	2.53	2.37	2.25	2.17	2.10	2.04	1.99	1.92	1.84	1.75	1.70	1.65	1.59	1.53	1.47	1.39
120	3.92	3.07	2.68	2.45	2.29	2.17	2.09	2.02	1.96	1.91	1.83	1.75	1.66	1.61	1.55	1.50	1.43	1.35	1.25
∞	3.84	3.00	2.60	2.37	2.21	2.10	2.01	1.94	1.88	1.83	1.75	1.67	1.57	1.52	1.46	1.39	1.32	1.22	1.00

Degrees of Freedom for the Denominator (v_2)

$F_{0.95, v_1, v_2} = 1/F_{0.05, v_2, v_1}$.

(continued)

$$F_{0.25,v_1,v_2}$$

Degrees of Freedom for the Numerator (v_1)

v_2 \ v_1	1	2	3	4	5	6	7	8	9	10	12	15	20	24	30	40	60	120	∞
1	647.8	799.5	864.2	899.6	921.8	937.1	948.2	956.7	963.3	968.6	976.7	984.9	993.1	997.2	1001.0	1006.0	1010.0	1014.0	1018.0
2	38.51	39.00	39.17	39.25	39.30	39.33	39.36	39.37	39.39	39.40	39.41	39.43	39.45	39.46	39.46	39.47	39.48	39.49	39.50
3	17.44	16.04	15.44	15.10	14.88	14.73	14.62	14.54	14.47	14.42	14.34	14.25	14.17	14.12	14.08	14.04	13.99	13.95	13.90
4	12.22	10.65	9.98	9.60	9.36	9.20	9.07	8.98	8.90	8.84	8.75	8.66	8.56	8.51	8.46	8.41	8.36	8.31	8.26
5	10.01	8.43	7.76	7.39	7.15	6.98	6.85	6.76	6.68	6.62	6.52	6.43	6.33	6.28	6.23	6.18	6.12	6.07	6.02
6	8.81	7.26	6.60	6.23	5.99	5.82	5.70	5.60	5.52	5.46	5.37	5.27	5.17	5.12	5.07	5.01	4.96	4.90	4.85
7	8.07	6.54	5.89	5.52	5.29	5.12	4.99	4.90	4.82	4.76	4.67	4.57	4.47	4.42	4.36	4.31	4.25	4.20	4.14
8	7.57	6.06	5.42	5.05	4.82	4.65	4.53	4.43	4.36	4.30	4.20	4.10	4.00	3.95	3.89	3.84	3.78	3.73	3.67
9	7.21	5.71	5.08	4.72	4.48	4.32	4.20	4.10	4.03	3.96	3.87	3.77	3.67	3.61	3.56	3.51	3.45	3.39	3.33
10	6.94	5.46	4.83	4.47	4.24	4.07	3.95	3.85	3.78	3.72	3.62	3.52	3.42	3.37	3.31	3.26	3.20	3.14	3.08
11	6.72	5.26	4.63	4.28	4.04	3.88	3.76	3.66	3.59	3.53	3.43	3.33	3.23	3.17	3.12	3.06	3.00	2.94	2.88
12	6.55	5.10	4.47	4.12	3.89	3.73	3.61	3.51	3.44	3.37	3.28	3.18	3.07	3.02	2.96	2.91	2.85	2.79	2.72
13	6.41	4.97	4.35	4.00	3.77	3.60	3.48	3.39	3.31	3.25	3.15	3.05	2.95	2.89	2.84	2.78	2.72	2.66	2.60
14	6.30	4.86	4.24	3.89	3.66	3.50	3.38	3.29	3.21	3.15	3.05	2.95	2.84	2.79	2.73	2.67	2.61	2.55	2.49
15	6.20	4.77	4.15	3.80	3.58	3.41	3.29	3.20	3.12	3.06	2.96	2.86	2.76	2.70	2.64	2.59	2.52	2.46	2.40
16	6.12	4.69	4.08	3.73	3.50	3.34	3.22	3.12	3.05	2.99	2.89	2.79	2.68	2.63	2.57	2.51	2.45	2.38	2.32
17	6.04	4.62	4.01	3.66	3.44	3.28	3.16	3.06	2.98	2.92	2.82	2.72	2.62	2.56	2.50	2.44	2.38	2.32	2.25
18	5.98	4.56	3.95	3.61	3.38	3.22	3.10	3.01	2.93	2.87	2.77	2.67	2.56	2.50	2.44	2.38	2.32	2.26	2.19
19	5.92	4.51	3.90	3.56	3.33	3.17	3.05	2.96	2.88	2.82	2.72	2.62	2.51	2.45	2.39	2.33	2.27	2.20	2.13
20	5.87	4.46	3.86	3.51	3.29	3.13	3.01	2.91	2.84	2.77	2.68	2.57	2.46	2.41	2.35	2.29	2.22	2.16	2.09
21	5.83	4.42	3.82	3.48	3.25	3.09	2.97	2.87	2.80	2.73	2.64	2.53	2.42	2.37	2.31	2.25	2.18	2.11	2.04
22	5.79	4.38	3.78	3.44	3.22	3.05	2.93	2.84	2.76	2.70	2.60	2.50	2.39	2.33	2.27	2.21	2.14	2.08	2.00
23	5.75	4.35	3.75	3.41	3.18	3.02	2.90	2.81	2.73	2.67	2.57	2.47	2.36	2.30	2.24	2.18	2.11	2.04	1.97
24	5.72	4.32	3.72	3.38	3.15	2.99	2.87	2.78	2.70	2.64	2.54	2.44	2.33	2.27	2.21	2.15	2.08	2.01	1.94
25	5.69	4.29	3.69	3.35	3.13	2.97	2.85	2.75	2.68	2.61	2.51	2.41	2.30	2.24	2.18	2.12	2.05	1.98	1.91
26	5.66	4.27	3.67	3.33	3.10	2.94	2.82	2.73	2.65	2.59	2.49	2.39	2.28	2.22	2.16	2.09	2.03	1.95	1.88
27	5.63	4.24	3.65	3.31	3.08	2.92	2.80	2.71	2.63	2.57	2.47	2.36	2.25	2.19	2.13	2.07	2.00	1.93	1.85
28	5.61	4.22	3.63	3.29	3.06	2.90	2.78	2.69	2.61	2.55	2.45	2.34	2.23	2.17	2.11	2.05	1.98	1.91	1.83
29	5.59	4.20	3.61	3.27	3.04	2.88	2.76	2.67	2.59	2.53	2.43	2.32	2.21	2.15	2.09	2.03	1.96	1.89	1.81
30	5.57	4.18	3.59	3.25	3.03	2.87	2.75	2.65	2.57	2.51	2.41	2.31	2.20	2.14	2.07	2.01	1.94	1.87	1.79
40	5.42	4.05	3.46	3.13	2.90	2.74	2.62	2.53	2.45	2.39	2.29	2.18	2.07	2.01	1.94	1.88	1.80	1.72	1.64
60	5.29	3.93	3.34	3.01	2.79	2.63	2.51	2.41	2.33	2.27	2.17	2.06	1.94	1.88	1.82	1.74	1.67	1.58	1.48
120	5.15	3.80	3.23	2.89	2.67	2.52	2.39	2.30	2.22	2.16	2.05	1.94	1.82	1.76	1.69	1.61	1.53	1.43	1.31
∞	5.02	3.69	3.12	2.79	2.57	2.41	2.29	2.19	2.11	2.05	1.94	1.83	1.71	1.64	1.57	1.48	1.39	1.27	1.00

Degrees of Freedom for the Denominator (v_2)

(continued)

$F_{0.975,v_1,v_2} = 1/F_{0.25,v_2,v_1}$.

$$F_{0.25,\,\nu_1,\,\nu_2}$$

Degrees of Freedom for the Numerator (ν_1)

ν_2	1	2	3	4	5	6	7	8	9	10	12	15	20	24	30	40	60	120	∞
1	4052.0	4999.5	5403.0	5625.0	5764.0	5859.0	5928.0	5982.0	6022.0	6056.0	6106.0	6157.0	6209.0	6235.0	6261.0	6287.0	6313.0	6339.0	6366.0
2	98.50	99.00	99.17	99.25	99.30	99.33	99.36	99.37	99.39	99.40	99.42	99.43	99.45	99.46	99.47	99.47	99.48	99.49	99.50
3	34.12	30.82	29.46	28.71	28.24	27.91	27.67	27.49	27.35	27.23	27.05	26.87	26.69	26.60	26.50	26.41	26.32	26.22	26.13
4	21.20	18.00	16.69	15.98	15.52	15.21	14.98	14.80	14.66	14.55	14.37	14.20	14.02	13.93	13.84	13.75	13.65	13.56	13.46
5	16.26	13.27	12.06	11.39	10.97	10.67	10.46	10.29	10.16	10.05	9.89	9.72	9.55	9.47	9.38	9.29	9.20	9.11	9.02
6	13.75	10.92	9.78	9.15	8.75	8.47	8.26	8.10	7.98	7.87	7.72	7.56	7.40	7.31	7.23	7.14	7.06	6.97	6.88
7	12.25	9.55	8.45	7.85	7.46	7.19	6.99	6.84	6.72	6.62	6.47	6.31	6.16	6.07	5.99	5.91	5.82	5.74	5.65
8	11.26	8.65	7.59	7.01	6.63	6.37	6.18	6.03	5.91	5.81	5.67	5.52	5.36	5.28	5.20	5.12	5.03	4.95	4.86
9	10.56	8.02	6.99	6.42	6.06	5.80	5.61	5.47	5.35	5.26	5.11	4.96	4.81	4.73	4.65	4.57	4.48	4.40	4.31
10	10.04	7.56	6.55	5.99	5.64	5.39	5.20	5.06	4.94	4.85	4.71	4.56	4.41	4.33	4.25	4.17	4.08	4.00	3.91
11	9.65	7.21	6.22	5.67	5.32	5.07	4.89	4.74	4.63	4.54	4.40	4.25	4.10	4.02	3.94	3.86	3.78	3.69	3.60
12	9.33	6.93	5.95	5.41	5.06	4.82	4.64	4.50	4.39	4.30	4.16	4.01	3.86	3.78	3.70	3.62	3.54	3.45	3.36
13	9.07	6.70	5.74	5.21	4.86	4.62	4.44	4.30	4.19	4.10	3.96	3.82	3.66	3.59	3.51	3.43	3.34	3.25	3.17
14	8.86	6.51	5.56	5.04	4.69	4.46	4.28	4.14	4.03	3.94	3.80	3.66	3.51	3.43	3.35	3.27	3.18	3.09	3.00
15	8.68	6.36	5.42	4.89	4.56	4.32	4.14	4.00	3.89	3.80	3.67	3.52	3.37	3.29	3.21	3.13	3.05	2.96	2.87
16	8.53	6.23	5.29	4.77	4.44	4.20	4.03	3.89	3.78	3.69	3.55	3.41	3.26	3.18	3.10	3.02	2.93	2.84	2.75
17	8.40	6.11	5.18	4.67	4.34	4.10	3.93	3.79	3.68	3.59	3.46	3.31	3.16	3.08	3.00	2.92	2.83	2.75	2.65
18	8.29	6.01	5.09	4.58	4.25	4.01	3.84	3.71	3.60	3.51	3.37	3.23	3.08	3.00	2.92	2.84	2.75	2.66	2.57
19	8.18	5.93	5.01	4.50	4.17	3.94	3.77	3.63	3.52	3.43	3.30	3.15	3.00	2.92	2.84	2.76	2.67	2.58	2.49
20	8.10	5.85	4.94	4.43	4.10	3.87	3.70	3.56	3.46	3.37	3.23	3.09	2.94	2.86	2.78	2.69	2.61	2.52	2.42
21	8.02	5.78	4.87	4.37	4.04	3.81	3.64	3.51	3.40	3.31	3.17	3.03	2.88	2.80	2.72	2.64	2.55	2.46	2.36
22	7.95	5.72	4.82	4.31	3.99	3.76	3.59	3.45	3.35	3.26	3.12	2.98	2.83	2.75	2.67	2.58	2.50	2.40	2.31
23	7.88	5.66	4.76	4.26	3.94	3.71	3.54	3.41	3.30	3.21	3.07	2.93	2.78	2.70	2.62	2.54	2.45	2.35	2.26
24	7.82	5.61	4.72	4.22	3.90	3.67	3.50	3.36	3.26	3.17	3.03	2.89	2.74	2.66	2.58	2.49	2.40	2.31	2.21
25	7.77	5.57	4.68	4.18	3.85	3.63	3.46	3.32	3.22	3.13	2.99	2.85	2.70	2.62	2.54	2.45	2.36	2.27	2.17
26	7.72	5.53	4.64	4.14	3.82	3.59	3.42	3.29	3.18	3.09	2.96	2.81	2.66	2.58	2.50	2.42	2.33	2.23	2.13
27	7.68	5.49	4.60	4.11	3.78	3.56	3.39	3.26	3.15	3.06	2.93	2.78	2.63	2.55	2.47	2.38	2.29	2.20	2.10
28	7.64	5.45	4.57	4.07	3.75	3.53	3.36	3.23	3.12	3.03	2.90	2.75	2.60	2.52	2.44	2.35	2.26	2.17	2.06
29	7.60	5.42	4.54	4.04	3.73	3.50	3.33	3.20	3.09	3.00	2.87	2.73	2.57	2.49	2.41	2.33	2.23	2.14	2.03
30	7.56	5.39	4.51	4.02	3.70	3.47	3.30	3.17	3.07	2.98	2.84	2.70	2.55	2.47	2.39	2.30	2.21	2.11	2.01
40	7.31	5.18	4.31	3.83	3.51	3.29	3.12	2.99	2.89	2.80	2.66	2.52	2.37	2.29	2.20	2.11	2.02	1.92	1.80
60	7.08	4.98	4.13	3.65	3.34	3.12	2.95	2.82	2.72	2.63	2.50	2.35	2.20	2.12	2.03	1.94	1.84	1.73	1.60
120	6.85	4.79	3.95	3.48	3.17	2.96	2.79	2.66	2.56	2.47	2.34	2.19	2.03	1.95	1.86	1.76	1.66	1.53	1.38
∞	6.63	4.61	3.78	3.32	3.02	2.80	2.64	2.51	2.41	2.32	2.18	2.04	1.88	1.79	1.70	1.59	1.47	1.32	1.00

Degrees of Freedom for the Denominator (ν_2)

$$F_{0.99,\,\nu_1,\,\nu_2} = 1/F_{0.01,\,\nu_2,\,\nu_1}.$$

Index